AF352792

Crop Disease Detection Using Remote Sensing Image Analysis

Crop Disease Detection Using Remote Sensing Image Analysis

Editor

Xanthoula Eirini Pantazi

MDPI • Basel • Beijing • Wuhan • Barcelona • Belgrade • Manchester • Tokyo • Cluj • Tianjin

Editor
Xanthoula Eirini Pantazi
Aristotle University of
Thessaloniki, Faculty of
Agriculture
Greece

Editorial Office
MDPI
St. Alban-Anlage 66
4052 Basel, Switzerland

This is a reprint of articles from the Special Issue published online in the open access journal *Remote Sensing* (ISSN 2072-4292) (available at: https://www.mdpi.com/journal/remotesensing/special_issues/crop_disease).

For citation purposes, cite each article independently as indicated on the article page online and as indicated below:

LastName, A.A.; LastName, B.B.; LastName, C.C. Article Title. *Journal Name* **Year**, *Volume Number*, Page Range.

ISBN 978-3-0365-5605-5 (Hbk)
ISBN 978-3-0365-5606-2 (PDF)

Contents

About the Editor

Xanthoula Eirini Pantazi

Xanthoula Eirini Pantazi, PhD is Assistant Professor at Faculty of Agriculture, Aristotle University of Thessaloniki, Greece. She holds a PhD in Biosystems Engineering and is an expert in bio-inspired computational systems and data mining. Her research interests include precision farming, plant stress detection, IoT, big data, robotics, sensor fusion, machine learning, deep learning and non-destructive sensing of bio-material and crop protection. She has been involved in several EU projects and ERANET and has developed apps for FIWARE future Internet applications based on Neural Network application. She has designed a DSS system for post-harvest quality assessment in organic crops. Her recent research interests consider Bioinformatics applications in field phenotyping with autonomous platforms (e.g. UAV, robots) and application of active learning in a number of sectors from condition monitoring, yield prediction, crop status determination, weed species recognition as well as post-harvest quality determination. She is co-author of the research monograph book "Intelligent Data Mining and Fusion Systems in Agriculture" (https://www.elsevier.com/books/intelligent-data-mining-and-fusion-systems-in-agriculture/pantazi/978-0-12-814391-9).

Preface to "Crop Disease Detection Using Remote Sensing Image Analysis"

The current book offers to advanced students and entry-level professionals in agricultural science and engineering, an in-depth overview of the novel approaches that employ remote sensing techniques for crop disease detection. This overview highlights how the improvement of crop disease monitoring and management can enhance global food security through addressing key disease management practices towards an enhanced crop production with lower environmental footprint under changing climate conditions.

Xanthoula Eirini Pantazi
Editor

Article

Assessment of Different Object Detectors for the Maturity Level Classification of Broccoli Crops Using UAV Imagery

Vasilis Psiroukis [1,*], Borja Espejo-Garcia [1], Andreas Chitos [1], Athanasios Dedousis [1], Konstantinos Karantzalos [2] and Spyros Fountas [1]

[1] Laboratory of Agricultural Engineering, Department of Natural Resources Management & Agricultural Engineering, School of Environment and Agricultural Engineering, Agricultural University of Athens, 11855 Athens, Greece; borjaeg@aua.gr (B.E.-G.); andreas.chitos@aua.gr (A.C.); adedousis@sarantisestate.gr (A.D.); sfountas@aua.gr (S.F.)

[2] Laboratory of Remote Sensing, Department of Topography, School of Rural, Surveying and Geoinformatics Engineering, National Technical University of Athens, 10682 Athens, Greece; karank@central.ntua.gr

[*] Correspondence: vpsiroukis@aua.gr

Citation: Psiroukis, V.; Espejo-Garcia, B.; Chitos, A.; Dedousis, A.; Karantzalos, K.; Fountas, S. Assessment of Different Object Detectors for the Maturity Level Classification of Broccoli Crops Using UAV Imagery. *Remote Sens.* **2022**, *14*, 731. https://doi.org/10.3390/rs14030731

Academic Editor: Xanthoula Eirini Pantazi

Received: 14 January 2022
Accepted: 2 February 2022
Published: 4 February 2022

Abstract: Broccoli is an example of a high-value crop that requires delicate handling throughout the growing season and during its post-harvesting treatment. As broccoli heads can be easily damaged, they are still harvested by hand. Moreover, human scouting is required to initially identify the field segments where several broccoli plants have reached the desired maturity level, such that they can be harvested while they are in the optimal condition. The aim of this study was to automate this process using state-of-the-art Object Detection architectures trained on georeferenced orthomosaic-derived RGB images captured from low-altitude UAV flights, and to assess their capacity to effectively detect and classify broccoli heads based on their maturity level. The results revealed that the object detection approach for automated maturity classification achieved comparable results to physical scouting overall, especially for the two best-performing architectures, namely Faster R-CNN and CenterNet. Their respective performances were consistently over 80% mAP@50 and 70% mAP@75 when using three levels of maturity, and even higher when simplifying the use case into a two-class problem, exceeding 91% and 83%, respectively. At the same time, geometrical transformations for data augmentations reported improvements, while colour distortions were counterproductive. The best-performing architecture and the trained model could be tested as a prototype in real-time UAV detections in order to assist in on-field broccoli maturity detection.

Keywords: object detection; UAV images; maturity detection; efficientdet; retinanet; centernet; deep learning; precision agriculture; broccoli

1. Introduction

The Brassicaceae are a family of flowering plants which are widely known for the multiple health benefits associated with their consumption [1]. Broccoli (*Brassica oleracea L. var. italica Plenck.*) is one of the most popular crops of this family, the global production of which reached 25 million tons in 2020 (Faostat, 2020). Approximately 10% of this quantity is produced within Europe (Eurostat, 2020, Faostat, 2020). Organic broccoli is an example of a high-value crop that requires delicate handling throughout the growing season and during its post-harvesting handling. In conventional farming, broccoli is mainly harvested mechanically, as the produce is typically intended for the process market (deep freezing). In the case of organic broccoli, as heads can be easily damaged, resulting in visible stains, it is still harvested 'on sight' by hand using handheld knives because it is targeted towards the fresh market. On top of that, this allows for a very strict time window of "optimal maturity" when the high-end quality broccoli heads should be harvested, before they remain exposed for too long in high-humidity conditions and become susceptible to fungal infections and quality degradation. Even slight delays from this time window can result in major losses in the final production (Figure 1). However, manual harvesting is a very laborious task, not only for the process of harvesting itself, but for the scouting required to initially identify the field segments where several broccoli plants have reached this maturity level. Moreover, the scouting process is performed on foot, as agricultural vehicles cruising across the fields result in soil compaction, which is highly undesirable in horticulture, especially in the case of organic systems [2].

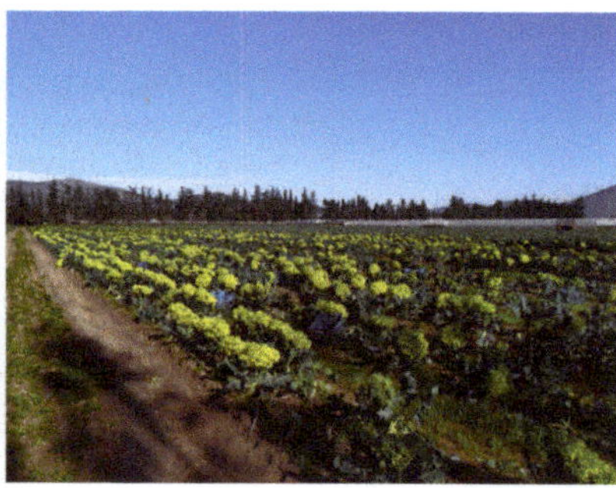

Figure 1. Examples of broccoli fields in full bloom, representing yield losses due to quality degradation.

This case creates a very interesting challenge. First of all, the scouting process can be automated using machine learning, drastically increasing the overall efficiency and reducing the human effort required. At the same time, UAVs can act as a double-benefit factor. They can easily supervise large areas rapidly, whilst diminishing any potential soil-compaction problems. There is a growing need for automated horticultural operations due to increasing uncertainty in the reliability of labour, and to allow for more targeted, data-driven harvesting [3]. To this end, images captured from Unmanned Aerial Vehicles (UAV) could be used to replace the labour work done for field observation. UAVs are widely used in precision agriculture for image capturing and the detection of specific conditions in the field [4]. Compared to the time-consuming work that should be performed by a group of people to find a potential problem in a crop, UAVs could quickly provide a high-resolution image of the field. The produced image, combined with image vision techniques, could output the potential problem/condition that would need to be dealt with [5].

Object detection is a primary field of computer vision, determining the location of certain objects in the image, and then classifying those objects [6]. Initially, the first methods used to address this problem consisted of two stages: (1) the Feature Extraction stage, in which different areas in the image are identified using sliding windows of different sizes, and (2) the Classification stage, in which the classes of the objects detected are estimated. A common method for the implementation of image classification is the sliding window approach, where the classifier runs at evenly spaced locations over the entire image. Object detection algorithms are evaluated based on the speed and the accuracy that are demonstrated, but their optimisation in both factors could be a very challenging task [6].

Object detection techniques apply classifiers or localizers to perform detections on images at multiple locations and scales. Recent approaches like the R-CNN and its variations use region proposal methods to generate, initially, several potential bounding boxes across the image, and then run the classifier on these proposed boxes. During training with those approaches, after every classification, post-processing steps are used to refine the bounding boxes, usually by increasing the score of the best-performed bounding boxes and decreasing the worse ones, ultimately eliminating potential duplicate detections [7]. Faster R-CNN [8] is one of the most widely used two-stage object detectors. The first stage uses a region proposal network, which is an attention mechanism developed as an alternative to the earlier sliding window-based approaches. In the second stage, bounding box regression and object classification are performed. Faster R-CNN is fairly well-recognized as a successful architecture for object detection, but it is not the only meta-architecture which is able to reach state-of-the-art results [9]. On the other hand, single-shot detectors, such as SSD [10] and RetinaNet [11], integrate the entire object detection process into a single neural network to generate each bounding box prediction.

Automation in agriculture presents a more challenging situation compared to industrial automation due to field conditions and the outdoor environment in general [12,13]. Fundamentally, most tasks demand a high accuracy of crop detection and localization, as they are both critical components for any automated task in agriculture [14]. The fact that there is a constant downward trend of the available agricultural labour force [15] also adds to this problem, and makes the automation of several production aspects a necessity. Accurate crop detection and classification are essential for several applications [5], including crop/fruit counting and yield estimation. Crop detection is often the preliminary step, followed by the classification operation, such as the quantification of the infestation level through the identification of disease symptoms [16], or as per the subject of the present paper, maturity detection for the automation of crop surveying. At the same time, it is the single most crucial component for automated real-time actuation tasks, such as automated targeted spraying applications or robotic harvesting.

Focusing on horticultural crops, automation in growth-stage identification has been an open challenge for multiple decades due to the very nature of the crops, which in their majority are high-value and demand timely interventions to maintain top-cut yield quality. Therefore, different approaches have been implemented to achieve the automated mapping of crop growth across larger fields and to assist harvesting, either by correlating the images' frequency bands with broccoli head sizes for maturity detection [17–19] or to combine image analysis techniques and neural networks to identify broccoli quality parameters [20].

As developments in computer vision allowed the research to move from simple image analysis frameworks to more complex and automated pipelines, the interest shifted towards Artificial Intelligence. Commercial RGB cameras and machine learning algorithms can provide affordable and versatile solutions for crop detection. Computer vision systems based on deep CNN [21] are immune to variations in illumination and large interclass variability [22], both of which have posed challenges in agricultural imaging in the past, thus achieving the robust recognition of the targets in open-field conditions. Recent research [23–25] have shown that the Faster R-CNN (region-based convolutional neural network) architecture [8,26] or different YOLO model versions [27–29] can produce accurate results for a large set of horticultural crops and fruit orchards. Moreover, a comparison of different computer vision techniques, such as object detection and object segmentation [30], has provided significant improvements in crop detection. In recent horticultural research literature, several studies have also focused on the localization of broccoli heads, without any evaluation of their maturity, by implementing deep learning techniques [31–34].

The objective of this study was to compare state-of-the-art object detection architectures and data augmentation techniques, and to assess their potential in the maturity classification of open-field broccoli crops, using a high-Ground Sampling Distance (GSD) RGB image dataset collected from low-altitude UAV flights. The best-performing architecture and the trained model could be tested as a prototype in real-time UAV detections in order to assist in on-field broccoli maturity detection.

2. Materials and Methods

2.1. Experimental Location

The experiment took place in Marathon, Greece (42 km north from Athens), in a commercial organic vegetable production unit (Megafarm Gaia mas). This region is specifically known for its horticultural production, being the main vegetable provider for Athens. The timing of the data acquisition flights was specifically designed to be performed a few hours prior to the first wave of selective harvesting. This was desired for two reasons: (1) to ensure that the entire field was intact (no broccoli heads were harvested), maximising the sample density in every image and the generated field orthomosaic; and (2) to make sure that individual plants of different maturity levels were present across the field, as it was at the very start of the harvesting season. The selected experimental parcel was located in the south-west part of the production unit, and occupied an area of approximately 1 ha. The segment on which the ground truth targets were deployed covered slightly more than half of it (Figures 2 and 3).

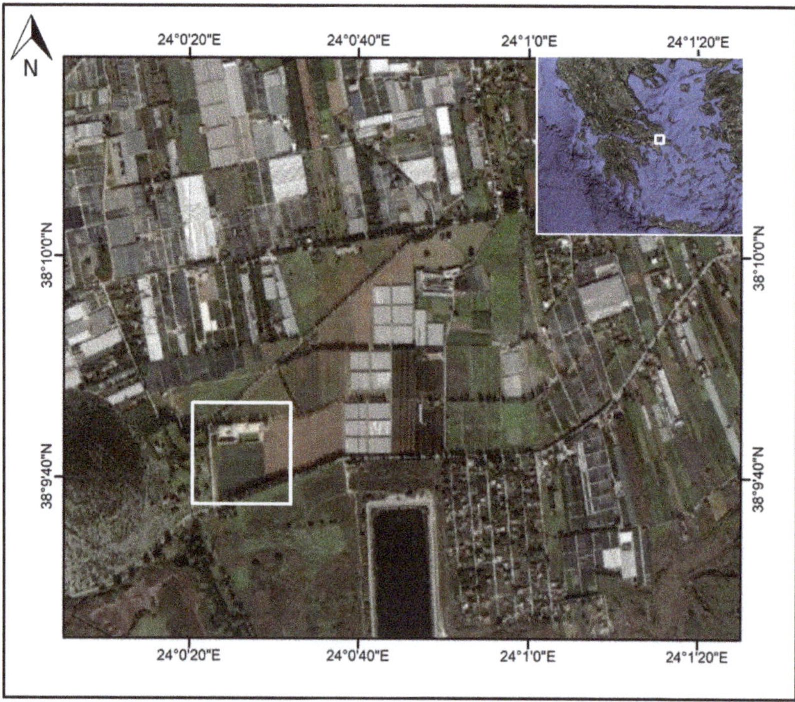

Figure 2. The experimental location in Marathon, Greece.

Figure 3. The field segment where the experiment took place.

2.2. Data Acquisition

The data collection was performed by a custom quadcopter drone equipped with a 20-megapixel (4096 × 2160 resolution) CMOS mechanical shutter RGB camera, which was used to generate the low-altitude aerial imagery dataset. The images were captured and saved in the 16-bit JPG format, accompanied by the georeferencing metadata of each image. A total of 3 flight missions were executed during the time window between 11:00 and 13:00 (when the solar elevation angle was greater than 45°), in order to avoid drastic deviations in solar illumination between the flights. The flights were executed with similar flight parameters across all of the flights, operating at a fixed altitude of 10 m AGL ±0.5 m (GPS vertical hovering error), which is considered to be consistent towards the ground because the field was levelled by a flattening cultivation roller approximately 4 months prior to the measurements. The sensor was set to capture images at a fixed interval of 2 sec/capture, and therefore the flight plans were designed around this parameter. The frontal and lateral overlaps were both selected to be 80%, resulting in a cruising speed of approximately 1.1 m/s. Finally, the orientation of the flight lines was selected to be parallel to the orientation of the planting rows. The generated flight plan is also presented below (Figure 4), and was executed three times consecutively by the UAV.

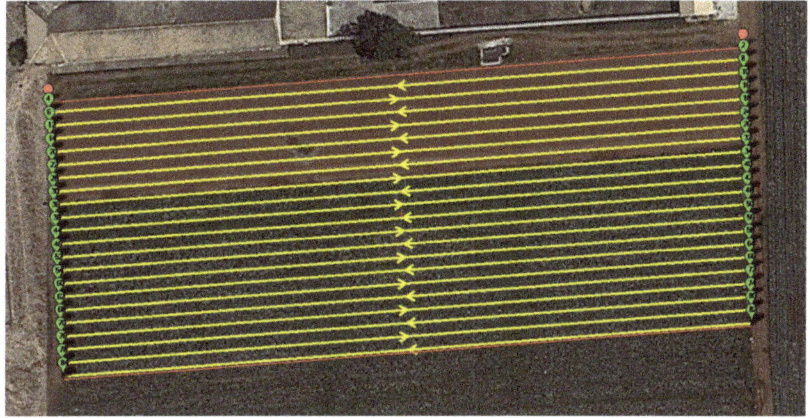

Figure 4. The flight mission executed by the UAV.

Before the start of the data collection flights, a total of 45 ground truth targets were deployed and stabilised across the field, in order to support the annotation stage (described in the following section). The targets indicated the maturity level of the selected broccoli crops, as they were categorised by an expert agronomist who participated in the process. The human expert indicated a total of 15 broccoli heads of 3 different broccoli maturity classes. These classes ranged from 1 to 3, with class 1 representing immature crops that would not be harvested for at least the following 15 days, class 2 representing heads that are estimated to reach harvesting level within a week, and finally, class 3, which contained exclusively "ready to harvest" heads. Due to the high humidity of the air near the surface and the constantly wet soil, the targets were enveloped inside transparent plastic cases in order to protect them from decomposing, as they would remain on the field for approximately two hours. The cases had been tested in the university campus to verify that the labels (numbering) remained visible in the UAV imagery. In case a ground truth label was covered heavily by the surrounding leaves, a bright-yellow point-like object was also placed nearby as a pointer. Examples of the deployed targets are presented below (Figure 5).

Figure 5. The ground truth targets for each plant's maturity class and their respective pointers (when deployed, in high plant coverage areas). (**a**) Maturity class 1; (**b**) maturity class 2; (**c,d**) maturity class 3.

2.3. Data Pre-Processing

Following the field data acquisition by the UAV and the imaging components, data pre-processing was performed in remote computational units. The first step of the data processing phase was to generate the orthomosaic maps from each flight. For this process, the photogrammetric software Pix4D Mapper (Pix4D SA) was used, generating a total of three (3) RGB orthomosaics, with a final GSD of 0.25 cm. In the following step, the single best orthomosaic was selected in order to ensure that the final dataset that was going to be presented to the machine did not contain duplicates of the same crops, as this would increase the initial bias of the experiment. After close inspection, the mosaic of the second flight was selected, as it produced an orthomosaic of slightly higher quality (less blurry spots and zero holes), potentially indicating that the flight conditions were better during that time window, which enabled the UAV to perform its flight in a more optimal way with fewer disruptions.

Once the mosaic was selected, the next step was to create the dataset that would be fed to the models. As the generated mosaic was georeferenced, an initial crop with a vector layer was performed in a GIS (QGIS 3.10) to eliminate the majority of the black, zero-valued pixels that were created during the mosaicking process (the exported mosaic is written in a minimum-bounding-box method, surrounding the mosaic map with black pixels to create a rectangle, were all of the bands are assigned a zero value for the pixels which did not contain any data). This cropping served another purpose, as the next pre-processing step involved "cutting" the mosaics into smaller images so that they could be used as an input for the models. This was easily performed using a script written in Python that iterated the entire mosaic and then copied the first X number of pixels in one direction and Y pixels in the other direction for all of the bands of the initial mosaic. In our case, the desired image dimensions were 500×500 pixels, and therefore the step of each loop (one for each axis, as the rectangular mosaic is scanned) was set to 500, to result in a dataset of rectangular RGB images with a uniform resolution.

The final step of the pre-processing involved object labelling on individual images. In this phase, the generated dataset was imported to the Computer Vision Annotation Tool (CVAT), and the images were annotated using the ground truth labels as a basis (Figure 4). Finally, the annotations of the dataset were exported in the PASCAL Visual Object Classes (VOC) format [35], as this format of bounding box annotations is required by Tensorflow. PASCAL VOC is an XML annotation format that requires a pixel-positioning encoding, meaning that once drawn, each annotation file is exported in the form of a text file containing the sequence of the four coordinates of each bounding box within the image. The annotations consisted of rectangular boxes assigned with the respective maturity class of each broccoli head they contained. The final dataset contained a total of 288 images with over 640 annotations, where most of the bounding boxes presented a squared shape (Figure 6).

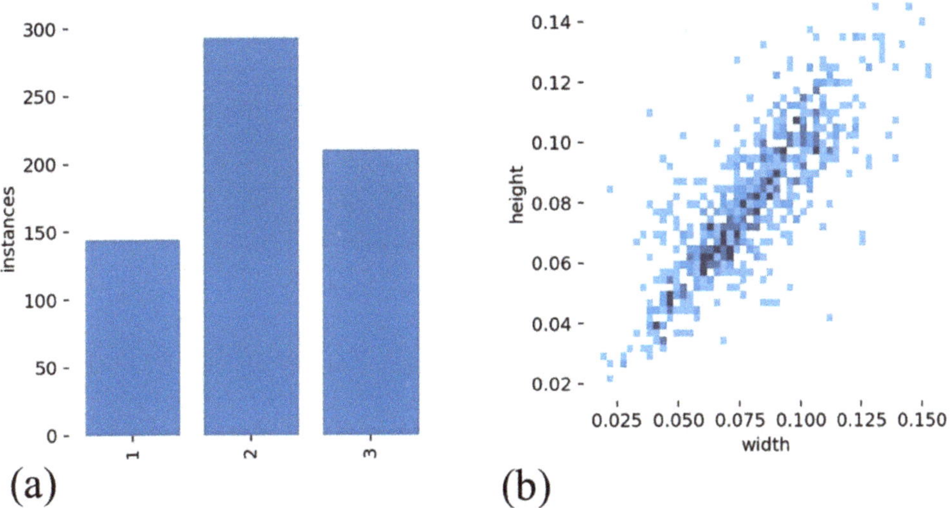

Figure 6. The dataset class representation (**a**) and annotation size plot (**b**).

2.4. Object Detection Pipelines

In order to create robust object detectors which are still reliable on test images, usually, it is necessary to use a training pipeline where implicit regularization, such as data augmentation, is applied during the training process without constraining the model capacity. Data augmentation in computer vision is a set of techniques performed on the original images to increase the number of images seen by the model while training. As studied in several works [36,37], having the right data augmentation pipeline is critical in order for agricultural computer vision systems to work effectively. As shown in Figure 7, the two particular augmentation transforms that seem to have the most impact are (i) geometrical transformations and (ii) colour distortions. In this work, in the geometrical augmentation approach, horizontal and vertical flipping, and random cropping with resizing were implemented. With regard to the colour distortion, slight modifications in the brightness, contrast, hue and saturation were implemented. All of the data augmentations were executed with a probability of 50%.

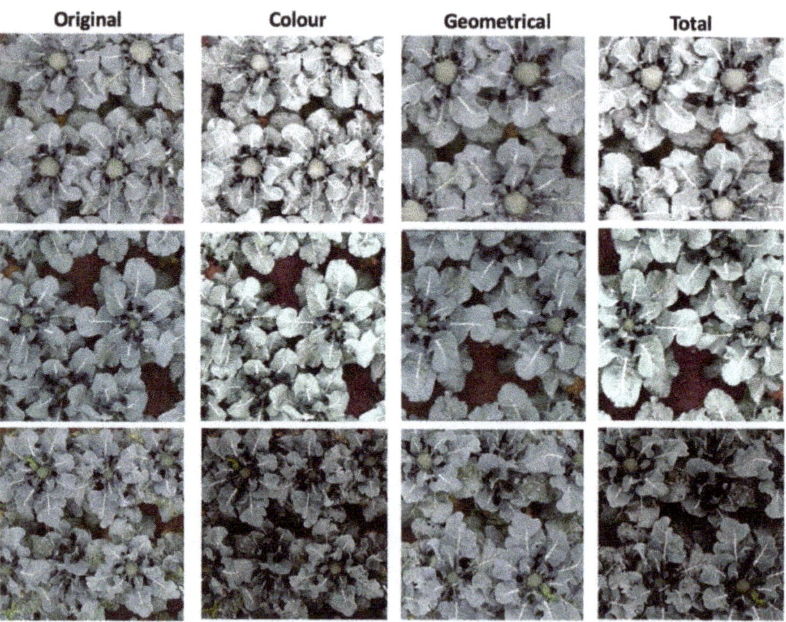

Figure 7. Example of the original images, and after applying the augmentation transformations.

Several object detectors have arisen in the last few years, with each of them having different advantages and disadvantages. Some of them are faster and less accurate, while others have higher performances but use more computational resources, which sometimes are not suitable depending on the deployment platform. In this work, five different object detectors were used: Faster R-CNN [8], SSD [10], CenterNet [38], RetinaNet [11] and EfficientDet-D1 [39]. Table 1 shows the specific detectors evaluated in this work. Except for the input size, the value of which was the closest to the original image size (500 × 500), the most promising ones were selected after some early experiments. It can be observed that ResNet-152 [22] is used as the backbone in two object detectors (Faster R-CNN and RetinaNet). Furthermore, HourGlass-104 [40], MobileNet-V2 [41] and EfficientNet-D1 [39] were evaluated. On the other hand, three of them use a Feature-Pyramid-Network (FPN), which is supposed to provide better performance in the detection of small objects. The reason for selecting these architectures is that different detection "families" are represented. For instance, SSD (e.g.: RetinaNet) and two-shot (Faster R-CNN) were compared in order to verify that the second one may lead to better performances than the basic approach (SSD and MobileNet-V2). However, the architectural improvements of RetinaNet (e.g., focal loss) could invert this assumption. On the other hand, the inference time was out of the scope of this paper, in which real-time detection is not discussed; however, theoretically, SSD could lead to faster inferences. Additionally, with these detectors, the anchorless approach was contrasted against the traditional anchor-based detection. Again, the most important theoretical gain could be the inference time, which is not discussed; however, related works presented the promising performances of CenterNet, which could be the chosen detector for future deployment on the field. Table 1 also reports the mAP (mean Average Precision) obtained on the COCO dataset. From this performance, it can be estimated that CenterNet and EfficientDet-D1 will be the best detectors, while SSD (MobileNet) and Faster R-CNN will be the least promising ones.

Table 1. The selected object detection architectures used for the experiment.

Object Detector	Backbone	Input Size	FPN	COCO mAP
Faster R-CNN	ResNet-152	640×640	No	32.4
SSD	MobileNet-V2	640×640	Yes	28.2
RetinaNet	ResNet-152	640×640	Yes	35.4
EfficientDet-D1	EfficientNet-D1	640×640	Yes	38.4
CenterNet	HG-104	512×512	No	41.9

When developing an object detection pipeline, it is important to fine-tune different hyper-parameters in order to select the ones that better fit a specific dataset. This means that different datasets could lead to different hyper-parameter configurations in every detector. Table 2 presents the evaluated hyper-parameter space used in this work to choose the most promising ones after 5 runs with different train-validations-test splits (see Table 3). The selected configurations were executed for 5 additional runs in order to complete the experimental trials and extract some statistics.

Table 2. Hyper-parameters evaluated for each detector.

Hyper-Parameter	Values
Optimizer	{Adam, SGD}
Learning Rate (LR)	{0.05, 0.01, 0.001}
Batch Size	{4, 8}
Warmup steps	{100, 500, 1000}
IoU Threshold	{0.1, 0.25, 0.5}
Max. Detections	{10, 50, 100}

Table 3. Hyper-parameters selected for the final experiments.

Object Detector	Optimizer	LR	Batch Size	Warmup Steps	IoU Threshold	Max. Detections
Faster R-CNN	SGD	0.05	4	500	0.5	100
SSD	SGD	0.01	8	1000	0.25	100
RetinaNet	SGD	0.01	8	100	0.25	50
EfficientDet-D1	SGD	0.04	4	500	0.5	50
CenterNet	Adam	0.001	8	1000	-	-

Two different optimizers were evaluated: Adam and SGD. All of the detectors performed better with SGD except CenterNet, which obtained its best performances with Adam. Related to the optimizer, the learning rate (LR) also played a major role. Adam worked better with the smallest evaluated LR (0.001), while SGD obtained the best performance with higher LRs (0.05 and 0.01). Two additional hyper-parameters that completely changed the training behavior were the warmup steps and the batch size. The warmup is the period of the training where the LR starts small and smoothly increases until it reaches the selected LR (0.05, 0.01 or 0.001). Every detector needed a different combination to obtain their best performance, but it is important to remark that the use of less than 500 steps (around 10 epochs) for warming obtained poor performances. Finally, all of the detectors, except CenterNet, which uses an anchorless approach, needed to run the Non-Maximum-Suppression (NMS) algorithm to remove redundant detections of the same object. Two important values configure this algorithm: the ratio to consider that two predicted bounding boxes point to the same object (the IoU threshold in the tables) and the maximum number of detections allowed (the Max. detections in the tables). As can be observed, again, all of the detectors found a different combination as the most promising one, which are presented in Table 3. Finally, an early stopping technique to avoid overfitting was implemented. Specifically, if the difference between the training and validation performances is greater than 5% for 10 epochs, the training process stops. All of the detectors were trained for a maximum of 50 epochs.

In this work, the experiments were carried out with GeForce RTX 3090 GPU under Ubuntu 18.04. For the software, Tensorflow 2.6.0 was used to implement the object detector pipelines. Early experiments to gain knowledge on the most promising detectors were implemented through parallel tasks in an HPC cluster (ARIS infrastructure https://hpc.grnet.gr/en/, accessed date: 13 January 2022) with 2 NVIDIA V100 GPU cards.

2.5. Evaluation Metrics

Because comparing the results of different architectures is not trivial, several benchmarks were developed and updated over the last few years for detection challenges, and different researchers may evaluate their model's skills on different benchmarks. In this paper, the evaluation method for the broccoli maturity detection task was based on the Microsoft COCO: Common Objects in Context dataset [42], which is probably the most commonly used dataset for object detection in images. Like COCO, the results of the broccoli maturity detection were reported using the Average Precision (AP). Precision is defined as the number of true positives divided by the sum of true positives (TP) and false positives (FP), while AP is the precision averaged across all of the unique recall levels. Because the calculation of AP only involves one class and, in object detection, there are usually several classes (3 in this paper), the mean Average Precision (mAP) is defined as the mean of the AP across all classes. In order to decide what a TP is, the Intersection over Union (IoU) threshold was used. IoU is defined as the area of the intersection divided by the area of the union of a predicted bounding box. For example, Figure 8 shows different IoUs in the same image and ground truth (red box) by varying the prediction (yellow box). If an IoU > 0.5 is configured, only the third image will contain a TP, while the other two will contain an FP. On the other hand, if an IoU > 0.4 is used, the central image will also count as a TP. In case multiple predictions correspond to the same ground truth, only the one with the highest IoU counts as a TP, while the remaining are considered FPs. Specifically, COCO reports the AP for two detection thresholds: IoU > 0.5 (traditional) and IoU > 0.75 (strict), which are the same thresholds used in this work. Additionally, the mAP with small objects (area less than 32 × 32 pixels) was also reported.

Figure 8. The ground truth (red box) and prediction (yellow box) determine the IoU score, which will be used to decide whether the prediction is correct or not. (**a**) Ground truth and predicted bounding box with IoU of 0; (**b**) ground truth and predicted bounding box with IoU of 0.42; and (**c**) ground truth and predicted bounding box with IoU of 0.98.

3. Results

The experiment results were obtained by averaging 10 different trials for each individual architecture. In each case, a stratified split was performed, with 70% of the samples being used for training, 10% of the samples being used for validation, and 20% of the samples being used for testing. Besides the performances on the test set, all of the tables include the mAP@50 on the training set, in order to illustrate the chances of overfitting.

3.1. Three-Class Approach

The performance of each architecture without using any type of data augmentation is shown in Table 4. As can be observed, Faster R-CNN (ResNet) is the object detector that showed the highest mAP@50 on average; however, it was not the case for mAP@75 and the performance with small objects, for which CenterNet performed better. SSD-MobileNet, RetinaNet and EfficientDet-D1 all demonstrated a lower performance, especially in the detection of small objects.

Table 4. The performance of each architecture without any data augmentation techniques. In the parentheses, the training performance is presented for mAP@50. The bold numbers correspond to the best performances.

Object Detector	mAP@50	mAP@75	mAP@small
Faster R-CNN (ResNet)	**82.6 $\pm$ 4.13** **(86.3)**	71.1 $\pm$ 4.01	54.66 $\pm$ 9.16
CenterNet	80.43 $\pm$ 2.46 (79.15)	**73.24 $\pm$ 2.81**	**57.05 $\pm$ 8.66**
SSD (MobileNet)	78.89 $\pm$ 3.69 (82.13)	69.1 $\pm$ 3.84	44.85 $\pm$ 8.75
RetinaNet	78.36 $\pm$ 1.84 (81.48)	62.24 $\pm$ 3.3	48.09 $\pm$ 6.01
EfficientDet-D1	77.68 $\pm$ 3.48 (79.16)	65.94 $\pm$ 4.22	48.85 $\pm$ 7.9

In the following step, a single form of augmentation was used at a time, and all of the architectures were evaluated in a similar process. Table 5 shows the performances when using geometrical augmentations. The performances improved in all of the cases. Faster R-CNN (ResNet) and CenterNet obtained the best results in mAP@50, mAP@75 and mAP@small. However, in this case, Faster R-CNN (ResNet) ranked first in the detection of small objects. Table 6 shows the performances when using only colour augmentations. Similarly to the previous configurations, Faster R-CNN (ResNet) and CenterNet obtained the best performances. The bold numbers correspond to the best performances.

Table 5. The performance of each architecture with only geometric augmentation active. In the parentheses, the training performance is presented for mAP@50.

Object Detector	mAP@50	mAP@75	mAP@small
Faster R-CNN (ResNet)	84.19 $\pm$ 2.96 (83.12)	73.6 $\pm$ 3.17	60.09 $\pm$ 7.78
CenterNet	83.68 $\pm$ 1.17 (85.18)	75.59 $\pm$ 1.47	59.39 $\pm$ 4.31
SSD (MobileNet)	82.81 $\pm$ 2.2 (81.36)	73.92 $\pm$ 2.1	59.17 $\pm$ 7.87
EfficientDet-D1	82.76 $\pm$ 2.56 (82.45)	71.91 $\pm$ 3.2	57.58 $\pm$ 7.89
RetinaNet	79.53 $\pm$ 3.21 (78.95)	65.44 $\pm$ 6.59	55.75 $\pm$ 5.8

Table 6. The performance of each architecture with only colour augmentation active. In the parentheses, the training performance is presented for mAP@50. The bold numbers correspond to the best performances.

Object Detector	mAP@50	mAP@75	mAP@small
Faster R-CNN (ResNet)	**83.17 ± 2.72** **(84.16)**	73.52 ± 2.09	**58.42 ± 6**
CenterNet	80.82 ± 2.7 (81.33)	**77.17 ± 2.82**	55.75 ± 6.21
EfficientDet-D1	79 ± 2.01 (80.47)	65.71 ± 2.97	52.44 ± 7.27
RetinaNet	77.45 ± 2.52 (76.74)	63.24 ± 3.91	46.18 ± 12.16
SSD (MobileNet)	77.38 ± 2.7 (76.49)	68.44 ± 3.2	45.4 ± 5.66

Table 7 shows the performances when using both colour and geometrical augmentations at the same time. In general, all of the architectures improved their mAP@50 except for Faster R-CNN (ResNet), the performance of which remained similar. As in the previous experiments, CenterNet was the best detector according to mAP@75 (besides mAP@50), confirming its superiority when giving accurate locations.

Table 7. The performance of each architecture with both augmentation types (geometrical and colour) active. In the parentheses, the training performance is presented for mAP@50. The bold numbers correspond to the best performances.

Object Detector	mAP@50	mAP@75	mAP@small
CenterNet	**83.52 ± 1.57** **(85.98)**	**79.53 ± 1.54**	59.73 ± 6.92
RetinaNet	83.28 ± 1.83 (84.17)	71.6 ± 2.15	**61.22 ± 5.21**
EfficientDet-D1	83.02 ± 1.44 (82.63)	72.73 ± 3.22	58.24 ± 5.29
SSD (MobileNet)	82.56 ± 3.18 (81.92)	72.14 ± 3.27	59.3 ± 7.51
Faster R-CNN (ResNet)	82.52 ± 4.32 (85.17)	70.96 ± 5.92	56.04 ± 8.17

Figure 9 depicts a box plot summarizing the performance of the detectors across all of the data augmentations. As can be inferred from the previous tables, Faster RCNN and CenterNet are the most consistent detectors at mAP@50 and mAP@75. However, both architectures presented a different behavior at the same time. On the one hand, Faster RCNN was able to obtain the highest performances with a higher variance. On the other hand, CenterNet did not reach the maximum, but showed more consistent performance. The other detectors (SSD, EfficientDet-D1 and RetinaNet) performed worse on average, but in some specific experiments, they were able to reach higher mAPs than CenterNet.

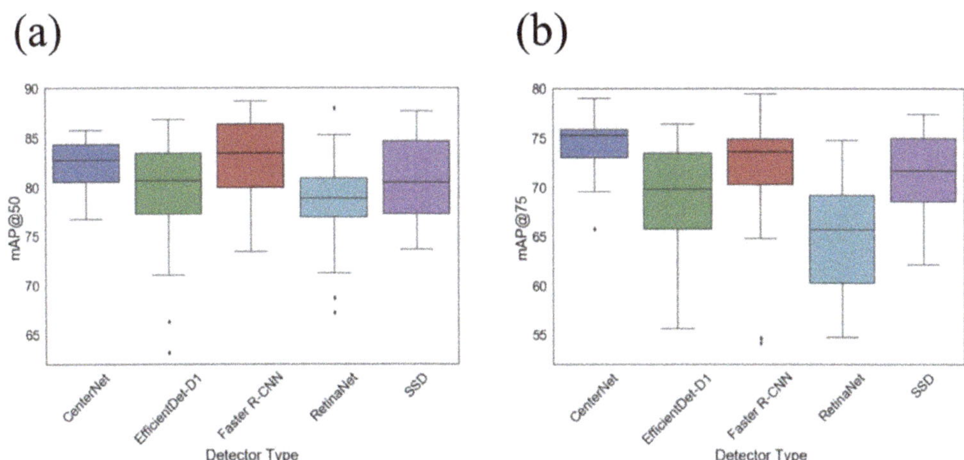

Figure 9. Box plot showing the performance of the different object detectors. (**a**) mAP@50; (**b**) mAP@75.

In order to provide a better description of how the data augmentation stage performs in the detection pipelines, Figure 10 depicts how each augmentation method has worked across all of the experiments and detectors. It can be observed that, in general, geometrical augmentation and complete augmentation show the higher median; however, with mAP@50, complete augmentation shows a more dispersed distribution, making it a less reliable augmentation. On the other hand, with mAP@75, geometrical augmentation obtains a more clear superiority over the rest of the detectors. Finally, Figure 10 also shows that all of the types of augmentations can reach the maximum performances (around 87.5%) with mAP@50, and close to maximum—obtained by geometrical augmentation—with mAP@75.

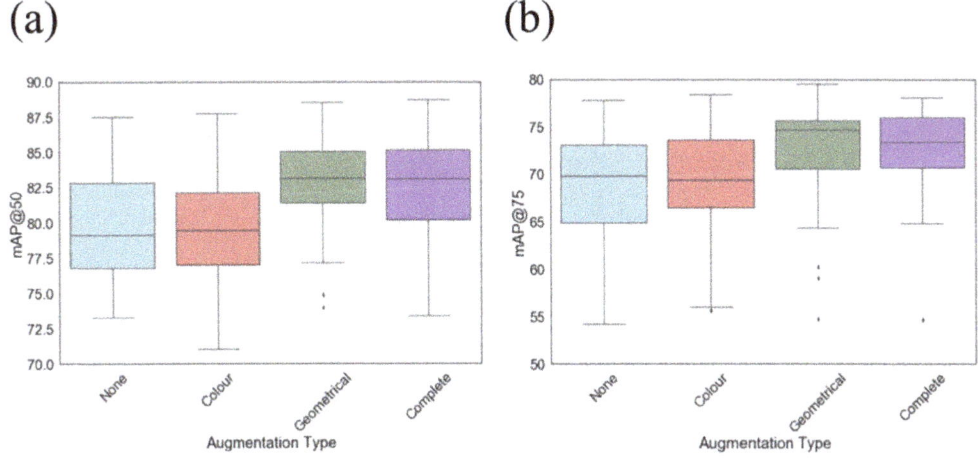

Figure 10. Box plot for image performance with and without augmentation transforms active. (**a**) mAP@50; (**b**) mAP@75.

3.2. Two-Class Approach

Regarding the performance for each individual maturity class, as observed in our results, the highest performing class was maturity class 3, followed by class 2, and finally class 1. Most class-3 broccoli heads were both detected successfully and classified correctly. In the case of lower maturity classes (1 and 2), despite the fact that the broccoli heads were distinguishable and thus correctly detected as objects, they would often be mixed between them in the classification step, as the architectures could not always tell them apart, and thus misclassified them. An example of this instance is presented in Figure 11, in which the single class-3 broccoli head present in the image was detected and classified correctly, while two class-2 heads, although properly detected, were misclassified as class 1.

Figure 11. Examples of (**a**) broccoli head detections and (**b**) ground truth labels using the three-class approach.

These observations led us to speculate that the architectures would potentially yield even better results and demonstrate a stronger performance if the problem presented was a simplified "ready to harvest" and "not ready yet" two-class problem. In order to test this hypothesis, the two best- and the two worst-performing pipelines (based on both data augmentation and architecture) were selected and evaluated on the two-class version of the dataset, where maturity classes 1 and 2 were merged into a single class, while class 3 was kept intact. Table 8 presents the performances of the four selected pipelines once the problem was simplified from three classes to two classes. As can be observed, all of the performances improved while the ranking of the pipelines was the expected. Finally, a similar detection example using the two-class approach, on the same image as Figure 11, is shown in Figure 12.

Table 8. The performance of the selected architectures and augmentation types on the two-class dataset. In the parentheses, the training performance is presented for mAP@50. The bold numbers correspond to the best performances.

Pipeline	mAP@50	mAP@75	mAP@small
Geometrical + CenterNet	**83.52 ± 1.57** **(84.79)**	**79.53 ± 1.54**	59.73 ± 6.92
Geometrical + Faster R-CNN (ResNet)	83.28 ± 1.83 (82.85)	71.6 ± 2.15	**61.22 ± 5.21**
Colour + SSD (MobileNet)	83.02 ± 1.44 (86.19)	72.73 ± 3.22	58.24 ± 5.29
Colour + RetinaNet	82.56 ± 3.18 (82.50)	72.14 ± 3.27	59.3 ± 7.51

Figure 12. Examples of (**a**) broccoli head detections and (**b**) ground truth labels using the two-class approach.

4. Discussion

The results have shown that, across all of the experiments, Faster R-CNN and CenterNet were the best-performing architectures for the task of broccoli maturity detection (Tables 4–7, and Figure 9). Their performances were, on average, always over 80% for mAP@50 and 70% for mAP@75 in the three-class problem, regardless of the hyperparameter configuration and augmentation techniques, which were solid performances considering the open-field nature of the dataset. The same pattern was encountered in the detection of small objects. Nonetheless, it is important to note that the differences were too small with high standard deviations, making it difficult to claim the best detector with small objects. According to Table 1, the high performance of CenterNet was expected due to its highest mAP on the COCO dataset in comparison to the other detectors. However, Faster R-CNN was, in theory, less promising than more recent architectures such as EfficientDet-D1 and RetinaNet. Additionally, Faster R-CNN and RetinaNet shared the same backbone, which made the comparison closer to the single-shot against two-shot detectors. In this case, the two-shot approach surpassed both SSD methods. This result highlights the fact that the evaluation of old architectures on new domains is always a worthy task, and opens the question of whether SSD may need deeper backbone architectures (e.g., ResNet 164) in order to consistently overcome the two-shot approach. This experiment could be carried out because the inference time in SSD is lower, and it could be increased by a deeper backbone to match the Faster R-CNN's times and performances. However, this approach could have a limit because embedded systems where the detector would be deployed have hardware limitations, and smaller architectures such as MobileNet would be preferable. However, according to the poor results shown in this work by MobileNet V2, the version 3 (both small and large) architectures should be evaluated to throw some light on this problem.

Another important aspect of this research was to evaluate different data augmentation approaches due to the relevance that this technique currently has in related literature. Without any form of augmentation, the best-performing architectures were Faster R-CNN (ResNet) for mAP@50 and CenterNet for mAP@75, achieving over 82% and 73%, respectively (Table 4). However, as it was hypothesized, not using augmentations did not obtain the best performances. Moreover, as can be observed in the presented tables, augmentation made the difference between training and testing lower, which can be translated into lower chances of overfitting. Focusing on the most promising augmentation techniques, geometrical techniques with and without the combination of colour distortions led to the highest performances overall (see Figure 10). By using only geometrical augmentation, Faster R-CNN (ResNet) managed to score the highest mAP@50 across all of the configurations and architectures (84.19%). On the other hand, it was quite remarkable that colour distortion by itself decreased the performance. For example, SSD-MobileNet decreased its mAP@50 from 78.89% (non-augmentation) and 82.82% (geometrical augmentation) to 77.38% (colour augmentation). Finally, with both colour and geometrical augmentations active, a surprising result was the improvement of RetinaNet, which not only obtained the second-best performance across all of the architectures but also achieved the highest mAP@small across all of the architectures and augmentation configurations. In addition, Faster R-CNN (ResNet) performed worse than the use of a single form of augmentation (colour and geometrical) separately.

The improvement of the geometrical augmentation is aligned with the results presented in [31,43], although the specific transformations were different. As discussed by these authors, these geometrically transformed images are similar to broccoli from the test set because the test images also contain broccolis of different sizes, scales, and positions. As a result, these transformed images allowed the neural network to better generalize, and to detect the broccolis in the test images with a higher mAP. However, with the colour transformations, the transformed images were less similar to the broccolis of the test set. Some unrealistic dark or bright images could appear, and even changes in textures that differed from the textural patterns learned while training could result in a lower mAP (See Figure 7). Moreover, this can lead to the conclusion that the use of standard augmentations, which usually work in popular datasets like COCO, could not work in some specific domains like agriculture and maturity level classification, where colour plays an important role. Another relevant remark for data augmentation is that all of the types of augmentation were able to reach performances of around 87.5% in some specific train–test splits and pipeline configurations. This reinforces the need to run several experiments with different splits in order to deeply understand the general behaviour of a specific detection pipeline.

For the two-class approach, the performance was significantly higher for the tested detectors, indicating that the merging of the two lower-maturity classes into one, and essentially only separating them from ready-to-harvest plants, was successful. This is another logical outcome considering that the "boundaries" between the classes were an approximation method for field labelling, and not an actual measurable metric. As a result, some level of confusion might occur even for a human observer tasked to separate the instances of these two classes. From an agronomic point of view, quantifying maturity in the early stages holds some value regarding the early planning of upcoming harvesting operations. Nevertheless, the most important aspect of maturity detection is the identification of the already-mature crops in order to avoid their prolonged exposure and delayed harvesting, which might lead to quality degradation. Additionally, for the simplified two-class problem, larger broccoli heads naturally cover more space in the image and are, therefore, more distinguishable in the first place. It can be inferred that the detection of class 3 (ready-to-harvest) outperforms the other two because the model can more easily detect them despite the fact that it is not the largest class (See Figure 6).

Focusing on the performances, the two (2) best-performing detectors both achieved over 91% mAP@50 and 83% mAP@75. These detectors were again CenterNet and Faster R-CNN, both with geometrical data augmentation. On the other hand, SSD and RetinaNet showed lower performances, but still maintained a significant improvement compared to their three-class problem counterpart. It is important to remark that, in the same way in which it happened in the three-class version of the problem, CenterNet made a larger difference when checking the mAP@75. This opens the question of whether the use of box centers instead of the box coordinates in the learning of the broccoli patterns is the only factor to boost localization accuracy. This could be discussed because Faster R-CNN, which is anchor-based, was also able to obtain high mAP@75 performances, but the variance of its results decreased the average severely.

One possible limitation of the presented work is the dataset size (288 images with 640 annotations) because it could be the cause of the overfitting. However, according to the results presented, the mAP@50 performances on the training and testing sets were quite similar. This means that, after 10 runs, the detectors did not overfit the training data, and they were able to generalize on the unseen test set. Additionally, the use of early stopping based on the difference between the training and validation performances reinforced the reduction of overfitting chances. On the other hand, these results could be seen as promising, but the real-world conditions could lead to more complex and diverse images. Therefore, this presented research is an initial experimental run of an open research subject, which is planned to be extended further (using larger data volumes and various learning techniques) in order to ensure its suitability in production.

Regarding a more technical aspect, the present study has been an opportunity to obtain knowledge regarding the proper planning and deployment of such experiments. First of all, during the orthomosaicing process, it is a common phenomenon that the flight lines around the perimeter of the flight plan, which often also reflect the boundaries of the experimental field, are the ones with the lowest values of overlap. The reason is that this area is only scanned once throughout the entire flight, such that the surrounding areas are only captured in the images of a single flight line. As a result, poor mosaicking quality can easily occur, and if most of the targets are placed in the perimeter, the entire mission is at risk of being abortive. On the other hand, the deployment of ground truth targets on the perimeter of the field is the easiest, as they are easily accessible, not covered by vegetation, and the perimeter's lower humidity level can potentially increase the time until the targets become soggy or covered by water droplets due to humidity, if they are not protected properly. Thus, the distribution of the ground truth targets is a factor that should always be considered. Therefore, the targets should be properly scattered towards the middle of the experimental area, and ideally across a large area. Additionally, in order to ensure their survivability in high-humidity conditions, waterproof materials should be used for the targets, and ideally should be placed on a slight incline in order to avoid the formation of droplets on their surface that would decrease their visibility.

Finally, as the timeframe during which data collection can be performed for this specific type of experiments is very strict, the utmost attention should be given to the minimization of as many risks as possible before committing to the field visit. One of the most unpredictable factors for all UAV missions is the weather. Certain UAVs have the capacity to perform flights under harsher conditions, while known thresholds are always a safety switch for the pilots to potentially cancel high-risk missions in time if they judge that the conditions do not allow for a safe flight. Weather forecasts can mitigate this risk to a certain extent by giving the pilots enough time to adjust/select the appropriate fleet for each mission based on the conditions they expect to encounter, although drastic changes—especially in wind speed and direction—are anything but rare.

5. Conclusions

The generated UAV dataset with the use of object-detection techniques has managed to automate the procedure of monitoring and detecting the maturity level of broccoli in open-field conditions. The implementation of the developed methodology can drastically reduce labour and increase the efficiency in scouting operations, while ensuring effective yield quality through optimal harvest timing, eliminating potential fungal infection and quality degradation issues. The results of the experiments have clearly indicated that the models were able to perform very well for the task of automated maturity detection. All of the experimental iterations maintained high mAP@50 and mAP@75 values of over 80% and 70%. The results showed that, in general, Faster R-CNN and CenterNet were the best broccoli maturity detectors. Moreover, geometrical transformations for data augmentations reported improvements, while colour distortions were counterproductive. Specifically, RetinaNet displayed a significant improvement in performance with the use of augmentations.

Finally, the utmost caution should be taken regarding the numerous parameters that are involved in the design of each flight plan, because they are decisive factors in the success of low-altitude missions. At the same time, the technical flight-related difficulties and limitations similar to the ones encountered during this experiment will hopefully serve the potential future researchers who will continue this research in the same domain, or transfer this knowledge to their respective field. In future experiments, we aim to collect additional imagery using both different acquisition parameters and sensing devices in order to validate and expand our approach. Furthermore, different machine learning techniques, such as semi-supervised learning approaches, capsule networks [44] and transfer learning from backbones previously trained on agricultural datasets (i.e., domain transfer) are already under experimentation in similar trials, while the integration of the entire pipeline in a real-time system is also currently being tested.

Author Contributions: Conceptualization, V.P., A.C. and S.F.; methodology, V.P., B.E.-G.; K.K. and S.F.; software, V.P., B.E.-G. and A.C.; validation, A.D., K.K. and S.F.; investigation, V.P., B.E.-G. and A.D.; resources, A.D.; data curation, A.C.; visualization, V.P, and B.E.-G.; supervision, K.K and S.F. All authors have read and agreed to the published version of the manuscript.

Funding: This research received no external funding.

Institutional Review Board Statement: Not applicable.

Informed Consent Statement: Not applicable.

Data Availability Statement: The data presented in this study are available on request from the corresponding author. The data are not publicly available due to privacy of the data collection location.

Acknowledgments: This work was supported by computational time granted from the National Infrastructures for Research and Technology S.A. (GRNET S.A.) in the National HPC facility-ARIS under project ID pr010012_gpu.

Conflicts of Interest: The authors declare no conflict of interest.

References

1. Latte, K.P.; Appel, K.E.; Lampen, A. Health benefits and possible risks of broccoli–an overview. *Food Chem. Toxicol.* **2011**, *49*, 3287–3309. [CrossRef] [PubMed]
2. Soane, B.D.; van Ouwerkerk, C. Chapter 1—Soil Compaction Problems in World Agriculture. In *Developments in Agricultural Engineering*; Elsevier: Amsterdam, The Netherlands, 1994; Volume 11, pp. 1–21. ISSN 0167-4137. ISBN 9780444882868. [CrossRef]
3. Bechar, A.; Vigneault, C. Agricultural robots for field operations: Concepts and components. *Biosyst. Eng.* **2016**, *149*, 94–111. [CrossRef]
4. Zhao, W.; Yamada, W.; Li, T.; Digman, M.; Runge, T. Augmenting Crop Detection for Precision Agriculture with Deep Visual Transfer Learning—A Case Study of Bale Detection. *Remote Sens.* **2020**, *13*, 23. [CrossRef]
5. Fan, Z.; Lu, J.; Gong, M.; Xie, H.; Goodman, E. Automatic Tobacco Plant Detection in UAV Images via Deep Neural Networks. *IEEE J. Sel. Top. Appl. Earth Obs. Remote Sens.* **2021**, *11*, 876–887. [CrossRef]

6. Felzenszwalb, P.F.; Girshick, R.B.; McAllester, D.; Ramanan, D. Object detection with discriminatively trained part based models. *IEEE Trans. Pattern Anal. Mach. Intell.* **2010**, *32*, 1627–1645. [CrossRef] [PubMed]

7. Girshick, R.B. Fast R-CNN. In Proceedings of the 2015 IEEE International Conference on Computer Vision (ICCV), Santiago, Chile, 7–13 December 2015. [CrossRef]

8. Ren, S.; He, K.; Girshick, R.B.; Sun, J. Faster R-CNN: Towards Real-Time Object Detection with Region Proposal Networks. *IEEE Trans. Pattern Anal. Mach. Intell.* **2015**, *39*, 1137–1149. [CrossRef] [PubMed]

9. Huang, J.; Rathod, V.; Sun, C.; Zhu, M.; Korattikara, A.; Fathi, A.; Fiscer, I.; Wojna, Z.; Song, Y.; Guadarrama, S.; et al. Speed/accuracy trade-offs for modern convolutional object detectors. In Proceedings of the IEEE Conference on Computer Vision and Pattern Recognition, Honolulu, HI, USA, 21–26 July 2017; pp. 7310–7311.

10. Liu, W.; Anguelov, D.; Erhan, D.; Szegedy, C.; Reed, S.E.; Fu, C.; Berg, A.C. SSD: Single Shot MultiBox Detector. In *Computer Vision—ECCV 2016*; Leibe, B., Matas, J., Sebe, N., Welling, M., Eds.; ECCV 2016; Lecture Notes in Computer Science; Springer: Cham, Switzerland, 2016; Volume 9905. [CrossRef]

11. Lin, G.S.; Tu, J.C.; Lin, J.Y. Keyword Detection Based on RetinaNet and Transfer Learning for Personal Information Protection in Document Images. *Appl. Sci.* **2021**, *11*, 9528. [CrossRef]

12. Oetomo, D.; Billingsley, J.; Reid, J.F. Agricultural robotics. *J. Field Robot.* **2009**, *26*, 501–503. [CrossRef]

13. Kirkpatrick, K. Technologizing Agriculture. In *Communications of the ACM*; Association for Computing Machinery: New York, NY, USA, 2019; Volume 62, pp. 14–16. [CrossRef]

14. Duckett, T.; Pearson, S.; Blackmore, S.; Grieve, B. Agricultural robotics: The future of robotic agriculture. CoRR, abs/1806.06762. *arXiv* **2018**, arXiv:1806.06762. Available online: http://arxiv.org/abs/1806.06762 (accessed on 20 December 2021).

15. Roser, M. Employment in agriculture. Our World in Data. 2019. Available online: https://ourworldindata.org/employment-in-agriculture (accessed on 20 December 2021).

16. Barbedo, J.G.A. Plant disease identification from individual lesions and spots using deep learning. *Biosyst. Eng.* **2019**, *180*, 96–107. [CrossRef]

17. Wilhoit, J.H.; Koslav, M.B.; Byler, R.K.; Vaughan, D.H. Broccoli head sizing using image texture analysis. *Trans. ASAE* **1990**, *33*, 1736–1740. [CrossRef]

18. Qui, W.; Shearer, S.A. Maturity assessment of broccoli using the discrete Fourier transform. *Trans. ASAE* **1992**, *35*, 2057–2062.

19. Shearer, S.A.; Burks, T.F.; Jones, P.T.; Qiu, W. One-dimensional image texture analysis for maturity assessment of broccoli. In Proceedings of the American Society of Agricultural Engineers, Kansas City, MO, USA, 19–22 June 1994.

20. Tu, K.; Ren, K.; Pan, L.; Li, H. A study of broccoli grading system based on machine vision and neural networks. In Proceedings of the 2007 International Conference on Mechatronics and Automation, Harbin, Heilongjiang, China, 5–8 August 2007; pp. 2332–2336.

21. LeCun, Y.; Bengio, Y.; Hinton, G. Deep learning. *Nature* **2015**, *521*, 436–444. [CrossRef] [PubMed]

22. He, K.; Zhang, X.; Ren, S.; Sun, J. Deep Residual Learning for Image Recognition. In Proceedings of the Conference on Computer Vision and Pattern Recognition (CVPR), Las Vegas, NV, USA, 27–30 June 2016. [CrossRef]

23. Bargoti, S.; Underwood, J. Deep fruit detection in orchards. In Proceedings of the 2017 IEEE International Conference on Robotics and Automation (ICRA), 29 May–3 June 2017; pp. 3626–3633. [CrossRef]

24. Sa, I.; Ge, Z.; Dayoub, F.; Upcroft, B.; Perez, T.; McCool, C. DeepFruits: A fruit detection system using deep neural networks. *Sensors* **2016**, *16*, 1222. [CrossRef] [PubMed]

25. Madeleine, S.; Bargoti, S.; Underwood, J. Image based mango fruit detection, localisation and yield estimation using multiple view geometry. *Sensors* **2016**, *16*, 1915.

26. García-Manso, A.; Gallardo-Caballero, R.; García-Orellana, C.J.; González-Velasco, H.M.; Macías-Macías, M. Towards selective and automatic harvesting of broccoli for agri-food industry. *Comput. Electron. Agric.* **2021**, *188*, 106263. [CrossRef]

27. Birrell, S.; Hughes, J.; Cai, J.Y.; Iida, F. A field-tested robotic harvesting system for iceberg lettuce. *J. Field Robot.* **2020**, *37*, 225–245. [CrossRef] [PubMed]

28. Junos, M.H.; Khairuddin, A.S.M.; Thannirmalai, S.; Dahari, M. Automatic detection of oil palm fruits from UAV images using an improved YOLO model. *Vis. Comput.* **2021**, 1–15. [CrossRef]

29. Mutha, S.A.; Shah, A.M.; Ahmed, M.Z. Maturity Detection of Tomatoes Using Deep Learning. *SN Comput. Sci.* **2021**, *2*, 441. [CrossRef]

30. Santos, T.T.; de Souza, L.L.; dos Santos, A.A.; Avila, S. Grape detection, segmentation, and tracking using deep neural networks and three-dimensional association. *Comput. Electron. Agric.* **2020**, *170*, 105247. [CrossRef]

31. Blok, P.M.; van Evert, F.K.; Tielen, A.P.; van Henten, E.J.; Kootstra, G. The effect of data augmentation and network simplification on the image-based detection of broccoli heads with Mask R-CNN. *J. Field Robot.* **2021**, *38*, 85–104. [CrossRef]

32. Le Louedec, J.; Montes, H.A.; Duckett, T.; Cielniak, G. Segmentation and detection from organised 3D point clouds: A case study in broccoli head detection. In Proceedings of the IEEE/CVF Conference on Computer Vision and Pattern Recognition Workshops, Seattle, WA, USA, 14–19 June 2020; pp. 64–65.

33. Bender, A.; Whelan, B.; Sukkarieh, S. A high-resolution, multimodal data set for agricultural robotics: A Ladybird's-eye view of Brassica. *J. Field Robot.* **2020**, *37*, 73–96. [CrossRef]

34. Zhou, C.; Hu, J.; Xu, Z.; Yue, J.; Ye, H.; Yang, G. A monitoring system for the segmentation and grading of broccoli head based on deep learning and neural networks. *Front. Plant Sci.* **2020**, *11*, 402. [CrossRef] [PubMed]

35. Everingham, M.; Gool, L.V.; Williams, C.K.; Winn, J.M.; Zisserman, A. The Pascal Visual Object Classes (VOC) Challenge. *Int. J. Comput. Vis.* **2009**, *88*, 303–338. [CrossRef]
36. Arsenovic, M.; Karanovic, M.; Sladojevic, S.; Anderla, A.; Stefanović, D. Solving Current Limitations of Deep Learning Based Approaches for Plant Disease Detection. *Symmetry* **2019**, *11*, 939. [CrossRef]
37. Zheng, Y.; Kong, J.; Jin, X.; Wang, X.; Su, T.; Zuo, M. CropDeep: The Crop Vision Dataset for Deep-Learning-Based Classification and Detection in Precision Agriculture. *Sensors* **2019**, *19*, 1058. [CrossRef]
38. Zhou, X.; Wang, D.; Krähenbühl, P. Objects as Points. *arXiv* **2019**. Available online: https://arxiv.org/abs/1904.07850 (accessed on 20 December 2021).
39. Tan, M.; Pang, R.; Le, Q.V. EfficientDet: Scalable and Efficient Object Detection. In Proceedings of the 2020 IEEE/CVF Conference on Computer Vision and Pattern Recognition (CVPR), Seattle, WA, USA, 13–19 June 2020; pp. 10778–10787.
40. Newell, A.; Yang, K.; Deng, J. Stacked Hourglass Networks for Human Pose Estimation. In Proceedings of the European Conference on Computer Vision (ECCV), Amsterdam, The Netherlands, 8–16 October 2016.
41. Sandler, M.; Howard, A.G.; Zhu, M.; Zhmoginov, A.; Chen, L. MobileNetV2: Inverted Residuals and Linear Bottlenecks. In Proceedings of the 2018 IEEE/CVF Conference on Computer Vision and Pattern Recognition, Salt Lake City, UT, USA, 18–23 June 2018; pp. 4510–4520.
42. Lin, T.; Maire, M.; Belongie, S.J.; Hays, J.; Perona, P.; Ramanan, D.; Dollár, P.; Zitnick, C.L. Microsoft COCO: Common Objects in Context. In Proceedings of the European Conference on Computer Vision (ECCV), Zurich, Switzerland, 5–12 September 2014.
43. Taylor, L.; Nitschke, G.S. Improving Deep Learning with Generic Data Augmentation. In Proceedings of the 2018 IEEE Symposium Series on Computational Intelligence (SSCI), Bengaluru, India, 18–21 November 2018; pp. 1542–1547.
44. Sabour, S.; Frosst, N.; Hinton, G.E. Dynamic Routing Between Capsules. *arXiv*. 2017. Available online: https://arxiv.org/abs/1710.09829 (accessed on 20 December 2021).

remote sensing

Article

Identification and Severity Monitoring of Maize Dwarf Mosaic Virus Infection Based on Hyperspectral Measurements

Lili Luo, Qingrui Chang *, Qi Wang and Yong Huang

College of Nature Resources and Environment, Northwest A&F University, Yangling 712100, China; lililuo@nwsuaf.edu.cn (L.L.); wangqieducation@nwsuaf.edu.cn (Q.W.); huangyong19920808@126.com (Y.H.)
* Correspondence: changqr@nwsuaf.edu.cn; Tel.: +86-1357-183-5969

Abstract: Prompt monitoring of maize dwarf mosaic virus (MDMV) is critical for the prevention and control of disease and to ensure high crop yield and quality. Here, we first analyzed the spectral differences between MDMV-infected red leaves and healthy leaves and constructed a sensitive index (SI) for measurements. Next, based on the characteristic bands (R_λ) associated with leaf anthocyanins (Anth), we determined vegetation indices (VI_s) commonly used in plant physiological and biochemical parameter inversion and established a vegetation index (VI_c) by utilizing the combination of two arbitrary bands following the construction principles of NDVI, DVI, RVI, and SAVI. Furthermore, we developed classification models based on linear discriminant analysis (LDA) and support vector machine (SVM) in order to distinguish the red leaves from healthy leaves. Finally, we performed UR, MLR, PLSR, PCR, and SVM simulations on Anth based on R_λ, VI_s, VI_c, and $R_\lambda + VI_s + VI_c$ and indirectly estimated the severity of MDMV infection based on the relationship between the reflection spectra and Anth. Distinct from those of the normal leaves, the spectra of red leaves showed strong reflectance characteristics at 640 nm, and SI increased with increasing Anth. Moreover, the accuracy of the two VI_c-based classification models was 100%, which is significantly higher than that of the VI_s and R_λ-based models. Among the Anth regression models, the accuracy of the MLR model based on $R_\lambda + VI_s + VI_c$ was the highest ($R^2_c = 0.85$; $R^2_v = 0.74$). The developed models could accurately identify MDMV and estimate the severity of its infection, laying the theoretical foundation for large-scale remote sensing-based monitoring of this virus in the future.

Keywords: plant disease; band selection; machine learning; anthocyanin; hyperspectral reflectance; linear discriminant analysis; precision crop protection

Citation: Luo, L.; Chang, Q.; Wang, Q.; Huang, Y. Identification and Severity Monitoring of Maize Dwarf Mosaic Virus Infection Based on Hyperspectral Measurements. *Remote Sens.* **2021**, *13*, 4560. https://doi.org/10.3390/rs13224560

Academic Editor: Xanthoula Eirini Pantazi

Received: 3 October 2021
Accepted: 9 November 2021
Published: 13 November 2021

Publisher's Note: MDPI stays neutral with regard to jurisdictional claims in published maps and institutional affiliations.

1. Introduction

As one of the most important food crops in the world, maize covered the largest area planted in China from 2015 to 2019, with more than 40 million hectares planted each year [1]. Therefore, guaranteeing healthy growths of maize is crucial for ensuring food security and achieving sustainable agricultural development worldwide [2]. However, maize dwarf mosaic virus (MDMV) infection adversely affects the growth and development of this crop, and it has become one of the major destructive diseases in the world, including China [3,4]. In 1962, MDMV was first detected in Ohio, USA, and had spread throughout the state by 1964, damaging 5 million corn plants in a dozen counties [5]. In 1965, Janson named the pathogen "maize dwarf mosaic virus" [6]. In 1968, MDMV was reported for the first time on a large scale in Xinxiang, Huixian, and other regions in the Henan Province of China, resulting in the loss of nearly 25 million kilograms of grain. In the 1980s, the disease was effectively prevented and controlled thanks to the promotion of specific resistant varieties and agronomic cultivation measures. However, since the 1990s, due to the increased acreage of MDMV-susceptible varieties, MDMV has become prevalent once again, occurring very frequently in China and causing substantial crop losses. Therefore, prompt and accurate identification of MDMV is crucial for proper field management in order to prevent and control disease spread, and regular yield assessments are essential to devise marketing plans [7–9].

MDMV infection can occur throughout the maize growth period. At the early stages of disease, many elliptical chlorotic spots or markings appear near the veins at the base of the heart lobe, arranged along the veins into intermittent strips of varying lengths. With further progression of the disease, wide chlorotic stripes formed on the leaves, particularly on the young ones [2]. After contracting the disease, the chlorophyll (Chl) content of leaves reduces, turning them yellow. In some cases, the symptoms start to develop from the tip and edge of the leaves, appearing as red-purple stripes; eventually, the entire leaf becomes red and dries. However, these red-purple streaks are mainly a result of color rendering by high concentrations of anthocyanins (Anth) in the infected leaves following Chl degradation [10,11]. As an important pigment, Anth confers all colors, except green, in plants and is sensitive to environmental and biological stresses [12–15]. In addition, pathogen infection can induce anthocyanin biosynthesis in plants, and the more severe the pathogen infection, the stronger the induction ability [16]. Ludmerszki et al. observed the fourth leaf of MDMV-infected maize and found that the observed leaf gradually turned red over time and that the content of anthocyanin in MDMV-infected leaves increased while the content of chlorophyll decreased by using fluorescence technology [17]. Singh and Sharma reported in 1998 that anthocyanins and phenols increase *Chkahao* resistance to a variety of common rice diseases (e.g., root rot, narrow brown spot, stem rot, false smut, bacterial leaf streaks, and bacterial leaf blight) and rice pests (e.g., stem borer, rice bug, green horned caterpillar, and rice skipper) [18]. Fasahat et al. reported in 2012 that a Malaysian colored rice, *Oryza Rufipogon*, containing anthocyanin pigment was highly resistant to bacterial leaf blight and brown plant hopper [19]. Therefore, we used red leaf Anth as a measure to indicate the severity of MDMV infection given the close association between Anth and plant disease.

Traditional MDMV monitoring methods include field observations and laboratory measurements, which are time-consuming and expensive. In addition, chemical methods are destructive, and they cannot reflect progressive changes in the disease status in the same leaf over time [20]. In contrast, remote sensing (RS) is a non-destructive technique for rapid monitoring at different scales. Moreover, owing to its high spatial resolution, hyperspectral technology can identify invisible symptoms reflecting the physiological status of plants at the initial stages of disease and has been widely used in crop pest and disease detection in recent years [21–23]. For instance, Mirik et al. classified Landsat 5 Thematic Mapper (TM) images of two cities in Texas from 2006 to 2008 by using the maximum likelihood method and achieved an overall classification accuracy of 89.47–99.07% for wheat streak mosaic virus [24]. Furthermore, Camino et al. coupled a spatial spread model with an RS-driven support vector for estimating the probability of *Xylella fastidiosa* (XF) infection and obtained highly accurate predictions of the spatial distribution of plant disease in almond trees [25]. Martins et al. conducted field surveys and small-format aerial photography (SFAP) with different cameras to obtain visible and near-infrared images and estimated the spatial distribution of ink disease in Northern Portugal during 1995–2004 by using a geostatistical method [26]. Liu et al. collected rice reflectance spectra in the field and laboratory and estimated the severity of brown rice spot disease based on the reflectance ratio [27]. These previous studies monitored plant diseases at different scales based on satellite, unmanned aerial vehicle (UAV), and near-ground platforms, achieving satisfactory results. Simultaneously, other plant diseases, including wheat stripe rust, rice spikelet rot disease, and maize stripe rust, have also been studied by using RS technology [28–30]. However, only Beverly et al. compared the spectra (400–2700) of healthy, MDMV-infected, and *Helminthosporium maydis*-infected maize leaves, laying the foundation for RS-based research on MDMV [31], and to our best knowledge, there have been no further studies on MDMV using RS technology. Therefore, RS-based research on MDMV is of paramount importance.

Furthermore, reliable physical and empirical models of leaf reflectance and a range of physiological parameters, such as leaf water content, leaf area, and pigment content (including Anth), have been established [32,33]. Therefore, we can indirectly assess the severity of MDMV infection by establishing the relationship between the reflectance spectrum and Anth. To this end, by analyzing the spectral characteristics of infected leaves through RS, we aimed to build a suitable model for the detection and monitoring of MDMV for minimizing its adverse effects and ensuring high crop quality and yield.

2. Materials and Methods

2.1. Study Area

The present study was conducted in an established research facility, which is part of the Northwest A&F University, located in the Shaanxi Province of China ($34°15'-34°20'$ N, $107°56'-108°7'$ E; 460 m a.s.l.). The region has warm, temperate, and continental monsoon climate. The study area was divided into 20 subplots, each with an area of 5.5×6 m^2, occupying a total area of 69×12 m^2. The maize cultivar "Dafeng 26" used in the experiment, and the cropping system was the rotation of winter wheat and summer maize. Five nitrogen (0, 45, 90, 135, and 180 kg N·ha^{-1}) and five phosphorus (0, 30, 60, 90, and 120 kg P$_2$O$_5$·ha^{-1}) application levels were set in the subplots. All fertilizers were applied at the time of sowing, and other management measures followed local practices. The location of the study area, plot setting and fertilization are shown in Figure 1.

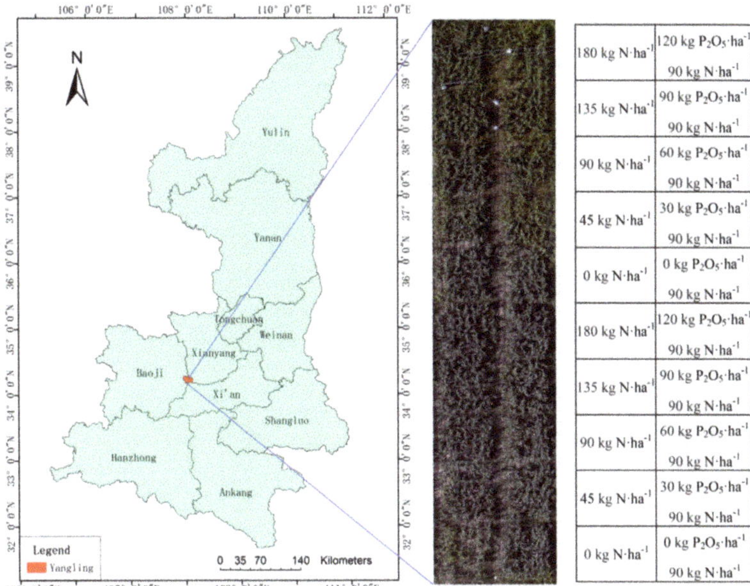

Figure 1. Location of study area and fertilization of plot.

2.2. Data Acquisition and Processing

Data collection and measurements were performed at the R2 blister stage period (14 September 2017); it is not only the peak season of MDMV but also an important period for yield estimation. Two healthy corn samples were collected from each plot, and destructive samples were obtained from the upper, middle, and lower layers (three pieces from each layer). Moreover, 360 healthy leaves and 72 MDMV-infected red leaves (hereinafter referred to as red leaves) were collected throughout the study area. All blades were placed in plastic bags and transported to the laboratory in an incubator. The collected red leaf samples are presented in Figure 2.

Figure 2. Maize plants and leaves infected with MDMV.

2.2.1. Anth Quantification

Leaf Anth was quantified with Dualex 4 (France), a new multifunctional blade measurement tool that calculates the absorbance of plants in the green region to obtain Anth ($\mu g \cdot cm^{-2}$). This tool can also accurately measure leaf Chl, surface flavonoid, and nitrogen content and is easy to use for real-time and non-destructive measurements. Each leaf was measured 10 times to obtain the representative value of Anth. The Anth measurement of healthy leaves was not required, except in the samples from the leaf vein; in infected leaves, Anth was only measured in parts that had turned red.

2.2.2. Hyperspectral Data Acquisition

The reflectance spectra of maize leaves were determined in an indoor measurement unit using a spectroradiometer (SVC HR~1024 i) with a spectral range of 350–2500 nm and a view of 25°. First, the reflectivity of the white reference plate was measured for spectral correction. Then, the leaves were placed into the clamp to measure 10 spectra of each leaf, and the average value was calculated as the actual spectrum. Finally, the spectral resolution was resampled to 1 nm and then a continuous smooth reflection spectrum was obtained by Savitzky–Golay smoothing. Considering that the wavelengths of plant pigments are concentrated in the visible and near-infrared regions and that the constituent wavelengths of vegetation indices commonly used for estimating plant physiological and biochemical parameters are within these region [34], we only studied the spectra at 400–1000 nm.

2.3. Analytical Methods

2.3.1. Definition of Sensitivity Index (SI)

We constructed an *SI* based on the ratio and difference algorithm to measure the spectral difference between healthy and red leaves. The closer *SI* is to 0, the smaller the spectral difference between the two leaves and vice versa. The specific formula is as follows:

$$SI = \frac{R_\lambda - R_h}{R_h} \tag{1}$$

where R_λ is the reflectance of the red leaf spectrum at wavelength λ, and R_h is the average reflectance of the healthy spectrum at wavelength λ.

2.3.2. Construction of Vegetation Indices with Two Arbitrary Bands

Vegetation indices have been widely used in the RS-based estimation of plant growth parameters. A VI constituting several bands can effectively minimize the errors associated with sensor specifications, atmosphere, and background differences, thus enhancing the description of the observation target [35,36]. However, due to the lack of disease specificity of these indicators, the quantification or identification of specific diseases based on common vegetation indices is currently impossible. Therefore, we combined different wavelengths to construct a VI (VI_c) for simplifying the spectral detection of plant diseases. Specifically, we used the normalized difference vegetation index (*NDVI*), ratio vegetation index (*RVI*), differential vegetation index (*DVI*), and soil-adjusted vegetation index (*SAVI*), defined as follows:

$$NDVI = (R_i - R_j) / (R_i + R_j) \tag{2}$$

$$RVI = R_i / R_j \tag{3}$$

$$DVI = R_i - R_j \tag{4}$$

$$SAVI = 1.5(R_i - R_j) / (R_i + R_j + 0.5) \tag{5}$$

where R_i and R_j are the reflectance at i and j nm over the entire reflectance spectrum.

2.3.3. Linear Discriminant Analysis (LDA) Classification Model

The LDA model uses a linear combination of features as the classification standard to project data from the higher-dimensional to the lower-dimensional space, while ensuring that the intra-class variance of each class after the projection is small but the mean difference between the classes is large. It can be used for both classification and dimension reduction and is better suited for the linear classification of smaller data volumes and fewer indicators. In the present study, 48 red leaf spectra and 240 healthy leaf spectra were randomly selected from all the spectra as the calibration set, and the remaining 24 red leaf spectra and 120 healthy leaf spectra were used as the validation set. The LDA model was built using The Unscrambler X 10.4.

2.3.4. Support Vector Machine (SVM) Classification Model

In principle, SVM uses a kernel function to project the spectral information of samples in higher-dimensional space, constructs the hyperplane with the largest classification interval, and then accurately identifies different types of samples. It is better suited for linearly indivisible sample data, specifically using relaxation variables and kernel functions. In the present study, C-SVM discriminant analysis was performed on data from the calibration and validation sets using The Unscrambler X 10.4. The kernel type was a radial basis function (RBF), and the penalty coefficient (C) was 1. We adopted a 10-fold cross validation during modeling in order to improve the stability of the classification model.

2.3.5. Regression Models

In general, hyperspectral data can be used to explore specific wavelengths and/or indices that are particularly useful for the assessment of plant and ecosystem variables [37]. These wavelengths and/or indices can be used to estimate plant variables based on multivariable statistical methods, such as multiple linear regression (MLR), principal component regression (PCR), partial least squares regression (PLSR), and SVM regression (SVMR) [38,39]. When introducing new independent variables into an MLR model, a collinearity check is essential on the substituted independent variables, and until then, no new variables can be introduced or no existing ones can be removed. The equation is simple and can effectively avoid collinearity between variables. PCR combines multiple features in a high-dimensional space into a few irrelevant principal components and contains most of the variation information in the original data, effectively reducing the amount of data and simplifying operations. PLSR combines the characteristics of MLR, canonical correlation analysis (CCA), and PCR; can avoid the multicollinearity problem; and offers advantages that classic regression methods do not. SVMR uses an inner product kernel function to replace nonlinear mapping in the higher-dimensional space, but it is still better suited when the feature dimension is larger than the sample number. Moreover, SVMR, based on the principle of minimizing the structural risk, avoids overlearning problems and shows high generalizability.

2.4. Evaluation of Precision

In order to compare the predictive performance of various spectral parameters and methods, we used statistical indicators, such as the coefficient of determination (R^2), root mean square error ($RMSE$), and relative error of prediction (REP), defined as follows:

$$R^2 = \sum_{i=1}^{n} (\hat{y}_i - \overline{y})^2 / \sum_{i=1}^{n} (y_i - \overline{y})^2 \tag{6}$$

$$RMSE = \sqrt{\sum_{i=1}^{n} (y_i - \hat{y}_i)^2 / n} \tag{7}$$

$$REP = \frac{100RMSE}{\overline{y}} \tag{8}$$

where y_i is the measured values, $\overline{y}$ is the average of the measured values, $\hat{y}_i$ is the predicted value, and n is the number of samples. The closer the value of R^2 is to 1, the smaller $RMSE$ and REP are, and the better the model accuracy. We applied 20-fold cross validation to assess the robustness of the estimation models.

2.5. Analytical Framework

Figure 3 presents the data, methods, and processing steps for the identification of MDMV and severity monitoring of its infection. First, the spectral differences between the healthy and infected leaves with different degrees of reddening were analyzed in order to explore the spectral response characteristics of MDMV-infected leaves, which were the basis for further analyses in the present study. Then, the LDA and SVM classification models for MDMV were constructed based on single-band spectral information (R_λ), two- or three-band VIs, and full-spectrum arbitrary two-band combined vegetation index (VI_c). Finally, high-precision Anth estimation models for healthy and red leaves were constructed based on three types of spectral parameters. The details are provided in Results.

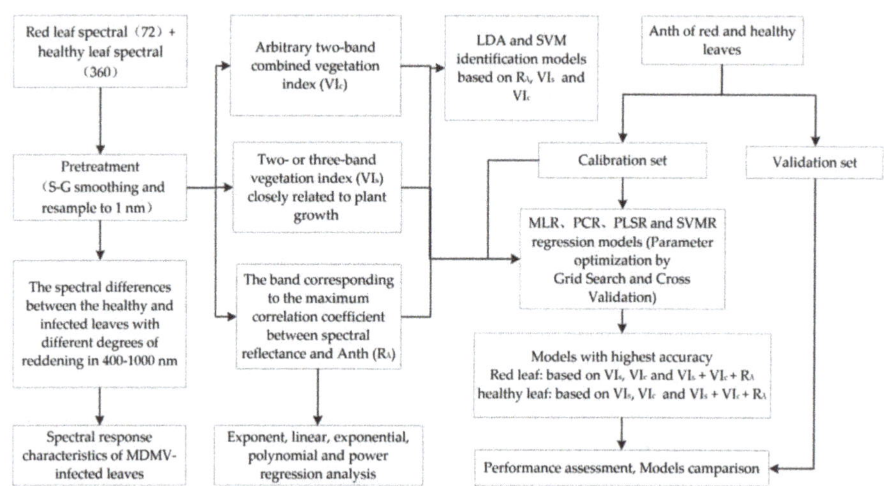

Figure 3. Analysis process of this study.

3. Results

3.1. Leaf Anth Statistics

The results of the statistical analysis are reported in Table 1. The range of Anth in red leaves was 0.04–0.76 µg·cm^{-2}, which was wider than that in healthy leaves (0.03–0.11 µg·cm^{-2}), although the Anth of two leaves overlapped. The average Anth of red leaves (0.19 µg·cm^{-2}) was much higher than that of healthy leaves (0.06 µg·cm^{-2}). SD, variance, and CV of healthy leaves were small, their Anth distribution was concentrated, and the spatial variability in values was low. In contrast, SD, variance, and CV of red leaves were large, their Anth distribution was scattered, and the spatial variability in values was high.

Table 1. Anth statistics of healthy and red leaves.

Type	N	Min (µg·cm^{-2})	Max (µg·cm^{-2})	Mean (µg·cm^{-2})	SD	Variance	CV (%)
healthy	360	0.03	0.11	0.06	0.02	3.24×10^{-4}	29.36
red	72	0.04	0.76	0.19	0.18	0.03	94.74

3.2. Characteristics of Reflectance Spectra

The spectral characteristics of plants are affected by their internal tissue structure, biochemical composition, and morphological features, which together constitute the biophysical and biochemical responses of plants to light. Figure 4a shows the spectra of the red and healthy leaves and SI of red leaves. Evidently, the variation in the spectral reflectance of healthy leaves is smaller than that of red leaves. In the 580–670 nm range, a new reflection peak appeared in the red leaf spectra, which was rather different from that in the healthy leaf spectra. The large SI values were mainly distributed in the 595–764 nm regions, and SI gradually increased with increases in Anth, reaching the maximum value exceeding three. At Anth values of 0.28, 0.58, and 0.68, SI showed obvious bimodal characteristics, and when the Anth value reached 0.75, the bimodal characteristic disappeared. The band corresponding to the maximum SI value appeared near 700 nm.

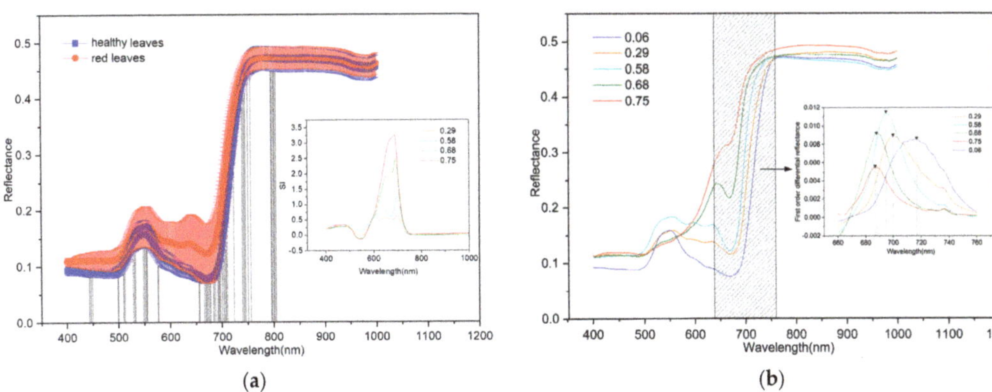

Figure 4. (a) Spectra of healthy and red leaves and SI of red leaves; (b) spectra and red edge of red leaves with different Anth.

Selected spectral characteristics of infected leaves with different degrees of reddening (Anth = 0.29, 0.58, 0.68, and 0.75 µg·cm^{-2}) compared with the spectral characteristics of healthy leaves (Anth = 0.06 µg·cm^{-2}) are presented in Figure 4b. The differences were mainly concentrated in the visible and near-infrared ranges (400–750 nm). The spectral curve of healthy leaves (Anth = 0.06 µg·cm^{-2}) showed an obvious absorption valley in the blue band at 450 nm and in the red band at 680 nm, as well as a strong reflection peak in the green band at 550 nm, which is consistent with the typical reflectance spectral characteristics of green plants. However, the spectral curve of red leaves exhibited obvious bimodal characteristics in the visible range (380–760 nm), with the left peak near 550 nm and the right peak near 640 nm. At Anth values of 0.29 and 0.58 µg·cm^{-2}, the reflectance of the left peak was higher than that of the right peak. However, at higher Anth values of 0.68 and 0.75 µg·cm^{-2}, the left peak disappeared, and the right peak increased sharply, forming a characteristic that is obviously different from the spectrum of healthy leaves. With the increase in Anth, the reflectance of the right peak continued to increase, whereas the absorption valley at 680 nm weakened until disappearing. The first-order differential of the original spectrum in the 640–760 nm range was calculated in Figure 4b. With an increase in Anth, the position of the red edge (wavelength corresponding to the maximum value of the first-order differential spectrum) moved in the short-wave direction; this trend is similar to that observed in cotton with different degrees of aphid damage [40]. As such, the higher the Anth is, the more severe the MDMV infection is, resulting in poor photosynthesis and low consumption of long-wave photons.

3.3. Correlation between Anth and Spectral Reflectance

The correlation between spectral reflectance and Anth of the two types of leaves was examined, as shown in Figure 5. The spectral reflectance of red leaves was negatively correlated with Anth in the 520–563 nm region but positively correlated with Anth in the 400–519 and 564–1000 nm regions. In the 400–449 and 587–753 nm regions, there was a strong correlation (P< 0.01), with the maximum correlation coefficient of 0.76 at 695 nm. The spectral reflectance of healthy leaves was positively correlated with Anth in the 400–1000 nm, and all correlations were strong (P< 0.01), with the maximum correlation coefficient of 0.68 at 554 nm.

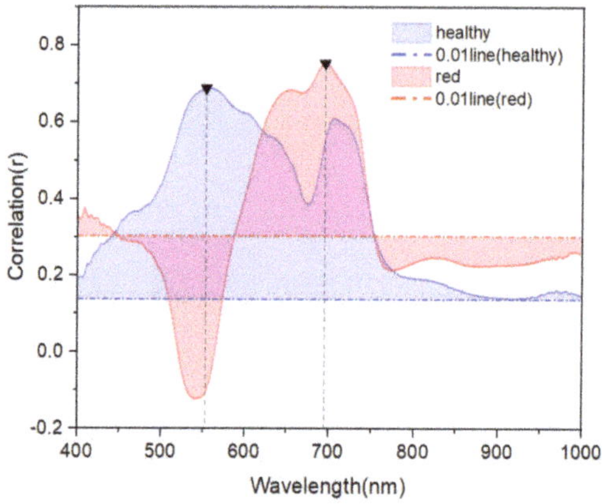

Figure 5. Correlation between Anth and spectral reflectance of healthy and red leaves.

In the 444–620 nm region, the correlation between the reflectance and Anth was stronger in healthy leaves than in red leaves, due to the strong reflection of Chl in this region, while anthocyanins had little effect on the reflectance of the spectrum in this region. The maximum difference in the correlation coefficient was recorded near 550 nm, coinciding with the position of the reflection peak of the healthy leaf spectrum shown in Figure 4b. In the 400–443, 621–748 and 776–1000 nm regions, the correlation between reflectance and Anth was stronger in red leaves than in healthy leaves due to the strong reflection of anthocyanins in this region. The maximum difference in the correlation coefficient was recorded near 680 nm, coinciding with the position of the right reflection peak of the red leaf spectrum shown in Figure 4b.

3.4. Vegetation Indices

3.4.1. Various Vegetation Indices

The vegetation index can reduce the influence of sensors and environment on the target through normalization and derivative processing as well as by improving data utilization efficiency through the use of several selected bands [41]. Numerous vegetation indices have been constructed to estimate the biophysical and biochemical properties of plants. In the present study, 12 vegetation indices with good correlation to Anth were selected, as shown in Table 2. Seven two-band vegetation indices and five three-band vegetation indices were included. The bands used in the vegetation indices are marked in Figure 4a. Specifically, the selected bands were mainly concentrated around 500–575, 670–725, 750, and 800 nm.

Table 2. Two-band and three-band vegetation indices.

VI_s	Equations	Bands	References	VI_s	Equations	Band	References
SR	R_{800}/R_{680}	2	[42]	SIPI	$(R_{680} - R_{500})/R_{750}$	3	[43]
NDVI	$(R_{800} - R_{670})/(R_{800} + R_{670})$	2	[44]	GNDVI	$(R_{800} - R_{550})/(R_{800} + R_{550})$	2	[45]
TVI	$0.5(120(R_{750} - R_{550}) - 200(R_{670} - R_{550}))$	3	[46]	MCARI	$R_{700}((R_{700} - R_{670}) - 0.2(R_{700} - R_{550}))/R_{670}$	3	[47]
CHLI	$R_{700}/(R_{700} + R_{710} - 1)$	2	[48]	MCARI1	$1.2(2.5(R_{800} - R_{670}) - 1.3(R_{800} - R_{550}))$	3	[49]
VRI	R_{740}/R_{720}	2	[50]	PSRI	$(R_{680} - R_{500})/R_{750}$	3	[51]
PRI	$(R_{531} - R_{570})/(R_{531} + R_{570})$	2	[52]	RNDVI	$(R_{750} - R_{705})/(R_{750} + R_{705})$	2	[53]

The correlations between the vegetation indices and Anth are shown in Figure 6. Among the two-band VI_s, CHLI (700,710), VRI (740,720), and RNDVI (750,705) showed a strong correlation with the Anth of healthy and red leaves. Among the three-band vegetation indices, PSRI was strongly correlated with the Anth of red leaves, and MCARI was strongly correlated with the Anth of healthy leaves. There were significant correlations between the vegetation indices and Anth in healthy leaves (P< 0.01), with GNDVI achieving the highest correlation coefficient of −0.73. In red leaves, CHLI achieves the highest correlation coefficient of −0.75.

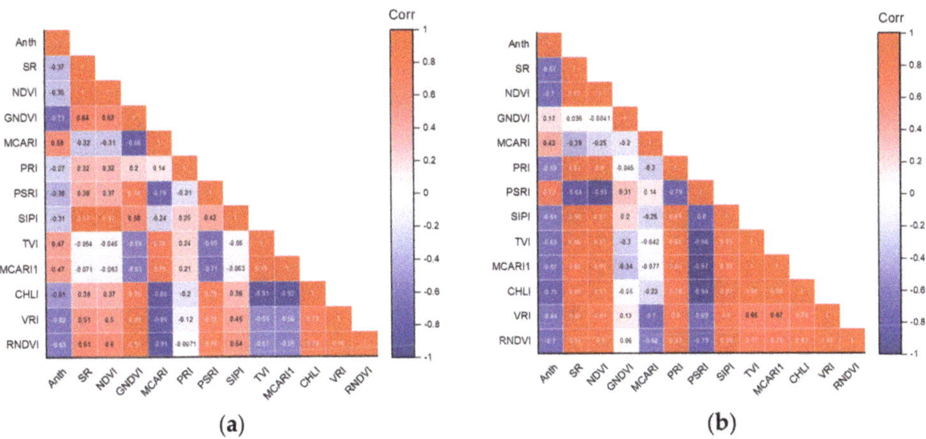

Figure 6. (a) Correlation analysis of healthy leaf VI_s; (b) correlation analysis of red leaf VI_s.

3.4.2. VI_c Based on Two Arbitrary Bands

The coefficients of determination (R^2) between plant-specific variables, and VI_c can reflect the predictive power of two independent band combinations. Figure 7 shows the contour maps of R^2 between Anth and NDVI, RVI, DVI, and SAVI using all combinations of two wavebands at i and j nm, which are very useful for selecting effective bandwidths and various combinations of wavelengths. For healthy leaves, the most significant areas noted around $NDVI_h$, RVI_h, DVI_h, and $SAVI_h$ were (R_{573}, R_{507}), (R_{572}, R_{507}), (R_{547}, R_{522}), and (R_{547}, R_{518}), respectively, and DVI_h was the most significant VI_c with an R^2 of 0.65. The contour map of $NDVI_h$ was very similar to that of RVI_h, and their significant bands and maximum R^2 values were close. For red leaves, the most significant areas noted around $NDVI_r$, RVI_r, DVI_r, and $SAVI_r$ were (R_{689}, R_{461}), (R_{460}, R_{692}), (R_{690}, R_{656}), and (R_{685}, R_{667}), respectively, and DVI_r was the most significant VI_c with an R^2 of 0.68. With the exception of RVI_h and RVI_r, the other vegetation indices showed symmetric spatial distribution patterns of R^2.

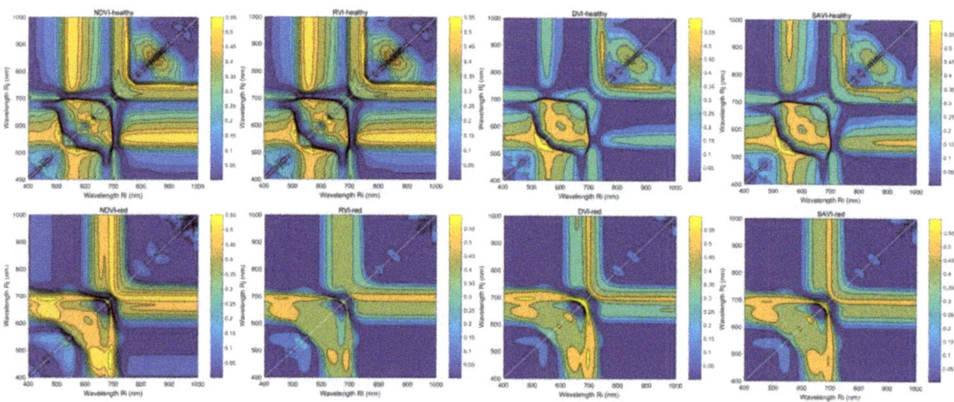

Figure 7. Contour maps of R^2 between Anth and VI_c of healthy and red leaves.

3.5. MDMV Identification

The results of the LDA and SVM classification models constructed based on R_λ, VI_s, and VI_c are shown in Table 3. All models could identify healthy leaves, and the models primarily differed in terms of their identification accuracy of red (i.e., infected) leaves. Both LDA and SVM classification models based on R_λ showed low accuracy, and the number of red leaves identified by the LDA-R_λ model was even lower than zero. Therefore, R_λ is not suitable for the RS-based identification of MDMV-infected leaves. Based on VI_s, the SVM model performed better than the LDA model, and the recognition accuracy of the calibration and validation sets exceeded 75% for the former. This is because the SVM algorithm has high classification accuracy and generalizability when the sample size is small, and it performs better in solving classification problems with high-dimensional features [54,55]. In addition, the RBF kernel function used by the SVM algorithm can be adjusted by using a grid search to obtain a better classification model [56,57]. The accuracy of both LDA and SVM classification models based on VI_c was 100%, which is significantly higher than that of models based on R_λ and VI_s and shows obvious superiority in the RS-based detection of MDMV. This is because the input parameter VI_c is constructed based on the spectrum of normal leaf and red leaf at any two bands in the 400–1000 nm region, which has its uniqueness. Moreover, in the VI_c contour maps of the red leaves, the most significant areas noted around $NDVI_h$, RVI_h, DVI_h, and $SAVI_h$ were (R573, R507), (R572, R507), (R547, R522), and (R547, R518). In the VI_c contour maps of the healthy leaves, the most significant areas around $NDVI_r$, RVI_r, DVI_r, and $SAVI_r$ were (R689, R461), (R460, R692), (R690, R656), and (R685, R667). These bands are located in the wavelength region corresponding to the high SI value, and there is a large difference between the two leaves in this region. In addition, the difference was further enhanced by using difference and ratio calculation to construct the vegetation index, thus achieving the high-precision identification of MDMV-infected and healthy leaves.

Table 3. LDA and SVM classification models.

Models	Parameters	Calibration Set ($n_r = 48$, $n_h = 240$)			Validation Set ($n_r = 24$, $n_h = 120$)		
		Red	Healthy	Accuracy/%	Red	Healthy	Accuracy/%
LDA	R_λ	6	240	56.3	6	120	62.5
	VI_s	34	240	85.4	12	120	75.0
	VI_c	48	240	100.0	24	120	100.0
SVM	R_λ	0	240	50.0	0	120	50.0
	VI_s	36	240	87.5	16	120	83.3
	VI_c	48	240	100.0	24	120	100.0

Note: R_λ is the spectral reflectance of red leaves at 695 nm and healthy leaves at 554 nm; VI_s is the narrow-band vegetation indices; VI_c is the vegetation index constructed based on two arbitrary bands; n_r and n_h are the number of samples of red and healthy leaves, respectively.

3.6. Classic Regression Analysis Based on a Sensitive Band

In the present study, random stratified sampling was used to divide the dataset at a ratio of 2:1 for obtaining representative samples for calibration and validation. The statistics of the calibration and validation datasets are shown in Figure 8a. The statistics of the calibration and validation datasets were similar for red leaves (max = 0.76 and 0.73; min = 0.05 and 0.04; mean = 0.20 and 0.18; SD = 0.19% and 0.17%; and CV = 96.48% and 95.60%, respectively). For healthy leaves, the maximum values of the calibration and validation datasets were 0.11 and 0.10, respectively, and the remaining statistical parameters were identical between the two datasets. Furthermore, the distribution of calibration and validation data was consistent with that of all data.

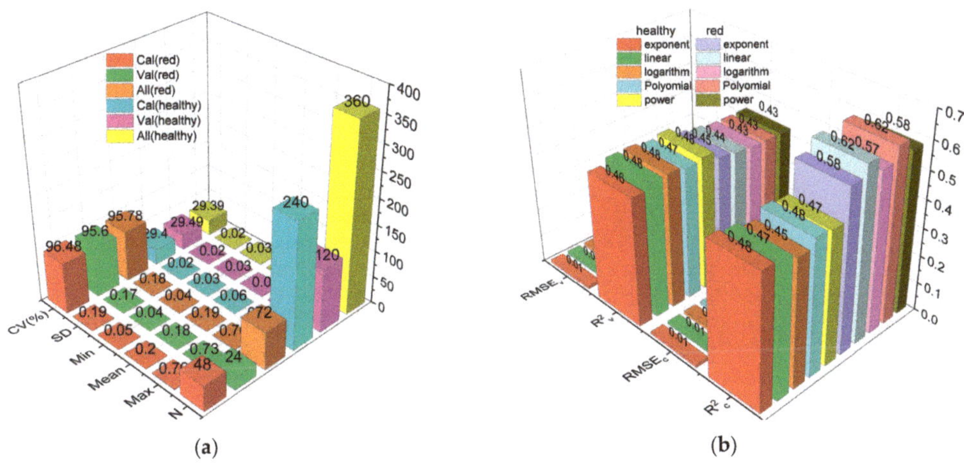

(a)   (b)

Figure 8. (**a**) Statistical analysis of calibration set and validation set; (**b**) precision comparison of UR models based on R_λ.

Exponent, linear, exponential, polynomial, and power regression models for Anth were built based on the spectral reflectance of red leaves at 665 nm (R_{695}) and healthy leaves at 554 nm (R_{554}). The R^2_c, R^2_v, $RMSE_c$, and $RMSE_v$ for each model are shown in Figure 8b. For the same sample, while the differences in $RMSE_c$ and $RMSE_v$ were small, R^2_c and R^2_v were significantly different among the various models. Among the models for healthy leaves, the exponential and polynomial models produced reliable results ($R^2_c = 0.48$, $R^2_v = 0.46$). Among the models for red leaves, the linear model produced the most reliable results ($R^2_c = 0.62$, $R^2_v = 0.44$). The R^2_c and R^2_v values of the univariate regression (UR) model for Anth based on R_λ were high ($p < 0.01$); however, model accuracy remained low, and it could not accurately estimate the anthocyanin content of leaves.

3.7. Anth Regression Models Based on VI_s, VI_c, and $VI_s + VI_c + R_\lambda$

The Anth estimation models for red and healthy leaves based on VI_s, VI_c, and $VI_s + VI_c + R_\lambda$ and constructed using MLR, PCR, PLSR, and SVMR are shown in Figure 9. The range of R^2_c and R^2_v values of the Anth models for red leaves was 0.63–0.85 and 0.57–0.74, respectively. Among the models based on VI_s, the PLSR model ($R^2_c = 0.73$, $R^2_v = 0.61$) showed the highest accuracy, followed by the MLR model. Among the VI_c-based models, the R^2_c of the MLR, PCA, and PLSR models was equal, and the difference in R^2_v was small. The SVMR model showed the highest accuracy ($R^2_c = 0.68$, $R^2_v = 0.62$). Among the models based on $VI_s + VI_c + R_\lambda$, the MLR model showed the highest R^2_c, followed by the SVMR model. However, there was overfitting in the SVMR model, with a small R^2_v; this can be attributed to the small number of red leaf spectra and a few outliers in the samples that affected the optimal classification hyperplane of the SVM. The most accurate Anth estimation model for red leaves was the MLR model based on $VI_s + VI_c + R_\lambda$ ($R^2_c = 0.85$, $R^2_v = 0.74$), which can be used for the quantitative estimation of Anth in red leaves as a measure of MDMV infection severity.

The range of R^2_c and R^2_v values in the Anth models for healthy leaves was 0.62–0.68 and 0.60–0.66, respectively. Among the models based on VI_s, the MLR model ($R^2_c = 0.67$, $R^2_v = 0.64$) showed the highest accuracy. Among the models based on VI_c, the R^2_c and R^2_v of the MLR, PCA, and PLSR models were identical ($R^2_c = 0.65$, $R^2_v = 0.64$), and the modeling method showed little effect on accuracy. Among the models based on $VI_s + VI_c + R_\lambda$, the SVMR model showed the highest accuracy ($R^2_c = 0.68$, $R^2_v = 0.66$), which can be used to accurately estimate the Anth of healthy leaves for promptly monitoring the health status of maize.

In the distribution diagram of the measured and predicted values of red leaves, most points with small Anth were concentrated near the 1:1 line. However, the points with large Anth were distributed far from the 1:1 line, and the red leaf models showed satisfactory predictive performance for the small values of Anth but poor performance for large values. First of all, most of the leaf samples collected in this study were mildly infected with MDMV, and their Anth content was small. Only a few samples were seriously infected with MDMV and had high Anth. Moreover, MDMV-infected leaves were randomly collected in the entire study area, which had nothing to do with the fertilization situation in the plot and also resulted in the uneven distribution of sample data. As shown in Figure 9a, the model input parameter VI_s includes a vegetation index with good effects in previous studies on the biophysical and biochemical parameters of healthy leaves (with low Anth content). It is not reconstructed for MDMV-infected leaves so that the model has a poor fitting effect on the high value of Anth. In the models of Figure 9b, only the reflectivity of 400–1000 nm was considered in the construction of the input parameter VI_c, which did not make full use of all the spectral information that could be detected by the spectroradiometer. Therefore, the ability of these models to estimate Anth is limited. The model in Figure 9c uses $R_\lambda + VI_s + VI_c$ as input to estimate Anth in red leaves. Compared with the model in Figure 9a,b, the accuracy is improved to some extent, but the good fitting effect is still a low Anth.

In the distribution diagram of the measured and predicted values of healthy leaves, all points were evenly distributed on both sides of the 1:1 line, and the healthy leaf models showed a better fit to the Anth values than the red leaf models. This is because healthy leaves were collected in plots with different fertilization; Anth data were evenly distributed and showed no obvious aggregation. Therefore, the models in Figure 9d–f had good fitting effects on both high and low values. MDMV-infected leaves were mainly distributed in the plots with $0 \text{ kg } P_2O_5 \cdot ha^{-1} + 90 \text{ kg } N \cdot ha^{-1}$, $60 \text{ kg } P_2O_5 \cdot ha^{-1} + 90 \text{ kgN} \cdot ha^{-1}$, $90 \text{ kg } P_2O_5 \cdot ha^{-1} + 90 \text{ kg } N \cdot ha^{-1}$, and $120 \text{ kg } P_2O_5 \cdot ha^{-1} + 90 \text{kg } N \cdot ha^{-1}$. It can be easily observed that the spatial distribution of MDMV-infected maize has a high degree of aggregation, but there is no obvious rule with the fertilization situation in the plots. Meanwhile, the results also ruled out the possibility that phosphorus deficiency was responsible for the reddening of leaves in the study area.

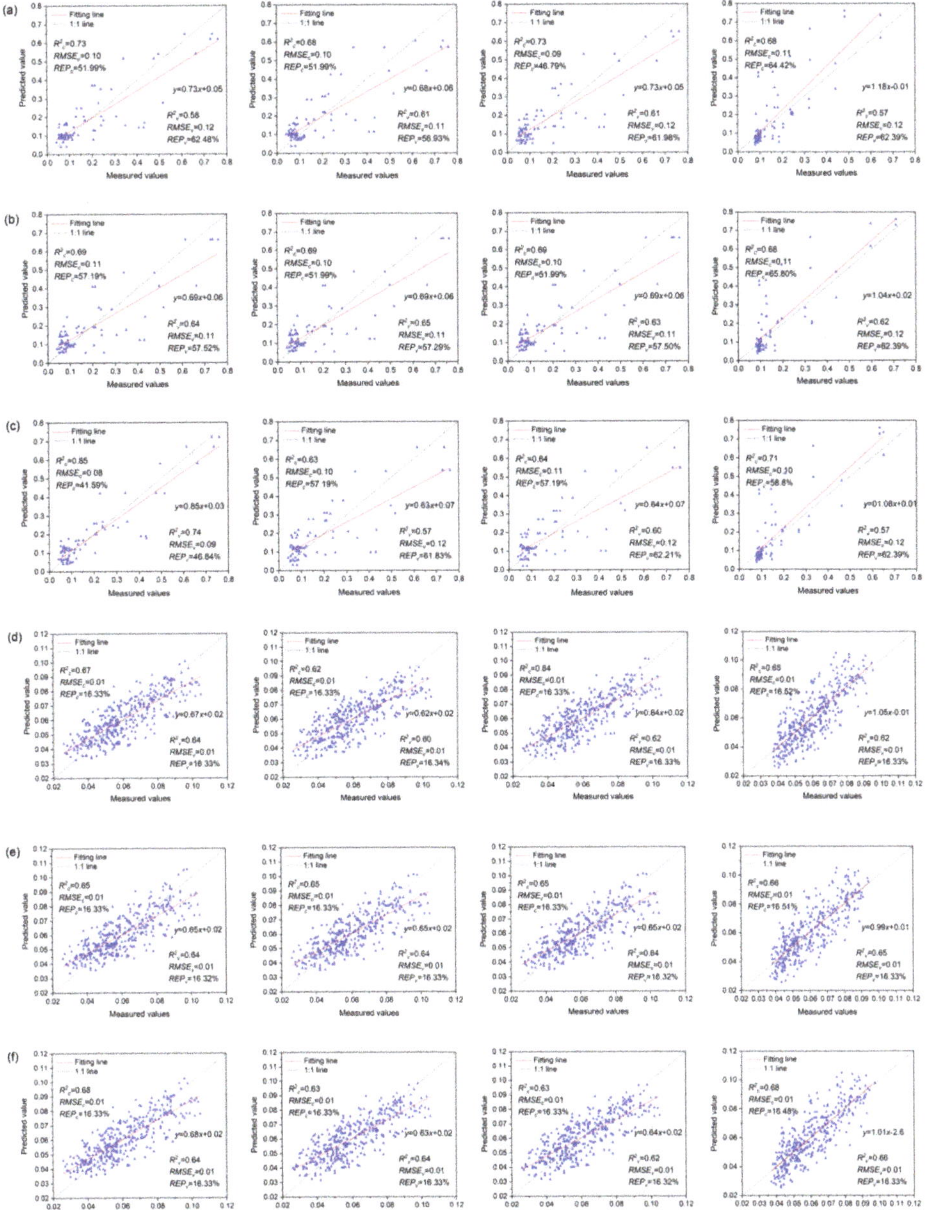

Figure 9. Anth estimation models of red and healthy leaves. (**a–c**) represent MLR, PCR, PLSR, and SVMR models of red leaves Anth based on VI_s, VI_c, and $R_\lambda + VI_s + VI_c$, respectively. (**d–f**) show MLR, PCR, PLSR, and SVMR estimation models of healthy leaf Anth based on VI_s, VI_c, and $R_\lambda + VI_s + VI_c$, respectively.

4. Discussion

4.1. Link between Spectral Reflectance and Plant Disease

Multispectral or hyperspectral RS technologies based on near-ground, low-altitude UAV and satellite platforms offer multiple opportunities to improve the productivity of agricultural production systems and provide an automatic and objective alternative to the visual assessment of plant diseases [58–60]. The utility of RS techniques in the field of plant disease detection has been well documented, and the potential of spectral sensors in the detection of fungal diseases has been proven [61–63]. In the present study, at Anth values below 0.58, the difference in the reflectance characteristics between red and healthy leaves in the visible range was small, and the spectral characteristics were similar; this result is consistent with the lack of significant differences in the spectra of MDMV-infected and healthy leaves of corn exhibiting mild mosaic symptoms in a study by Beverly [30]. At the Anth values exceeding 0.58, however, the spectra of MDMV-infected and healthy leaves differed significantly in both visible and near-infrared regions, which is also consistent with Beverly's speculation that the near-infrared region is critical for studying MDMV [30].

However, in order to effectively use spectral reflectance measurements for disease detection, the key is to identify the most important spectral wavelengths that are closely linked to a particular disease. Only a few bands of the reflectance spectrum are of interest depending on the type of disease and the range of application. In the present study, bands closely linked to MDMV infection were mainly concentrated at 611–743 nm, with the maximum correlation coefficient of 0.76.

4.2. Application of Vegetation Indices Based on Two Arbitrary Bands

Vegetation indices have been commonly used for analyzing and detecting changes in plant physiology and biochemistry. These indices, based on information at specific wavelengths, have been developed to reflect diverse plant parameters, such as pigment content, water content, and leaf area. Moreover, vegetation indices can be used as a measure of plant diseases. However, the quantitative analysis or identification of a specific disease based on the commonly used vegetation indices is not possible at present due the lack of disease specificity of the available indices. Therefore, we combined different wavelengths to construct vegetation indices (VI_c) for simplifying disease detection by using spectral sensors, since each disease influences a spectral signature in a characteristic manner. In the present study, compared with the commonly used vegetation indices, the newly developed VI_c showed improved classification accuracy for the red leaf spectrum (Table 3) and enhanced monitoring accuracy for MDMV severity (Figure 9). Inoue et al. [21] constructed a vegetation index based on two-band combinations for monitoring rice canopy nitrogen content and found that the RSI (D_{740}, D_{522}) model constructed based on the first-order differential spectra at 740 and 522 nm showed better performance. Mahlein et al. [64] created contour maps of correlation coefficients for the disease severity of *Cercospora* leaf spot, rust, and powdery mildew in beet based on the NDVI of two arbitrary bands to identify and monitor plant diseases. The strongly correlated bands used for NDVI in their study were mainly concentrated at 500–500 nm, close to the optimal bands for $NDVI_r$ based on VI_c in the present study.

4.3. RS-Based Identification of MDMV-Infected Leaves

In the present study, among the LDA and SVM models, the classification model based on VI_c showed the highest accuracy, followed by the models based on VI_s and R_λ. Additionally, in the classification models based on VI_s, SVM performed better than LDA, which is consistent with the trends reported by Shi et al. in the identification of wheat stripe rust, by Shang et al. in the classification of Australian virgin forest species, and by Calderón et al. in the early identification of olive *Verticillium* wilt [65–67]. SVM classifiers are constructed based on the threshold discriminant rules, which map samples to an appropriate feature space and can solve nonlinear and small-sample classification problems well. Both LDA and SVM classification models based on R_λ showed low accuracy mainly because R_λ offers very little spectral information for accurately identifying the target.

4.4. Application of Machine Learning Algorithms in Precision Agriculture

The major RS-based estimation methods in plant physiology and biochemistry include the physical radiative transfer models and empirical statistical models [68–70]. The physical model simulates the reflectivity of a leaf blade based on a limited number of variables according to different mathematical and physical principles. The greatest advantages of this approach are that it is based on the radiative transmission model of electromagnetic waves and the ecological theory of vegetation and that it is not affected by vegetation type and other factors. However, despite being powerful, these models require a large amount of local perception data related to biotic and abiotic factors for calibration [71], thus limiting their applicability in the large-scale monitoring of crop growth [72].

Most empirical statistical models are constructed based on vegetation indices, and they are relatively simple and diverse in structure. However, these models are susceptible to vegetation type, light conditions, and canopy structure and are sensitive to soil background; thus, the universality of these models is poor. Among the empirical statistical models, the machine learning models have the advantage of realizing the high-precision prediction of leaf pigments by analyzing the relationship between leaf nutrient drivers and pigment content, without relying on specific crop parameters. Moreover, with the advantages of RS technology, some machine learning algorithms can assess crop growth at a regional scale from satellite images. However, the inversion results obtained by linear models are typically not reliable, and nonlinear machine learning algorithms, such as MLR, PLSR, PCA, SVM, and RF, can analyze complex relationships between vegetation indices and multiple factors of crop growth [73,74]. In the present study, the MLR, PCR, PLSR, and SVMR models based on machine learning algorithms showed significant differences in their accuracy with different modeling parameters; however, the predictive performance of the four models was satisfactory, as evidenced by the high $R^2{}_c$ and $R^2{}_v$ values, and their accuracy was significantly higher than that of the simple regression model.

Rapid and large-scale monitoring of crop diseases can effectively reduce the pressure on plant protectors and is an important means to prevent and control diseases and to ensure food health. In addition, such efforts have made significant contributions in eliminating hunger, ensuring global food security, improving nutrition, and achieving sustainable development goals for food proposed by the United Nations [75].

5. Conclusions

The key to the effective identification of MDMV based on spectral reflectance is to find the appropriate wavelength that is closely linked to the disease. In the present study, we first analyzed the spectral differences between red and healthy leaves and determined the band region of the observed spectral difference between the two types of leaves. Next, by comparing the spectra of leaves with different severities of MDMV infection (as indicated by different Anth values), we obtained the variation rule of the reflectance spectra for MDMV infection severity and the band with the strongest correlation with Anth (R_λ). However, single-band spectra are not sufficient for fully characterizing plants. Therefore, in order to detect plant growth statuses, we selected 12 vegetation indices that are closely correlated with Anth for MDMV monitoring. Following the construction principle of NDVI, RVI, DVI, and SAVI, we constructed VI_c based on the combination two arbitrary bands. Furthermore, we constructed LDA and SVM classification models for MDMV based on R_λ, VI_s, and VI_c and identified a classification model that could accurately distinguish red leaves from the healthy ones. In addition, we constructed the Anth regression model based on three spectral parameters in order to accurately assess the severity of MDMV infection. The major conclusions are as follows:

(1) The spectral differences between red and healthy leaves were mainly concentrated in the 493–764 nm region, and the maximum difference was recorded near 700 nm.

(2) The red leaf spectrum showed bimodal characteristics in the visible range. With the aggravation of the disease (i.e., increase in Anth), the reflectance of the left peak of the spectrum (550 nm) gradually decreased until it disappeared. Simultaneously, the reflectance of the right peak increased gradually, and the absorption characteristics near 680 nm disappeared. With worsening MDMV infection, the position of the red edge of the reflectance spectrum appeared as a "blue shift."

(3) The LAD and SVM models constructed based on VI_c performed better in recognizing MDMV, with the classification accuracy of 100%, followed by the models based on VI_s; the models based on R_λ showed the poorest classification accuracy.

(4) The MLR model based on $R_\lambda + VI_s + VI_c$ ($R^2_c = 0.85$, $R^2_v = 0.74$) was the best for monitoring the severity of MDMV infection, while the SVMR model based on $R_\lambda + VI_s + VI_c$ ($R^2_c = 0.68$, $R^2_v = 0.66$) was the best for the estimation of Anth in healthy maize leaves.

Author Contributions: Conceptualization, L.L.; methodology, L.L.; software, L.L.; validation, L.L., Q.C.; formal analysis, L.L.; investigation, L.L.; resources, Q.C.; data curation, L.L., Q.W. and Y.H.; writing—original draft preparation, L.L.; writing—review and editing, L.L..; visualization, L.L.; supervision, Q.C.; project administration, Q.C.; funding acquisition, Q.C. All authors have read and agreed to the published version of the manuscript.

Funding: This research was funded by the National High Technology Research and Development Program of China (863 Program), grant number 2013AA102401-2.

Institutional Review Board Statement: This study not involving humans.

Informed Consent Statement: This study not involving humans.

Data Availability Statement: Data sharing is not application to this article.

Acknowledgments: We would like to thank all the students in Chang's team for collecting the data for us, as well as the managing editors and anonymous reviewers for their constructive comments, which greatly improved the quality of this paper.

Conflicts of Interest: The authors declare no conflict of interest.

References

1. Food and Agricultural Organizations of the United Nations. Available online: https://www.fao.org/faostat/zh/#data/QCL (accessed on 3 November 2021).
2. Kang, Y.; Ozdogan, M.; Zhu, X.; Ye, Z.; Hain, C.; Anderson, M. Comparative assessment of environmental variables and machine learning algorithms for maize yield prediction in the US Midwest. *Environ. Res. Lett.* **2020**, *15*, 064005. [CrossRef]
3. Jiang, J.; Zhou, X. Maize dwarf mosaic disease in different regions of China is caused by Sugarcane mosaic virus. *Arch. Virol.* **2002**, *147*, 2437–2443. [CrossRef] [PubMed]
4. Ali, F.; Yan, J. Disease resistance in maize and the role of molecular breeding in defending against global threat. *J. Integr. Plant Biol.* **2012**, *54*, 134–151. [CrossRef] [PubMed]
5. Williams, L.E.; Alexander, L.J. Maize dwarf mosaic, a new Corn disease. *Phytopathology* **1965**, *55*, 802–804.
6. Janson, B.F.; Williams, L.E.; Findley, W.R.; Dollinger, E.J.; Ellett, C.W. Maize dwarf mosaic: New corn virus disease in Ohio. *Res. Circ. Ohio Agric. Exp. Station* **1965**, *460*, 16.
7. Johnson, D.M. An assessment of pre- and within-season remotely sensed variables for forecasting corn and soybean yields in the United States. *Remote Sens. Environ.* **2014**, *141*, 116–128. [CrossRef]
8. Jiang, H.; Hu, H.; Zhong, R.; Xu, J.; Lin, T. A deep learning approach to conflating heterogeneous geospatial data for corn yield estimation: A case study of the US Corn Belt at the county level. *Glob. Chang. Biol.* **2020**, *26*, 1754–1766. [CrossRef]
9. Guan, K.; Wu, J.; Kimball, J.S.; Anderson, M.C.; Frolking, S.; Li, B.; Hain, C.R.; .Lobell, D.B. The shared and unique values of optical, fluorescence, thermal and microwave satellite data for estimating large-scale crop yields. *Remote Sens. Environ.* **2017**, *199*, 333–349. [CrossRef]
10. Novak, A.B.; Short, F.T. Leaf reddening in the seagrass Thalassia testudinum in relation to anthocyanins, seagrass physiology and morphology, and plant protection. *Mar. Biol.* **2011**, *158*, 1403–1416. [CrossRef]
11. Brefort, T.; Tanaka, S.; Neidig, N.; Doehlemann, G.; Vincon, V.; Kahmann, R. Characterization of the largest effector gene cluster of Ustilago maydis. *PLoS Pathog.* **2014**, *10*, e1003866. [CrossRef]
12. Wang, L.Z.; Hu, Y.B.; Zhang, H.H.; Xu, N.; Sun, G.Y. Photoprotective mechanisms of leaf anthocyanins: Research progress. *Chin. J. Appl. Ecol.* **2012**, *23*, 835–841.

13. Landi, M.; Tattini, M.; Gould, K.S. Multiple functional roles of anthocyanins in plant-environment interactions. *Environ. Exp. Bot.* **2015**, *119*, 4–17. [CrossRef]

14. Pietrini, F.; Iannelli, M.A.; Massacci, A.; Iannelli, M.; Iannelli, M. Anthocyanin accumulation in the illuminated surface of maize leaves enhances protection from photo-inhibitory risks at low temperature, without further limitation to photosynthesis. *Plant Cell Environ.* **2008**, *13*, 1529–1546. [CrossRef]

15. Merzlyak, M.N.; Chivkunova, O.B.; Solovchenko, A.E.; Razi, N.K. Light absorption by anthocyanins in juvenile, stressed, and senescing leaves. *J. Exp. Bot.* **2008**, *59*, 3903–3911. [CrossRef]

16. Goławska, S.; Łukasik, I. Antifeedant activity of luteolin and genistein against the pea aphid, Acyrthosiphon pisum. *J. Pest Sci.* **2012**, *85*, 443–450. [CrossRef] [PubMed]

17. Ludmerszki, E. The beneficial effects of S-methyl-methionine in maize in the case of Maize dwarf mosaic virus infection. *Acta Biol. Szeged.* **2011**, *55*, 109–112.

18. Singh, M.; Sharma, P. *Rice Germplasms of Manipur: Varietal Description and Cataloguing*; Plant Breeding Technical Report No. 1; Department of Plant Breeding and Genetics, Central Agriculture University: Imphal Manipur, India, 1998; Volume 1, pp. 75–88.

19. Fasahat, P.; Muhammad, K.; Abdullah, A.; Ratnam, W. Proximate nutritional composition and antioxidant properties of 'Oryza rufipogon', a wild rice collected from Malaysia compared to cultivated rice, MR219. *Aust. J. Crop Sci.* **2012**, *6*, 1502–1507.

20. Gitelson, A.A.; Keydan, G.P.; Merzlyak, M.N. Three-band model for noninvasive estimation of chlorophyll, carotenoids, and anthocyanin contents in higher plant leaves. *Geophys. Res. Lett.* **2006**, *33*, 431–433. [CrossRef]

21. Haboudane, D.; Miller, J.R.; Pattey, E.; Zarco-Tejada, P.J.; Strachan, I.B. Hyperspectral vegetation indices and novel algorithms for predicting green LAI of crop canopies: Modeling and validation in the context of precision agriculture. *Remote Sens. Environ.* **2004**, *90*, 337–352. [CrossRef]

22. Inoue, Y.; Sakaiya, E.; Zhu, Y.; Takahashi, W. Diagnostic mapping of canopy nitrogen content in rice based on hyperspectral measurements. *Remote Sens. Environ.* **2012**, *126*, 210–221. [CrossRef]

23. Bongiovanni, R.; Lowenberg-DeBoer, J. Precision agriculture and sustainability. *Precis. Agric.* **2004**, *5*, 359–387. [CrossRef]

24. Mirik, M.; Agrilife, T.; Vernon, F.; Price, J.A.; Agrilife, T. Satellite Remote Sensing of Wheat Infected by Wheat streak mosaic virus. *Plant Dis.* **2011**, *95*, 4–12. [CrossRef]

25. Camino, C.; Calderón, R.; Parnell, S.; Dierkes, H.; Beck, P. Detection of Xylella fastidiosa in almond orchards by synergic use of an epidemic spread model and remotely sensed plant traits. *Remote Sens. Environ.* **2021**, *260*, 112420. [CrossRef] [PubMed]

26. Martins, L.; Castro, J.; Macedo, W.; Marques, C.; Abreu, C. Assessment of the spread of chestnut ink disease using remote sensing and geostatistical methods. *Eur. J. Plant Pathol.* **2007**, *119*, 159–164. [CrossRef]

27. Liu, Z.Y.; Huang, J.F.; Tao, R.X. Characterizing and Estimating Fungal Disease Severity of Rice Brown Spot with Hyperspectral Reflectance Data. *Rice Sci.* **2008**, *15*, 232–242. [CrossRef]

28. Feng, W.; Shen, W.; He, L.; Duan, J.; Guo, B.; Li, Y.; Wang, C.; Guo, T. Improved remote sensing detection of wheat powdery mildew using dual-green vegetation indices. *Precis. Agric.* **2016**, *17*, 608–627. [CrossRef]

29. Dhau, I.; Adam, E.; Mutanga, O.; Ayisi, K.K. Detecting the severity of maize streak virus infestations in maize crop using in situ hyperspectral data. *Trans. R. Soc. S. Afr.* **2018**, *73*, 8–15. [CrossRef]

30. Huang, S.; Wang, L.; Liu, L.; Liu, E.; Hou, E.; Xiao, D.; Fan, Z. Isolation, identification and biological characters of pathogens of rice spikelet rot disease. *Chin. J. Rice Sci.* **2012**, *26*, 341–350.

31. Ausmus, B.S.; Hilty, J.W. Reflectance studies of healthy, maize dwarf mosaic virus-infected, and Helminthosporium maydis-infected corn leaves. *Remote Sens. Environ.* **1971**, *2*, 77–81. [CrossRef]

32. Croft, H.; Chen, J.M.; Zhang, Y.; Simic, A. Modelling leaf chlorophyll content in broadleaf and needle leaf canopies from ground, CASI, Landsat TM 5 and MERIS reflectance data. *Remote Sens. Environ.* **2013**, *133*, 128–140. [CrossRef]

33. Yi, Q.; Wang, F.; Bao, A.; Jiapaer, G. Leaf and canopy water content estimation in cotton using hyperspectral indices and radiative transfer models. *Int. J. Appl. Earth Obs. Geoinf.* **2014**, *33*, 67–75. [CrossRef]

34. Ferrer, M.; Echeverría, G.; Pereyra, G.; Gonzalez-Neves, G.; Pan, D.; Mirás-Avalos, J.M. Mapping vineyard vigor using airborne remote sensing: Relations with yield, berry composition and sanitary status under humid climate conditions. *Precis. Agric.* **2020**, *21*, 178–197. [CrossRef]

35. Huete, A.R. A soil-adjusted vegetation index (SAVI). *Remote Sens. Environ.* **1988**, *25*, 295–309. [CrossRef]

36. Ren, H.; Zhou, G.; Zhang, F. Using negative soil adjustment factor in soil-adjusted vegetation index (SAVI) for aboveground living biomass estimation in arid grasslands. *Remote Sens. Environ.* **2018**, *209*, 439–445. [CrossRef]

37. Feng, W.; Yao, X.; Zhu, Y.; Tian, Y.; Cao, W. Monitoring leaf nitrogen status with hyperspectral reflectance in wheat. *Eur. J. Agron.* **2008**, *28*, 394–404. [CrossRef]

38. Yuan, H.; Yang, G.; Li, C.; Wang, Y.; Liu, J.; Yu, H.; Feng, H.; Xu, B.; Zhao, X.; Yang, X. Retrieving soybean leaf area index from unmanned aerial vehicle hyperspectral remote sensing: Analysis of RF, ANN, and SVM regression models. *Remote Sens.* **2017**, *9*, 309. [CrossRef]

39. Zhai, Y.; Cui, L.; Zhou, X.; Gao, Y.; Fei, T.; Gao, W. Estimation of nitrogen, phosphorus, and potassium contents in the leaves of different plants using laboratory-based visible and near-infrared reflectance spectroscopy: Comparison of partial least-square regression and support vector machine regression methods. *Int. J. Remote Sens.* **2013**, *34*, 2502–2518.

40. Wei, G.; Hong, B.Q.; Heng, Q.Z.; Juan, J.Z.; Peng, C.P.; Ze, L.L. Cotton Aphid Damage Monitoring Using UAV Hyperspectral Data Based on Derivative of Ratio Spectroscopy. *Spectrosc. Spectr. Anal.* **2021**, *41*, 1543–1550.

41. Jiang, Z.; Huete, A.R.; Didan, K.; Miura, T. Development of a two-band enhanced vegetation index without a blue band. *Remote Sens. Environ.* **2008**, *112*, 3833–3845. [CrossRef]
42. Jordan, C.F. Derivation of leaf-area index from quality of light on the forest floor. *Ecology* **1969**, *50*, 663–666. [CrossRef]
43. Penuelas, J.; Filella, I. Visible and near-infrared reflectance techniques for diagnosing plant physiological status. *Trends Plant Sci.* **1998**, *3*, 151–156. [CrossRef]
44. Rouse, J.W.; Haas, R.W.; Schell, J.A.; Deering, D.W.; Harlan, J.C. *Monitoring the Vernal Advancement and Retrogradation (Green Wave Effect) of Natural Vegetation*; NASA/GSFC: Greenbelt, MD, USA, 1974; Type III, Final Report; p. 371.
45. Gitelson, A.A.; Kaufman, Y.J.; Merzlyak, M.N. Use of a green channel in remote sensing of global vegetation from EOS-MODIS. *Remote Sens. Environ.* **1996**, *58*, 289–298. [CrossRef]
46. Barati, S.; Rayegani, B.; Saati, M.; Sharifi, A.; Nasri, M. Comparison the accuracies of different spectral indices for estimation of vegetation cover fraction in sparse vegetated areas. *Egypt. J. Remote Sens. Space Sci.* **2011**, *14*, 49–56. [CrossRef]
47. Daughtry, C.S.; Walthall, C.L.; Kim, M.S.; De Colstoun, E.B.; McMurtrey Iii, J.E. Estimating corn leaf chlorophyll concentration from leaf and canopy reflectance. *Remote Sens. Environ.* **2000**, *74*, 229–239. [CrossRef]
48. Gitelson, A.A.; Vina, A.; Ciganda, V.; Rundquist, D.C.; Arkebauer, T.J. Remote estimation of canopy chlorophyll content in crops. *Geophys. Res. Lett.* **2005**, *32*, L08403. [CrossRef]
49. Haboudane, D.; Miller, J.; Tremblay, N.; Pattey, E.; Vigneault, P. Estimation of leaf area index using ground spectral measurements over agriculture crops: Prediction capability assessment of optical indices. In Proceedings of the 20th ISPRS Congress: "Geo-Imagery Bridging Continents", Istanbul, Turkey, 12–23 July 2004.
50. Vogelmann, J.; Rock, B.; Moss, D. Red edge spectral measurements from sugar maple leaves. *Int. J. Remote Sens.* **1993**, *14*, 1563–1575. [CrossRef]
51. Merzlyak, M.N.; Gitelson, A.A.; Chivkunova, O.B.; Rakitin, V.Y. Non-destructive optical detection of pigment changes during leaf senescence and fruit ripening. *Physiol Plant.* **1999**, *106*, 135–141. [CrossRef]
52. Gamon, J.; Serrano, L.; Surfus, J. The photochemical reflectance index: An optical indicator of photosynthetic radiation use efficiency across species, functional types, and nutrient levels. *Oecologia* **1997**, *112*, 492–501. [CrossRef] [PubMed]
53. Gitelson, A.A.; Gritz, Y.; Merzlyak, M.N. Relationships between leaf chlorophyll content and spectral reflectance and algorithms for non-destructive chlorophyll assessment in higher plant leaves. *J. Plant Physiol.* **2003**, *160*, 271–282. [CrossRef] [PubMed]
54. Pal, M.; Foody, G.M. Feature selection for classification of hyperspectral data by SVM. *IEEE Trans. Geosci. Remote Sens.* **2010**, *48*, 2297–2307. [CrossRef]
55. Neumann, J.; Schnörr, C.; Steidl, G. Combined SVM-based feature selection and classification. *Mach. Learn.* **2005**, *61*, 129–150. [CrossRef]
56. Li, C.H.; Lin, C.T.; Kuo, B.C.; Chu, H.S. An automatic method for selecting the parameter of the RBF kernel function to support vector machines. In *2010 IEEE International Geoscience and Remote Sensing Symposium*; IEEE: Piscataway, NJ, USA, 2010.
57. Han, S.; Qu, B.C.; Meng, H. Parameter selection in SVM with RBF kernel function. In *World Automation Congress 2012*; IEEE: Piscataway, NJ, USA, 2012.
58. Johansen, K.; Morton, M.J.; Malbeteau, Y.M.; Aragon, B.; Al-Mashharawi, S.K.; Ziliani, M.G.; Angel, Y.; Fiene, G.M.; Negrão, S.S.; Mousa, M.A. Unmanned aerial vehicle-based phenotyping using morphometric and spectral analysis can quantify responses of wild tomato plants to salinity stress. *Front. Plant Sci.* **2019**, *10*, 370–407. [CrossRef] [PubMed]
59. Hassan, M.A.; Yang, M.; Rasheed, A.; Yang, G.; Reynolds, M.; Xia, X.; Xiao, Y.; He, Z.A. Rapid monitoring of NDVI across the wheat growth cycle for grain yield prediction using a multi-spectral UAV platform. *Plant Sci.* **2019**, *282*, 95–103. [CrossRef] [PubMed]
60. Thorp, K.R.; Wang, G.; Bronson, K.F.; Badaruddin, M.; Mon, J. Hyperspectral data mining to identify relevant canopy spectral features for estimating durum wheat growth, nitrogen status, and grain yield. *Comput. Electron. Agric.* **2017**, *136*, 1–12. [CrossRef]
61. Huang, W.; Guan, Q.; Luo, J.; Zhang, J.; Zhao, J.; Liang, D.; Huang, L.; Zhang, D. New optimized spectral indices for identifying and monitoring winter wheat diseases. *IEEE J. Sel. Top. Appl. Earth Obs. Remote Sens.* **2014**, *7*, 2516–2524. [CrossRef]
62. Mahlein, A.-K.; Steiner, U.; Dehne, H.W.; Oerke, E.C. Spectral signatures of sugar beet leaves for the detection and differentiation of diseases. *Precis. Agric.* **2010**, *11*, 413–431. [CrossRef]
63. Rumpf, T.; Mahlein, A.-K.; Steiner, U.; Oerke, E.C.; Dehne, H.W.; Plümer, L. Early detection and classification of plant diseases with support vector machines based on hyperspectral reflectance. *Comput. Electron. Agric.* **2010**, *74*, 91–99. [CrossRef]
64. Mahlein, A.K.; Rumpf, T.; Welke, P.; Dehne, H.W.; Plümer, L.; Steiner, U.; Oerke, E.C. Development of spectral indices for detecting and identifying plant diseases. *Remote Sens. Environ.* **2013**, *128*, 21–30. [CrossRef]
65. Calderón, R.; Navas-Cortés, J.A.; Zarco-Tejada, P.J. Early detection and quantification of Verticillium wilt in olive using hyperspectral and thermal imagery over large areas. *Remote Sens.* **2015**, *7*, 5584–5610. [CrossRef]
66. Shang, X.; Chisholm, L.A. Classification of Australian native forest species using hyperspectral remote sensing and machine-learning classification algorithms. *IEEE J. Sel. Top. Appl. Earth Obs. Remote Sens.* **2013**, *7*, 2481–2489. [CrossRef]
67. Shi, Y.; Huang, W.; Luo, J.; Huang, L.; Zhou, X. Detection and discrimination of pests and diseases in winter wheat based on spectral indices and kernel discriminant analysis. *Comput. Electron. Agric.* **2017**, *141*, 171–180. [CrossRef]
68. Khaki, S.; Wang, L.; Archontoulis, S.V. A cnn-rnn framework for crop yield prediction. *Front. Plant Sci.* **2020**, *10*, 1750. [CrossRef]

69. Yendrek, C.R.; Tomaz, T.; Montes, C.M.; Cao, Y.; Morse, A.M.; Brown, P.J.; McIntyre, L.M.; Leakey, A.D.; Ainsworth, E.A. High-throughput phenotyping of maize leaf physiological and biochemical traits using hyperspectral reflectance. *Plant Physiol.* **2017**, *173*, 614–626. [CrossRef] [PubMed]

70. Maimaitijiang, M.; Sagan, V.; Sidike, P.; Hartling, S.; Esposito, F.; Fritschi, F.B. Soybean yield prediction from UAV using multimodal data fusion and deep learning. *Remote Sens. Environ.* **2020**, *237*, 111599. [CrossRef]

71. Ma, Y.; Zhang, Z.; Kang, Y.; Özdoğan, M. Corn yield prediction and uncertainty analysis based on remotely sensed variables using a Bayesian neural network approach. *Remote Sens. Environ.* **2021**, *259*, 112408. [CrossRef]

72. Sakamoto, T.; Gitelson, A.A.; Arkebauer, T.J. MODIS-based corn grain yield estimation model incorporating crop phenology information. *Remote Sens. Environ.* **2013**, *131*, 215–231. [CrossRef]

73. Han, L.; Yang, G.; Dai, H.; Xu, B.; Yang, H.; Feng, H.; Li, Z.; Yang, X. Modeling maize above-ground biomass based on machine learning approaches using UAV remote-sensing data. *Plant Methods* **2019**, *15*, 10. [CrossRef] [PubMed]

74. Shah, S.H.; Angel, Y.; Houborg, R.; Ali, S.; McCabe, M.F. A random forest machine learning approach for the retrieval of leaf chlorophyll content in wheat. *Remote Sens.* **2019**, *11*, 920. [CrossRef]

75. Zheng, Q.; Ye, H.; Huang, W.; Dong, Y.; Jiang, H.; Wang, C.; Li, D.; Wang, L.; Chen, S. Integrating spectral information and meteorological data to monitor wheat yellow rust at a regional scale: A case study. *Remote Sens.* **2021**, *13*, 278. [CrossRef]

remote sensing

Article

Early Detection of Powdery Mildew Disease and Accurate Quantification of Its Severity Using Hyperspectral Images in Wheat

Imran Haider Khan [1,†], Haiyan Liu [1,†], Wei Li [1], Aizhong Cao [2], Xue Wang [1], Hongyan Liu [1], Tao Cheng [1], Yongchao Tian [1], Yan Zhu [1], Weixing Cao [1] and Xia Yao [1,*]

[1] National Engineering and Technology Center for Information Agriculture, Key Laboratory for Crop System Analysis and Decision Making, Ministry of Agriculture, Jiangsu Key Laboratory for Information Agriculture, Jiangsu Collaborative Innovation Center for Modern Crop Production, Nanjing Agricultural University, Nanjing 210095, China; 2018101174@njau.edu.cn (I.H.K.); 2018101173@njau.edu.cn (H.L.); 2020201096@stu.njau.edu.cn (W.L.); wangxue@njau.edu.cn (X.W.); 2016101049@njau.edu.cn (H.L.); tcheng@njau.edu.cn (T.C.); yctian@njau.edu.cn (Y.T.); yanzhu@njau.edu.cn (Y.Z.); caow@njau.edu.cn (W.C.)

[2] National Key Laboratory of Crop Genetics and Germplasm Enhancement, Cytogenetics Institute, Nanjing Agricultural University/JCIC-MCP, Nanjing 210095, China; caoaz@njau.edu.cn

* Correspondence: yaoxia@njau.edu.cn; Tel.: +86-25-84396565; Fax: +86-25-84396672

† Imran Haider Khan and Haiyan Liu are the co-first author.

Citation: Khan, I.H.; Liu, H.; Li, W.; Cao, A.; Wang, X.; Liu, H.; Cheng, T.; Tian, Y.; Zhu, Y.; Cao, W.; et al. Early Detection of Powdery Mildew Disease and Accurate Quantification of Its Severity Using Hyperspectral Images in Wheat. *Remote Sens.* **2021**, *13*, 3612. https://doi.org/10.3390/rs13183612

Academic Editor: Xanthoula Eirini Pantazi

Received: 12 August 2021
Accepted: 7 September 2021
Published: 10 September 2021

Abstract: Early detection of the crop disease using agricultural remote sensing is crucial as a precaution against its spread. However, the traditional method, relying on the disease symptoms, is lagging. Here, an early detection model using machine learning with hyperspectral images is presented. This study first extracted the normalized difference texture indices (NDTIs) and vegetation indices (VIs) to enhance the difference between healthy and powdery mildew wheat. Then, a partial least-squares linear discrimination analysis was applied to detect powdery mildew with the combined optimal features (i.e., VIs & NDTIs). Further, a regression model on the partial least-squares regression was developed to estimate disease severity (DS). The results show that the discriminant model with the combined VIs & NDTIs improved the ability for early identification of the infected leaves, with an overall accuracy value and Kappa coefficient over 82.35% and 0.56 respectively, and with inconspicuous symptoms which were difficult to identify as symptoms of the disease using the traditional method. Furthermore, the calibrated and validated DS estimation model reached good performance as the coefficient of determination (R^2) was over 0.748 and 0.722, respectively. Therefore, this methodology for detection, as well as the quantification model, is promising for early disease detection in crops.

Keywords: wheat powdery mildew; hyperspectral imaging; early; detect the crop disease; quantify the disease severity

1. Introduction

Wheat (Triticum aestivum) is an important food crop in China. The sustainable production of wheat is important for the stability of China's society and economy. However, wheat is often attacked by various diseases due to unfavorable environments and management conditions. Powdery mildew (PM) (*Blumeria graminis f.* sp. tritici) is one of the major diseases that seriously affect wheat yield and quality [1]. Currently, people use disease-resistant varieties [2,3] and spray pesticides [4] to control PM, but these are not optimal solutions, and will produce an increased cost and environmental pollution. The essential approach to combating PM is the early identification and quantitative assessment of disease severity (DS), thereby helping farmers to ensure timely use of fungicides [5].

Traditional methods to identify the disease mainly depend on visual inspection and laboratory tests. Farmers usually visually judge its presence in the field, but there are frequent mistakes. Laboratory tests include enzyme-linked immunosorbent assay (ELISA), immunofluorescence, fluorescence in situ hybridization (FISH), and polymerase chain reaction (PCR) methods, which are more accurate than visual inspection but usually lag and require precision instruments and strict operations, therefore, most are time consuming.

With the development of spectroscopy technologies, spectral identification methods (Mutka and Bart, 2015) have achieved remarkable results in the field of intelligent identification of plant diseases, such as RGB imagery and multi/hyperspectral technology [6]. A disease identification method using RGB imagery is easier to implement and does not require sophisticated chemical analysis instruments or expensive sensors, imaging equipment, etc. RGB imagery recognition manually extracts the recognizable features of each disease, such as color, shape, and texture features. These features can further be used to obtain a model for distinguishing disease types (Siricharoen) [7]. However, some factors seriously affect the application of visible imaging to plant disease including light condition, removal of the background, and other automatic image processing techniques (Barbedo) [8]. Multispectral images contain more band information compared with RGB images. Satellite imagery and images obtained from the combined use of a UAV with multispectral camera are two common ways to obtain multispectral image. This method seems to be one of the best methods for quickly and accurately detecting and mapping disease incidence over large areas [9]. However, it appears to be insufficient to utilize this combination of narrow bands for disease mapping at earlier stages of the disease's development [10].

Compared with the above methods, hyperspectral technology has a high cost and its complex data interpretation remains challenging [11]. However, its applications are mature and reliable, and have been widely used to monitor the physiological status of vegetation [12], drought stress [13], nutrient deficiencies [14], and disease stress [15]. The principle is that the interaction between plants and the electromagnetic spectrum relies on photosynthetic chlorophyll in the visible light band (VIS), leaf anatomy in the near-infrared band (NIR), and leaf water in the short-wave infrared band [16]. When wheat is subjected to PM stress, its spectral reflectance changes according to physiological and biochemical changes in its leaves, such as decreased chlorophyll content or destroyed cell structure [17]. When combined with aerial platforms, hyperspectral technology is very well-suited for field phenotyping due to its capability for the characterization of subtle details of diseases. Many studies monitoring plant diseases have been conducted using ground and aerial hyperspectral reflectance. For example, Shi et al. [18] proposed a spectral vegetation indices-based kernel discriminant approach for the detection and classification of yellow rust, aphid, and powdery mildew in winter wheat. Zheng et al. [19] developed optimal spectral indices with three-band combinations based on sensitive wavebands to detect yellow rust disease in wheat at different growth stages. However, by using only spectral information, those studies identified disease at the middle or late infection stages, thereby missing the critical time for prevention and control of the disease.

Hyperspectral imaging (HI) is a new nondestructive monitoring technology developed in recent years. It combines imaging with spectroscopic techniques to obtain both spectral and spatial information [20]. As well as the spectral information of HI, its spatial information offers an accurate and reliable resource to identify disease. Texture is a typical spatial information that refers to the recurring local patterns and permutation rules of pixels. The texture features can reflect changes in the intensity of pixels to distinguish and identify objects [21]. Analyzing texture information from images can detect foliar surface changes due to fungal disease. For example, Yang et al. [22] used gray histogram and a gray-level co-occurrence matrix to extract the texture features of rice blast leaves, establishing a stepwise discriminant model with a classification accuracy of more than 90%. Al-Saddik et al. [23] combined spectral and texture data to detect yellowness and esca in grapevines with an overall accuracy of 99% for both of the diseases. Although the above studies achieved some useful results, there is scant evidence regarding the detection of PM through the fusion of spectral and texture information, especially at the early stages of disease infection.

The main element of HI processing for the detection of disease at its early stages is feature engineering, which transforms the raw hyperspectral data into features suitable for modeling. Feature selection is one of the important methods of feature engineering for dimensionality reduction of hyperspectral data. Feature selection methods, such as stepwise discriminant analysis, select a subset of characteristic wavelengths that preserve certain information of the full spectrum. The selected wavelengths can be used to construct vegetation indices [24] and disease indices [25]. However, most of the feature-selection methods are based on a single dataset or a single model, which do not account for the impaction of the changes of sample or variable on variable selection. To solve these problems, introduced herein is sub-window permutation analysis (SPA) [26]. The main idea of SPA is to build hundreds of sub-models based on different sub-datasets by a random sampling of the total samples and variables, and then to test those models statistically to determine the importance of each variable. SPA is a simple and effective feature-selection method, which considers both the influence of the samples and interaction among the variables. The SPA method was first proposed for chemistry and life science analysis and has proved its advantage. Mo et al. [27] combined near infrared spectroscopy with SPA to select the important variables related to haloxyfop-p-methyl residue in edible oil from the whole band and established a PLS-LDA model with an accuracy of 94.74%. Sun et al. [28] used near infrared (NIR) spectroscopy and SPA to detect complex adulteration of camellia oil. In the field of disease detection, SPA was initiated in wheat PM and successfully extracted information from non-imaging hyperspectral data by Wang [29]. However, the point-by-point spectral characteristics of the entire wheat leaf cannot be analyzed due to the measurement characteristics of the analytical spectral device. In the early stage of wheat infection by powdery mildew, the disease spots are scattered and usually not evenly distributed. For this reason, the combination of spectral and imaging technology can provide a new perspective for the early diagnosis of wheat powdery mildew.

Here, the main objectives of this study were (I) to determine the sensitive features by SPA and combine spectral and texture analysis, (II) to develop a robust discriminant model to early detect wheat PM by PLS-LDA, and (III) to establish an accurate model to quantify disease severity of wheat PM by PLSR. This methodology is expected to provide a new technique to detect the disease. This study demonstrates the capability of hyperspectral imaging to determine occurrence and severity of PM infection at the early stage.

2. Materials and Methods

2.1. Experimental Site and Design

The experiment was carried out at the Pailou experiment station, Nanjing Agricultural University (32°1′N, 118°15′E) during the wheat growing season of 2017–2018. Nannong 0686 wheat with high susceptibility to disease was grown in greenhouses. We planted 24 pots (280 mm × 280 mm × 500 mm) with a density of ten seeds per pot. Eight pots were health for control, eight for inoculation at the jointing stage, and the residual eight pots were put separately in the greenhouse for inoculation at booting stage. At the two growth stages, the infected wheat strains were used to shake off spores to inoculate the wheat plant in the pot, then they were placed horizontally in the pot, and all of them were incubated in a controlled environment at a temperature of 25 °C and at 80% relative humidity for 72 h.

2.2. Definition of Disease Severity (DS)

The DS of wheat PM is defined as the percentage of the disease pustules portion relative to the total leaf area, which was recorded by visual assessment according to a protocol by Graeff et al. [30]. Because this method is subjective, the assessment was carried out by just one investigator to eliminate any bias due to individual differences. Apart from using DS to quantify the incidence of PM, the leaves are also classified into the categories of healthy leaves (DS less than 1%) and diseased leaves (DS more than 1%) for detection of wheat PM (Figure 1).

0%   1%   5%   10%   20%   40%   60%   80%   100%

Figure 1. Images of wheat leaves with different DS.

2.3. Acquisition and Pre-Processing of Hyperspectral Images

In this study, we used the GaiaField portable HI system, consisting of an imaging lens coupled with an imaging spectrometer (V10E; Specim, Oulu, Finland) and a CCD camera (C8484-05; Hamamatsu Photonics, Osaka, Japan) (Figure 2). The HI system also included two light sources of 150 W halogen lamps (Oriel Instruments, Stratford, CT, USA) angled at 45°, a computer, a dark box, and a tripod. The spectral data were recorded in the VIS-NIR range of 400–1000 nm with a spectral resolution of 2.8 nm. The specification of each hyperspectral image was 1392 × 1040 (spatial dimensions) × 256 (spectral bands). The distance between the leaves and the camera was set as 300 mm and the exposure time was set as 0.13 s to obtain clear and non-deformable images. The data were processed and analyzed using the ENVI 5.2 (Exelis Visual Information Solutions, Boulder, CO, USA) and MATLAB R2014a (The MathWorks Inc., Natick, MA, USA) software packages.

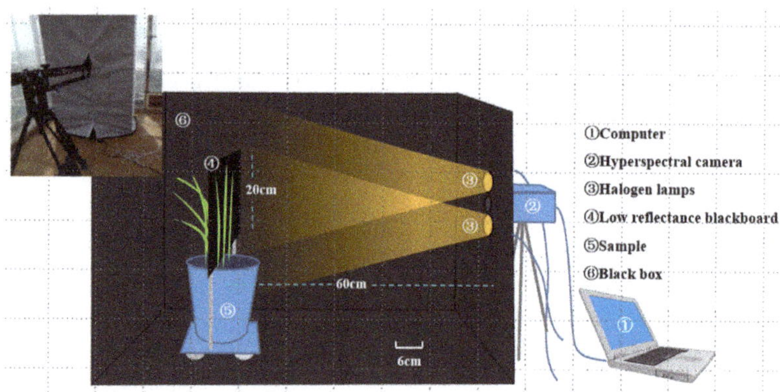

Figure 2. Test scenario of the hyperspectral imaging system.

The HI preprocessing steps included reflectance conversion, noise removal, and reflectance extraction. Because of the dark current and physical configuration of the imaging system, some bands with weaker light intensity contain excessive noise. Therefore, a raw hyperspectral image was calibrated to reduce the noise according to the following equation:

$$I_c = \frac{I_\Gamma - I_d}{I_w - I_d} \tag{1}$$

where I_c is the calibrated reflectance image, I_r is the raw hyperspectral image, I_d is the dark current image, and I_w is the white reference image; I_d is obtained by covering the camera lens with an opaque cap and turning off the light source, and I_w is acquired by imaging a polytetrafluoroethylene white panel with spectral reflectance of 99%.

After obtaining the calibrated spectral reflectance, to smooth and minimize the noise signals in the images, the noise of the hyperspectral image data was reduced by using the minimum noise fraction [31]. The average hyperspectral reflectance of the whole leaf was directly extracted from an image of a single leaf.

2.4. Selection of the Sensitive Feature

In this study, an SPA algorithm was applied to determine the sensitive wavelength, and then texture features of the corresponding sensitive wavelength were calculated for developing the normalized difference texture index (NDTI).

2.4.1. Construction of Texture Indices

Hyperspectral image has a gray image in each of its spectral bands, so there is a large amount of redundant information. Based on the gray-level co-occurrence matrix (GLCM) of the gray image in each selected spectral band, eight texture features were acquired for each gray image [32], as listed in Table 1. In addition to the texture features, a new texture-based index called the normalized difference texture index (NDTI) was proposed, which follows the conventional definition of normalized difference vegetation index. A new NDTI was constructed with all possible two-texture feature combinations at all sensitive wavelengths with eight GLCM-based texture features of each gray image. The texture index NDTI has been used to classify sun and shade leaves of crops and plant biomass estimation [33]. For example, six wavelengths were selected in this research with 8 texture features at each wavelength, thus 48 texture features were acquired. NDTI can be calculated with the combination of two different texture features from all 48 texture features. A total of 2256 combination pairs were obtained, and the specific formula used for NDTI calculation was NDTI = (T1 − T2)/(T1 + T2), where T1 and T2 are the texture features of the selected sensitive wavelength.

Table 1. Texture features based on gray-level co-occurrence matrix.

Number	Name	Equation	Description
1	Mean, MEA	$Mean = \sum_{i,j=1}^{G} iP(i,\ j)$	Reflects the average of grayscale
2	Variance, VAR	$Variance = \sum_{i=1}^{G}\sum_{j=1}^{G}(i-u)^2 P(i,j)$	Reflects the size of the grayscale change
3	Homogeneity, HOM	$Homogeneity = \sum_{i=1}^{G}\sum_{j=1}^{G}\frac{P(i,j)}{1+(i-j)^2}$	Reflects local homogeneity of texture
4	Contrast, CON	$Contrast = \sum_{i=1}^{G}\sum_{j=1}^{G}(i-j)^2 P(i,j)$	Reflects the clarity of the texture
5	Dissimilarity, DIS	$Dissimilaity = \sum_{i=1}^{G}\sum_{j=1}^{G} P(i,j)|i-j|$	Same as contrast, used to detect similarity
6	Entropy, ENT	$Entropy = -\sum_{i=1}^{G}\sum_{j=1}^{G} P(i,j)\log P(i,j)$	Measures the amount of information of an image
7	Second Moment, SEM	$Second\ Moment = \sum_{i=1}^{G}\sum_{j=1}^{G} P^2(i,j)$	Reflects the uniformity of the grayscale distribution of the image
8	Correlation, COR	$Correlation = \sum_{i=1}^{G}\sum_{j=1}^{G}\frac{(i-Mean_i)(j-Mean_j)P(i,j)}{\sqrt{Variance_i}\sqrt{Variance_j}}$	Reflects the extension of a gray value along a certain direction

Notes: $P(i,j) = V(i,j)/\sum_{i=0}^{G}\sum_{j=0}^{G} V(i,j)$ where $V(i,j)$ is the value in the cell at row of i and column of j and G is the number of rows or columns.

Here, the sensitive wavelengths for each texture feature were selected by SPA. SPA is a new statistics-based feature selection algorithm that combines a random-forest permutation method and model-based cluster analysis. The main aspects of SPA are the establishment of sub-models based on samples and variable sub-windows, and the subjection of parameters of interest to strict statistical analysis. The detailed implementation of SPA can be seen in Li et al. [26].

2.4.2. Selection of Vegetation Indices

When wheat is infected with PM the physiological and biochemical parameters of its leaves change accordingly [16], thereby changing its spectral reflectance. In total, 15 vegetation indices (VIs) frequently investigated in disease research were selected for this study, as listed in Table 2. And these 15 VIs are all highly related to the physiological and biochemical parameters.

Table 2. Vegetation indices used in this study.

	Definition	Equations	Reference
1.	Powdery mildew index, PMI	$(R515 - R698)/(R515 + R698) - 0.5 * R738$	[34]
2.	Modified simple ratio, MSR	$(R800/R670 - 1)/(R800/R670 + 1)^{1/2}$	[35]
3.	Photochemical reflectance index, PRI	$(R570 - R531)/(R570 + R531)$	[36]
4.	Photosynthetic radiation, PhRI	$(R550 - R531)/(R550 + R531)$	[36]
5.	Modified chlorophyll absorption ratio index, MCARI	$[(R701 - R671) - 0.2(R701 - R549)]/(R701/R671)$	[37]
6.	Anthocyanin reflectance index, ARI	$(R550)^{-1} - (R700)^{-1}$	[38]
7.	Structure independent pigment index, SIPI	$(R800 - R445)/(R800 - R680)$	[39]
8.	Normalized pigment chlorophyll ration index, NPCI	$(R680 - R430)/(R680 + R430)$	[39]
9.	Red-edge vegetation stress index, RVSI	$[(R712 + R752)/2] - R732$	[40]
10.	Narrow-band normalized difference vegetation index, NBNDVI	$(R850 - R680)/(R850 + R680)$	[41]
11.	Nitrogen reflectance index, NRI	$(R570 - R670)/(R570 + R670)$	[42]
12.	Triangular vegetation index, TVI	$0.5[120(R750 - R550) - 200(R670 - R550)]$	[43]
13.	Transformed chlorophyll absorption and reflectance index, TCARI	$3[(R700 - R670) - 0.2(R700 - R550)(R700/R670)]$	[44]
14.	Plant senescence reflectance index, PSRI	$(R680 - R500)/R750$	[45]
15.	Aphid index, AI	$(R740 - R887)/(R691 - R698)$	[46]

2.5. Development of the Recognition Model for Wheat Leaf Disease

The PLS-LDA method was used to classify the healthy and diseased wheat leaves with the sensitive features of vegetation and texture indices. PLS-LDA is effective in processing data such as small sample size, high dimensionality, and multicollinearity [39]. In the recognition model, the actual measured DS was quantitatively classified into two classes of healthy and diseased for discriminating PM infection. The model performance was evaluated using the overall accuracy (OA) derived from a confusion matrix and Kappa coefficient. An accurate model should have higher values of OA and Kappa coefficient.

2.6. Construction of the DS Estimation Model for Wheat Leaf Disease

Furthermore, partial least-squares regression (PLSR) was used to establish the regression relationships between the indices (VIs and NDTIs) and the DS for wheat leaves at different growth stages. As a mature multiple linear regression algorithm, PLSR is the simplest partial least-squares method, and has been widely used to analyze agricultural remote-sensing data [40]. In the severity model, the relationship between actual measured DS and indices was examined to evaluate the sensitivity of indices to PM at different growth stages. The performance of the wheat DS estimation model was evaluated by the coefficient of determination (R^2), the root mean square error ($RMSE$), and the relative root mean square error ($RRMSE$). R^2 can illustrate the correlation between VIs/NDTIs and DS. The robustness of models was assessed by both $RMSE$ and $RRMSE$. The higher R^2 and the lower $RMSE$ and $RRMSE$ values, the better the model performs. R^2, $RMSE$, and $RRMSE$ can be calculated as following equations:

$$R^2 = 1 - \frac{\sum_i \left(\hat{y}^i - y^i\right)^2}{\sum_i \left(\overline{y}^i - y^i\right)^2} \tag{2}$$

$$RMSE = \sqrt{\left(\frac{1}{M} \sum_1^M (y^i - \hat{y}^i)^2\right)} \tag{3}$$

$$RRMSE = \frac{RMSE}{\overline{y}^i} \tag{4}$$

where y^i was the observed DS, $\hat{y}^i$ was the estimated DS, $\overline{y}^i$ was the mean of DS observation, and M was the number of samples.

The calibration model was built using all data to achieve better performance, but the model may be overfitting. To avoid this phenomenon, 10-fold cross validation was adopted to assess the performance of models. In this process, the data were separated in 10 folds. The 9-fold data were used to build PLSR model. Then, the constructed model was used to predict the remaining 1-fold data. After 10 trials, all DS estimates can be obtained.

3. Results

3.1. Time-Series Variation of Spectral Reflectance

Figure 3 shows the spectral-reflectance changes of healthy and diseased leaves on different days after infection at the jointing stage. The average DS of the wheat leaves increased gradually over time. Compared with healthy leaves, the spectral reflectance of diseased leaves was higher in the VIS region but lower in the NIR region, which is consistent with previous studies [17]. Moreover, the spectral signature in the NIR region differed more obviously between normal and diseased leaves at late infection stage.

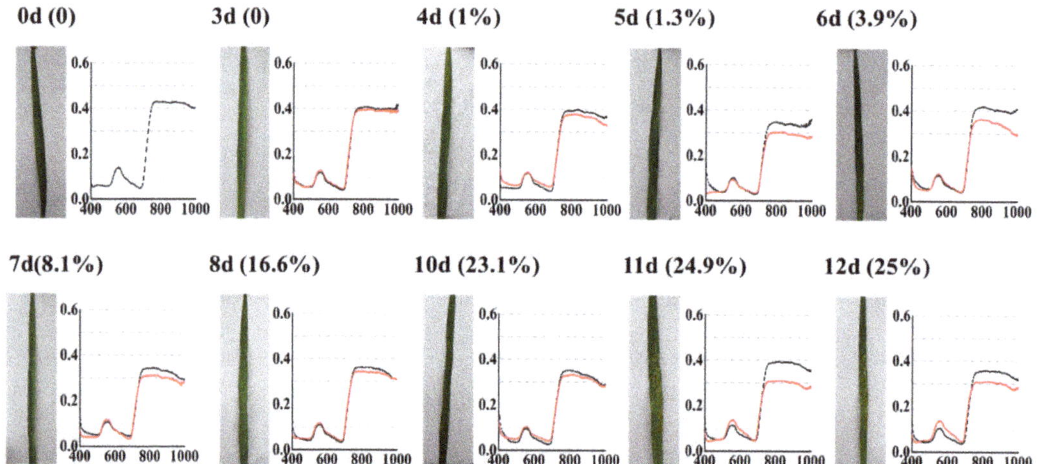

Figure 3. Reflectance changes of healthy (black line) and diseased (red line) wheat leaves after infection at the jointing stage. Note: X axis: Wavelength (nm); Y axis: reflectance; "12d (25%)" represents the 12th day after infection with an average DS of 25%.

3.2. Selection of the Sensitive Features

In this study, we used a new feature-selection algorithm SPA for selection of sensitive wavelengths for wheat PM, and then we calculated 8 texture features based on the selected sensitive wavelengths for developing the NDTIs. Meanwhile, sensitive VIs were selected and analyzed to compare with the performance of NDTIs. The results of feature selection are as the below.

3.2.1. Selection of Sensitive Wavebands

Due to the noisy spectral reflectance data around 400 and 1000 nm, the wavelengths used in this study were in the range of 437.2–976.2 nm. By using an SPA algorithm, we obtained the conditional synergetic score (COSS) values of each wavelength, indicating its sensitivity to PM, as shown in Figure 4A. The higher the COSS value, the more the waveband is sensitive to disease. Then we ranked the spectral bands in descending order according to their COSS values and selected the top 20 bands further for construction of PLS-LDA models. We successively added these bands to the PLS-LDA model one by one according to their COSS values in descending order and recorded their calibration and validation accuracies for selection of the best combination of sensitive wavebands. These calibration and leave-one-out cross validation (LOOCV) accuracies are shown in Figure 4B. Cross validation accuracies gradually increased and achieved highest coinciding point at the sixth waveband, hence, top six wavebands with higher COSS values were selected as sensitive wavelengths, which were 533.5, 659.2, 528.5, 661.7, 968.2, and 523.6 nm by descending COSS values.

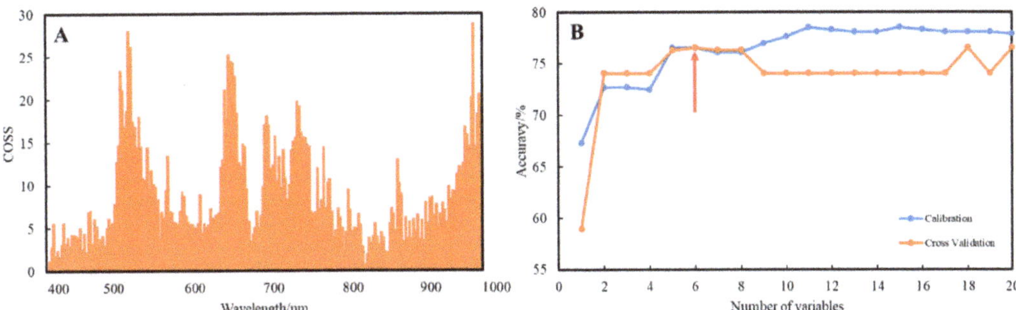

Figure 4. (**A**): COSS value computed by SPA for each waveband. (**B**): Calibration and cross validation classification accuracies at different number of variables based on PLS-LDA.

3.2.2. Selection of Optimal Vegetation Indices

The sensitive VIs were selected by correlation analysis between the VIs and DS of wheat PM in two growth stages. Figure 5 summarizes the sensitivity of 15 VIs to DS at each of the two individual growth stages and at both growth stages as a whole. Except for TVI, these selected VIs were significantly correlated to DS at both stages. The response of VIs to DS at different growth stages differed from that of both stages. Most of the VIs showed high correlation to DS at booting stage, while only a few showed good relationships with DS at the jointing stage. The response of the VIs in different growth stages is not the same. In total, six spectral vegetation indices (PMI, MSR, MCARI, ARI, SIPI, and PSRI) were significantly correlated with DS at each growth stage as well as for both stages as a whole. They were selected for further disease recognition regression and classification models.

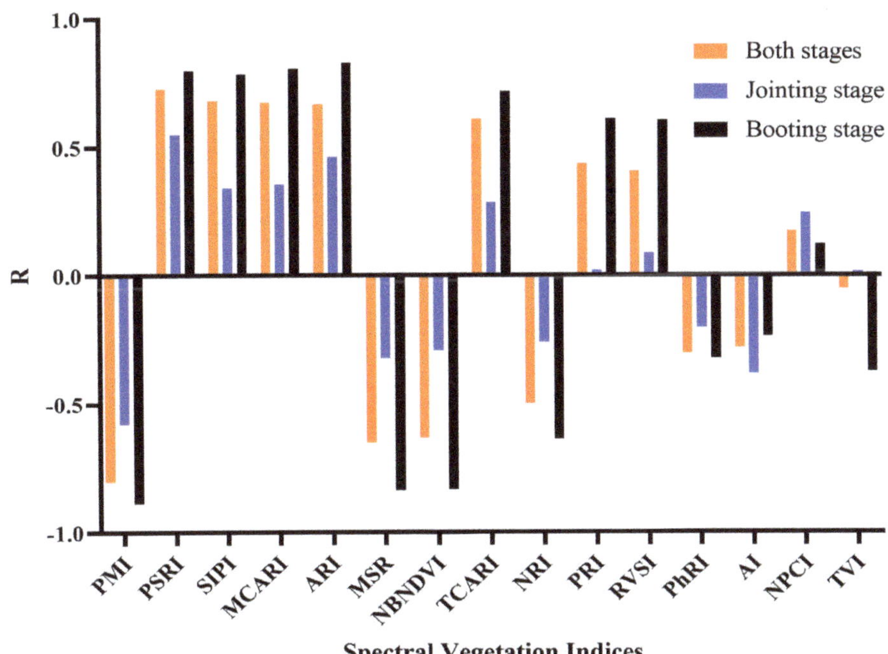

Figure 5. Response of spectral vegetation indices to DS at both and single growth stages.

3.2.3. Extraction of Texture Features

Because of the redundant information in hyperspectral images, here texture features of images at the six selected sensitive wavelengths were extracted, for the further development of NDTIs. Eight texture features for each spectral band and in total 48 texture features were adopted. Figure 6 shows one of the texture images at the 533.5 nm wavelength. Texture images at same wavelength are various due to different texture features.

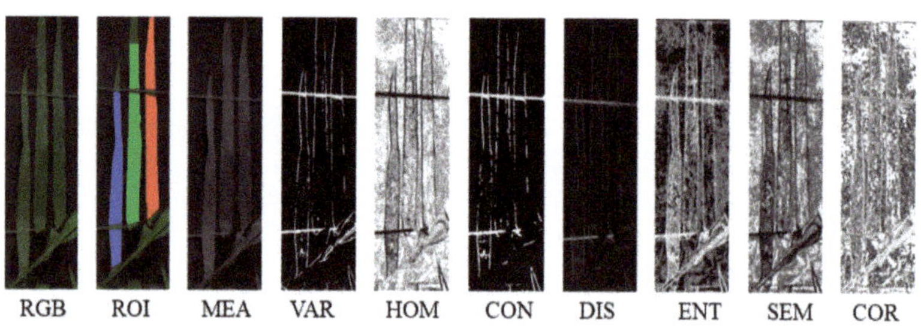

Figure 6. Texture images at wavelength of 533.5 nm. Note: MEA = mean; VAR = variance; HOM = homogeneity; CON = contrast; DIS = dissimilarity; ENT = entropy; SEM = second moment; COR = correlation).

The Pearson correlation coefficient r was used to examine the relationships between the texture features and the observed DS at different growth stages (Figure 7). The results showed that the texture features were poorly correlated with DS at the jointing stage but highly correlated at the booting stage. Whether it was at the jointing or the booting stage, 20 textures, namely homogeneity (HOM), dissimilarity (DIS), entropy (ENT), and second moment (SEM) at the wavelengths of 523.6, 528.5, 533.5, 659.2, and 661.7 nm were significantly correlated with DS.

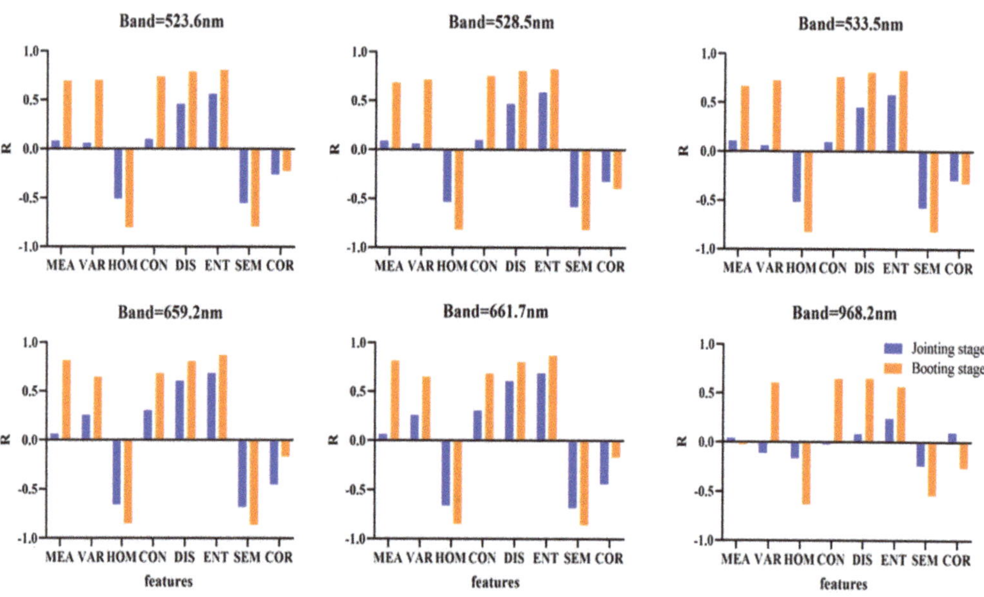

Figure 7. Pearson correlation analysis between texture features and DS.

3.2.4. Calculation of Normalized Difference Texture Indices (NDTIs)

NDTIs were constructed to improve the performance of texture analysis in disease recognition as the texture features did not exhibit consistent relationships to DS at two growth stages. Based on the aforementioned 48 texture features, 2256 NDTIs were realized by combining all two possible texture features, and the significance of those NDTIs were analyzed by correlation analysis. Among the 2256 NDTIs, only 10 NDTIs were sensitive to DS at different growth stages and throughout both growth stages, as listed in Figure 8. Ten best performing NDTIs were constructed by a combination of the following seven texture features, namely the second moment (SEM) at 659.2 nm and 661.7 nm and the homogeneity (HOM) at 523.6 nm, 528.5 nm, 533.5 nm, 659.2 nm, and 661.7 nm. Compared with texture features in Figure 7, the NDTIs are better correlated to disease severity. Thus, these NDTIs were selected for further construction of the disease detection model.

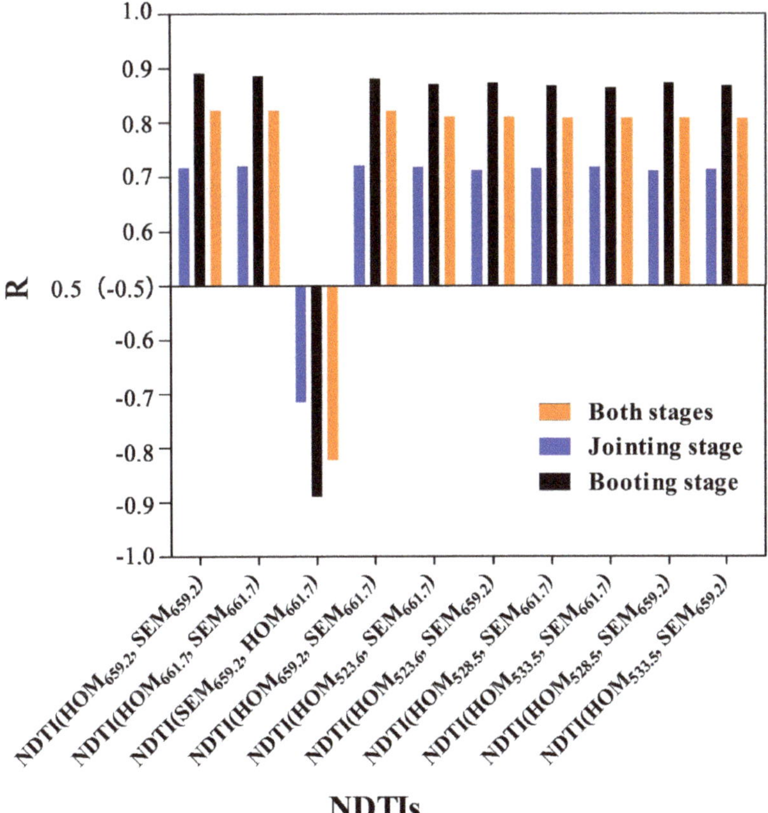

Figure 8. The correlation coefficient of the top 10 normalized difference texture indices (NDTI) which are highly correlated with DS.

3.3. PLS-LDA Model for Classifying the Healthy and Diseased Leaves

3.3.1. Evaluation of PLS-LDA Model Based on Different Selected Sensitive Features

The PLS-LDA model was built to evaluate the performance of different indices in terms of detecting wheat PM at different growth stages. The classification results were evaluated both for calibration and validation accuracies, as listed in Table 3. Regardless of growth stage, spectral vegetation indices achieved better classification accuracies than texture indices. However, the model based on the combination of VIs and NDTIs performed best in comparison with that of VIs and NDTIs when used alone as input features. The combination of VIs and NDTIs significantly increased the classification accuracy, yielding calibration accuracies of 75.34%, 80.72%, and 76.46% and validation accuracies of 74.88%, 78.93%, and 76.23% at jointing stage, booting stage, and throughout both stages, respectively. The classification accuracies at booting stage were superior, followed by that at both stages and jointing stage, respectively.

Table 3. Classification results of PLS-LDA model based on different features.

Dataset	Inputted Features	Features	Calibration Accuracy (%)			Validation Accuracy (%)		
			Healthy	Infected	Overall	Healthy	Infected	Overall
Both stages	VIs	6	76.49	78.26	77.13	74.19	74.00	74.22
	NDTIs	10	72.63	67.70	70.85	72.53	65.96	70.18
	VIs & NDTIs	16	77.17	75.16	76.46	77.62	73.72	76.23
Jointing stage	VIs	6	71.71	76.06	73.09	72.70	69.64	72.65
	NDTIs	10	73.68	73.24	73.54	73.47	67.80	72.23
	VIs & NDTIs	16	76.97	71.83	75.34	74.63	74.43	74.88
Booting stage	VIs	6	85.71	62.22	76.23	82.36	63.29	75.36
	NDTIs	10	75.19	63.33	70.40	77.42	63.36	70.30
	VIs & NDTIs	16	84.21	75.56	80.72	91.97	61.40	78.93

Note: VIs = vegetation indices; NDTIs = normalized difference texture indices; VIs & NDTIs = combination of VIs and NDTIs.

3.3.2. Classification of Healthy and Diseased Leaves at Early Stage after Inoculation

As demonstrated in Section 3.3.1 (Table 3), the model based on combination of VIs and NDTIs revealed better results. So, further for classification of healthy and diseased leaves as early as possible after inoculation, we established the PLS-LDA model based on combined features (VIs & NDTIs), on datasets of different days after inoculation (DAI) of each growth stage (Table 4). At the jointing stage, healthy and diseased leaves could be better distinguished on 6 DAI with an average DS of 3.9%, as the overall classification accuracy on this day approached more than 85% with a kappa value of 0.73 and the accuracies for healthy and diseased leaves were 82.35% and 100%, respectively. While, at booting stage, on 3 DAI the overall accuracy reached more than 85% with a kappa value of 0.73 and the accuracies for healthy and diseased leaves were recorded as 81.25% and 100%, respectively. The mean disease severity on this day was noted as 1.1%. Therefore, it can be concluded that wheat PM can be better diagnosed at an early-stage with a maximum of up to 3–6 DAI or at a mean DS of around 1–9%, by using combined (VIs & NDTIs) sensitive features as an input in classification model.

Table 4. PLS-LDA classification results of healthy and diseased leaves based on combined spectral and texture indices (VIs & NDTIs) at the jointing and booting stage.

		Classification Accuracies (%)							
Growth stage	DAI	4	5	6	7	8	10	11	12
	T (°C)	21.5	21	17.5	11	13	12	8.8	15.8
	DS/State	1%	1.3%	3.9%	8.1%	16.6%	23.1%	24.9%	25%
Jointing stage	Healthy	56.25	57.14	82.35.	87.5	84.62	87.5	100	100
	Diseased	100	100	100	87.5	100	100	90.91	90.91
	OA (%)	63.16	64.71	87.50	87.50	90.91	91.67	95.65	95.65
	Kappa	0.29	0.32	0.73	0.73	0.82	0.82	0.91	0.91
Growth stage	DAI	3	4	5	7	9	10	11	12
	T (°C)	27.7	25.8	31	31.5	30.8	32.4	29.3	33.2
	DS/State	1.1%	6.2%	9.6%	13.5%	25.6%	25.7%	31.9%	45.4%
Booting stage	Healthy	81.25	88.89	92.31	100	91.67	100	100	100
	Diseased	100	100	100	100	100	100	100	100
	OA	86.96	94.44	95.83	1	95.45	1	1	1
	Kappa	0.73	0.89	0.92	1	0.91	1	1	1

Note: DAI indicates days after inoculation, T is mean temperature of green house in degree centigrade, DS is mean disease severity, Healthy and Diseased indicate classification accuracies of inoculated and non-inoculated wheat leaves, OA is overall classification accuracy.

3.4. PLSR Model for Estimating the Disease Severity

For estimating the disease severity, the aforementioned six sensitive vegetation indices, 10 texture indices and 16 combined features (i.e., VIs & NDTIs) were used to construct the partial least square regression (PLSR) model with 10-fold cross-validation (Table 5). Across growth stages, the R^2 and *RMSE* results of booting stage were superior regardless of input features, followed by pooled data of both stages and then jointing stage, respectively. Based on input features, the model constructed by using VIs was comparatively better in the validation model than that of the model constructed by using NDTIs. However, the model constructed by using combined features (i.e., VIs & NDTIs) yielded significantly better results, both for calibration and validation, than the models separately based on VIs or NDTIs.

Table 5. Results of partial least-squares regression (PLSR) models based on different inputted features.

Growth Stage	Inputted Features	Number of Features	Calibration		Validation		
			R^2	*RMSE*	R^2	*RMSE*	*RRMSE*
Both stages	VIs	6	0.687	14.166	0.660	14.761	0.597
	NDTIs	10	0.694	14.001	0.649	14.992	0.606
	VIs & NDTIs	16	0.748	12.711	0.722	13.356	0.540
Jointing stage	VIs	6	0.527	15.166	0.431	16.636	0.924
	NDTIs	10	0.531	15.102	0.344	17.872	0.993
	VIs & NDTIs	16	0.619	13.624	0.532	15.162	0.842
Booting stage	VIs	6	0.831	9.990	0.792	11.511	0.384
	NDTIs	10	0.815	11.374	0.747	13.230	0.443
	VIs & NDTIs	16	0.855	10.060	0.818	11.320	0.377

Figure 9 shows DS estimation models based on combined features (i.e., VIs & NDTIs), by using the validation datasets of different growth stages. It can be observed that the measured DS is close to the predicted DS which lies along line of concordance (1:1 solid line), confirming the accuracy of the DS estimation model. The performance of the estimation model at booting dataset was superior, i.e., the R^2 values between measured and estimated DS was 0.818 and *RMSE* was 11.320, followed by pooled data of both stages (R^2 = 0.722, *RMSE* = 13.356). While the model performance at jointing stage was relatively poor (R^2 = 0.532, *RMSE* = 15.32) and the predicted DS was overestimated. The above results suggest that the PLSR model based on the combined features of VIs and NDTIs performed better for detecting wheat PM disease severity.

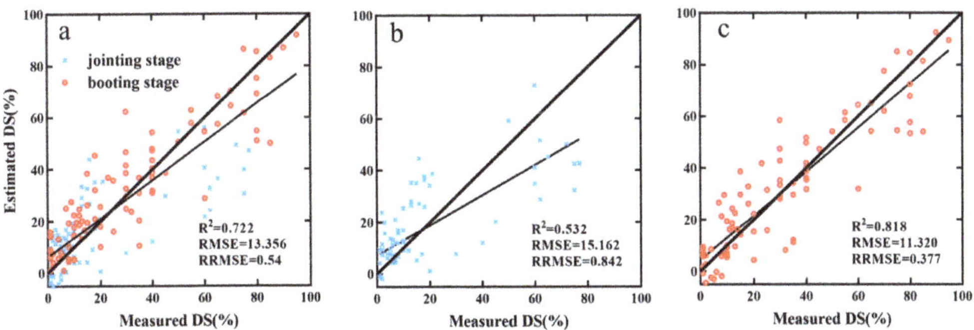

Figure 9. Comparison of measured and estimated wheat powdery mildew DS based on combined features of VIs & NDTIs, at both stages (**a**), jointing stage (**b**), and booting stage (**c**).

4. Discussion

4.1. Why the Selected Features Are Rational?

In this study, six sensitive spectral bands (533.5, 659.2, 528.5, 661.7, 968.2, and 523.6 nm) were selected by SPA method to reduce the hyperspectral dimensions, in agreement with previous studies revealing sensitive spectral regions to wheat PM [30]. Among these spectral bands, three wavelengths at around 550 nm are sensitive to photosynthetic pigments and have great potential for early disease monitoring, in accordance with the findings of Maimaitiyiming [47]. When a pathogen infects wheat, the physiological parameters respond accordingly, such as decreased chlorophyll content, which affects the spectral reflectance in the visible light region. As shown in Figure 10, the chlorophyll content of healthy and diseased leaves on different days after infection were analyzed by analysis of variance ($p < 0.05$). At the jointing stage, the chlorophyll content of healthy and diseased leaves differed significantly on day 8 after infection. At the booting stage, the chlorophyll content of healthy and diseased leaves differed significantly on day 6 after infection. Decreased chlorophyll content weakened the ability to absorb light, therefore resulting in higher reflectance in the VIS range.

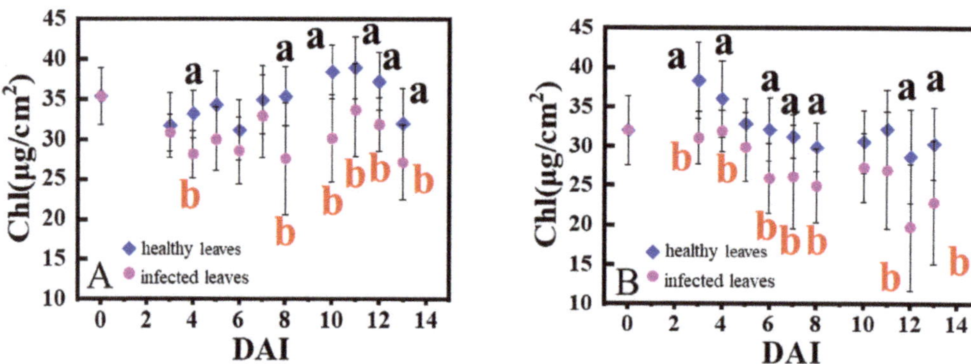

Figure 10. Changes and comparison of chlorophyll content between healthy leaves and diseased leaves infected at jointing stage (**A**) and booting stage (**B**) on the day after inoculation (DAI). Note: ab represents a difference that is significant ($p < 0.05$).

In the near-infrared range, the reflectance is affected by leaf anatomy, which corresponds to the sensitive wavelength at 968.2nm in this study. As shown in Figure 11, the chloroplast structure (marked by "Ch") in leaves with different DS on 13 DAI was observed using a transmission electron microscope. As the DS increased, the chloroplast in the wheat leaves gradually swelled in the cells. Cavities of varying sizes appeared surrounded by a single layer of membrane, and the chloroplast disintegrated eventually. The cells in the normal leaves form multiple scattering of light resulting in high reflectance, while powdery mildew affects the internal structure of leaves, thus leading to lower reflectance of infected leaves.

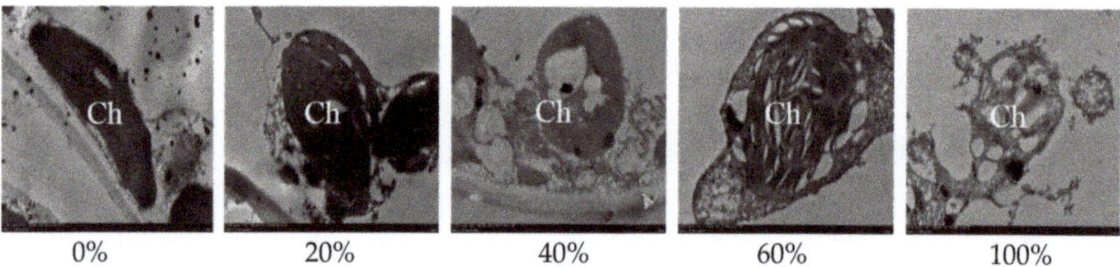

Figure 11. Chloroplast structure of wheat at different disease severity (DS) after infection.

For texture features, Xie et al. [48] first compared spectral and texture features for detecting different diseases on tomato leaves, and the results obtained by texture features (60.2%) were slightly worse than those obtained by reflectance (97.1%) in classification accuracy. Al-Saddik et al. [23] used hyperspectral images to detect yellowness and esca in grapevines by combining spectral and texture data with an overall accuracy of 99% for both of the diseases. The adoption of NDTI in disease detection of this study is inspired by Zheng [33], who combined NDTIs and VIs to estimate above ground crop biomass with R^2 of 0.78. However, the features selected in this study are specific to the wheat disease PM. When this methodology is used to monitor diseases on other crops, it would need to re-select the sensitive spectral features and corresponding texture features, as the symptoms (shape and color) of each disease are different.

4.2. What Is the Reason of Detection Performance at Varied Growth Stages?

The detection at the booting stage performed significantly better than that of jointing stage, which might be due to the different rates of disease propagation at two growth stages. Most of the DS values for diseased leaves obtained after infection at the jointing and booting stages were 1–30% and 1–50%, respectively. The factors which affect the development of wheat PM are environment conditions [49], variety resistance [3], and the number of pathogen spores [50]. The varieties, cultivation practices and the number of spores for inoculations used in the two growth stages were the same. Hence, the difference was mainly due to the different climatic conditions during the two inoculation periods. The inoculation interval between the two growth stages differed by 14 days. The inoculation time at the jointing stage was early spring, at this stage the temperature was still low and gradually rising, therefore the development of the disease was slow. At low temperature, the PM fungus tends to reproduce in small quantities and the physiological and biochemical characteristics of diseased plants may not be obvious. However, the disease developed rapidly at the booting stage because of the suitable environment conditions. It was the time of late spring and conditions of high temperature and humidity were optimal for disease development, consistent with the observations of Yao et al. [51]. Such disease favorable environmental conditions enhanced the rate of disease development at the booting stage, thus inducing obvious changes in the biophysical and biochemical parameters of plants which resulted in changed spectral responses [52]. Moreover, high lesion ratios on the leaf surfaces at booting stage changed the texture patterns of wheat leaves more severely thus, ensuring robust and accurate disease detection at booting stage.

4.3. How Early to Detect the Disease When the Combined Feature Is Applied?

This paper focuses on the methodology with regard to the feasibility of detecting the disease at its early stage or even before symptoms appear, at which it is difficult to acquire the weak signal due to the limited bands and course resolution. To the best of our knowledge, the present work is the first attempt to detect wheat PM by combining spectral vegetation indices and texture indices. Wheat PM lesions induce modification of color and surface properties of wheat leaves, and the texture information describes the intensity of the pixel's changes in an image. This phenomenon explains the robustness of the texture approach combined with the spectral approach in the better diagnosis of disease. Previous studies found that the spectral approach performed better when the DS is usually greater than 20%, with classification accuracy of about 87%. Meanwhile, the texture approach performed better in distinguishing DS of less than 20% with classification accuracy of 84%, but the accuracy by using a combination of texture and spectral features can reach up to 90% [53]. Therefore, combining spectral and texture information allows significantly better classification accuracy of diseased and healthy leaves and DS estimation at the early stage of wheat PM.

The early detection of disease on a daily basis after inoculation at different growth stages was carried out by combining the spectral and texture indices, and the classification accuracies of healthy and diseased leaves increased with increasing DS. On 6 DAI at jointing stage, the classification of healthy and diseased leaves was robust and stable, when the average DS was 3.9%. While at the booting stage, disease was recognized well on 3 DAI with an average DS of 1.1%. As the DS increased, the spectral and texture features changed, and the classification accuracies improved as the days passed after inoculation. However, at 9 DAI of booting stage, a slight decrease in classification accuracy was observed. These slight variations in classification accuracies were due to the unbalanced number of healthy and diseased leaves. At the booting stage on 9th DAI, most of the leaves were infected, so the model tended to predict the diseased one, resulting in decreased accuracy of healthy leaves. In another aspect, the variations of chlorophyll contents in leaves are also consistent with the model's performance for disease detection, as the model correctly identified disease on 6 and 3 DAI (Table 4) with an OA of over 85% and chlorophyll contents significantly changed between healthy and diseased leaves on 8 and 6 DAI (Figure 10), at jointing and booting stages, respectively. Therefore, our results well support the idea that HI is an efficient and reliable method for plant disease detection at early stage. It is worth mentioning that ground-based hyperspectral analysis is not suitable for a large area, and it is too expensive to be used by average farmers compared with UAV platforms. Multi-spectral UAV platforms can monitor disease in large areas with stable and high efficiency but perform poorly at the early disease infection stage. In the future, we believe there would be more hyperspectral equipment mounted on UAV systems, however currently there is only one kind of UHD185 in the community.

5. Conclusions

In this paper, spectral and texture analysis of hyperspectral imaging technique were used to discriminate wheat powdery mildew without obvious symptoms and estimate corresponding disease severity. This study indicates that the combination of spectral and texture approaches is the sensitive feature for disease detection when building a PLS-LDA classification model and PLSR DS estimation model, which offered significantly improved accuracies in detection and quantification of wheat PM disease. The model based on selected sensitive features identified the disease even before the initiation of significant variations in the physiological and biochemical parameters of leaves after disease. Therefore, it can provide the basis for early detection, preventing and controlling the plant disease worldwide. The stable performance of the spectral and texture indices enabled the early detection of disease with more than 85% overall classification accuracies of PLS-LDA model, especially at early infection stage of 3–6 DAI with a DS of 1–6%. The DS can be well estimated by PLSR model with the R^2 value of 0.818 at booting growth stage. However, the present study is conducted on leaf scale with relatively few trials, and only used limited number of diseased and healthy leaves for classification on a daily basis after inoculation. In the future, we would evaluate the feature and model with a greater number of trials and data, and hope to expand to field and canopy level.

Author Contributions: Conceptualization, X.Y., H.L. (Hongyan Liu); methodology, H.L. (Haiyan Liu), H.L. (Hongyan Liu) and I.H.K.; software, H.L. (Hongyan Liu); validation, H.L. (Haiyan Liu); formal analysis, I.H.K.; investigation, I.H.K. and H.L. (Hongyan Liu); resources, W.L., X.W.; writing—original draft preparation, H.L. (Hongyan Liu) and H.L. (Haiyan Liu); writing—review and editing, H.L. (Haiyan Liu), X.Y., A.C., W.L., T.C., Y.T., Y.Z., W.C., supervision, X.Y.; project administration, X.Y.; funding acquisition, X.Y. All authors have read and agreed to the published version of the manuscript.

Funding: This work was supported by grants from the National Key Research and Development Plan of China (2019YFE011721), National Natural Science Foundation of China (31971780), the Key Projects (Advanced Technology) of Jiangsu Province (BE 2019383), 333 Project of Jiangsu Province (JS333), Collaborative Innovation Center for Modern Crop Production co-sponsored by Province and Ministry(CIC-MCP), the Priority Academic Program Development of Jiangsu Higher Education Institutions (PAPD), and the 111 project (B16026).

Institutional Review Board Statement: Not applicable.

Informed Consent Statement: Not applicable.

Data Availability Statement: Not applicable.

Acknowledgments: The authors would like to thank the reviewers for recommendations which improved the manuscript.

Conflicts of Interest: The authors declare no conflict of interest.

References

1. Liu, L.Y.; Song, X.Y.; Li, C.J.; Qi, N.; Huang, W.J.; Wang, J.H. Monitoring and evaluation of the diseases of and yield winter wheat from multi-temporal remotely-sensed data. *Trans. CSAE* **2009**, *25*, 137–143.
2. Foulkes, M.J.; Paveley, N.D.; Worland, A.; Welham, S.J.; Thomas, J.; Snape, J.W. Major genetic changes in wheat with potential to affect disease tolerance. *Phytopathology* **2006**, *96*, 680–688. [CrossRef] [PubMed]
3. Bingham, I.J.; Walters, D.R.; Foulkes, M.J.; Paveley, N.D. Crop traits and the tolerance of wheat and barley to foliar disease. *Ann. Appl. Biol.* **2009**, *154*, 159–173. [CrossRef]
4. Qin, W.; Xue, X.; Zhang, S.; Gu, W.; Wang, B. Droplet deposition and efficiency of fungicides sprayed with small UAV against wheat powdery mildew. *Int. J. Agric. Biol. Eng.* **2018**, *11*, 27–32. [CrossRef]
5. Kuska, M.; Wahabzada, M.; Leucker, M.; Dehne, H.W.; Kersting, K.; Oerke, E.C. Hyperspectral phenotyping on the microscopic scale: Towards automated characterization of plant-pathogen interactions. *Plant Methods* **2015**, *11*, 28. [CrossRef]
6. Mutka, A.M.; Bart, R.S. Image-based phenotyping of plant disease symptoms. *Front. Plant Sci.* **2015**, *5*, 734. [CrossRef]
7. Siricharoen, P.; Scotney, B.; Morrow, P.; Parr, G. A lightweight mobile system for crop disease diagnosis. In *International Conference Image Analysis and Recognition. Lecture Notes in Computer Science*; Springer: Cham, Switzerland, 2016; Volume 9730, pp. 783–791.
8. Barbedo, J.G.A. A review on the main challenges in automatic plant disease identification based on visible range images. *Biosyst. Eng.* **2016**, *144*, 52–60. [CrossRef]
9. Mirik, M.; Jones, D.C.; Price, J.A.; Workneh, F.; Ansley, R.J.; Rush, C.M. Satellite remote sensing of wheat infected by wheat streak mosaic virus. *Plant Dis.* **2011**, *95*, 4–12. [CrossRef]
10. Zhang, D.; Zhou, X.; Zhang, J.; Lan, Y.; Xu, C.; Liang, D. Detection of rice sheath blight using an unmanned aerial system with high-resolution color and multispectral imaging. *PLoS ONE* **2018**, *13*, e0187470. [CrossRef]
11. Li, L.; Zhang, Q.; Huang, D. A review of imaging techniques for plant phenotyping. *Sensors* **2014**, *14*, 20078–20111. [CrossRef] [PubMed]
12. Zarco-Tejada, P.J.; Morales, A.; Testi, L.; Villalobos, F. Spatio-temporal patterns of chlorophyll fluorescence and physiological and structural indices acquired from hyperspectral imagery as compared with carbon fluxes measured with eddy covariance. *Remote. Sens. Environ.* **2013**, *133*, 102–115. [CrossRef]
13. Kim, D.M.; Zhang, H.; Zhou, H.; Du, T.; Wu, Q.; Mockler, T.C.; Berezin, M.Y. Highly sensitive image-derived indices of water-stressed plants using hyperspectral imaging in SWIR and histogram analysis. *Sci. Rep.* **2015**, *5*, 15919. [CrossRef] [PubMed]
14. Ye, X.; Abe, S.; Zhang, S. Estimation and mapping of nitrogen content in apple trees at leaf and canopy levels using hyperspectral imaging. *Precis. Agric.* **2020**, *21*, 198–225. [CrossRef]
15. Zhou, R.Q.; Jin, J.J.; Li, Q.M.; Su, Z.Z.; Yu, X.J.; Tang, Y.; Luo, S.M.; He, Y.; Li, X.L. Early detection of magnaporthe oryzae-infected barley leaves and lesion visualization based on hyperspectral imaging. *Front. Plant Sci.* **2019**, *9*, 1962. [CrossRef] [PubMed]
16. Shi, Y.; Huang, W.; González-Moreno, P.; Luke, B.; Dong, Y.; Zheng, Q.; Ma, H.; Liu, L. Wavelet-based rust spectral feature set (WRSFS): A novel spectral feature set based on continuous wavelet transformation for tracking progressive host–pathogen interaction of yellow rust on wheat. *Remote Sens.* **2018**, *10*, 525. [CrossRef]
17. Zhang, J.C.; Pu, R.L.; Wang, J.H.; Huang, W.J.; Yuan, L.; Luo, J.H. Detecting powdery mildew of winter wheat using leaf level hyperspectral measurements. *Comput. Electron. Agric.* **2012**, *85*, 13–23. [CrossRef]
18. Shi, Y.; Huang, W.; Luo, J.; Huang, L.; Zhou, X. Detection and discrimination of pests and diseases in winter wheat based on spectral indices and kernel discriminant analysis. *Comput. Electron. Agric.* **2017**, *141*, 171–180. [CrossRef]
19. Zheng, Q.; Huang, W.; Cui, X.; Dong, Y.; Shi, Y.; Ma, H.; Liu, L. Identification of wheat yellow rust using optimal three-band spectral indices in different growth stages. *Sensors* **2019**, *19*, 35. [CrossRef]
20. Gowen, A.A.; O'Donnell, C.P.; Cullen, P.J.; Downey, G.; Frias, J.M. Hyperspectral imaging—An emerging process analytical tool for food quality and safety control. *Trends Food Sci. Technol.* **2007**, *18*, 590–598. [CrossRef]
21. Eckert, S. Improved forest biomass and carbon estimations using texture measures from WorldView-2 satellite data. *Remote Sens.* **2012**, *4*, 810–829. [CrossRef]
22. Yang, Y. *The Key Diagnosis Technology of Rice Blast Based on Hyperspectral Image*; Zhejiang University: Hangzhou, China, 2012.
23. Al-Saddik, H.; Laybros, A.; Billiot, B.; Cointault, F. Using image texture and spectral reflectance analysis to detect yellowness and esca in grapevines at leaf-level. *Remote Sens.* **2018**, *10*, 618. [CrossRef]
24. Lu, J.; Zhou, M.; Gao, Y.; Jiang, H. Using hyperspectral imaging to discriminate yellow leaf curl disease in tomato leaves. *Precis. Agric.* **2018**, *19*, 379–394. [CrossRef]
25. Mahlein, A.K.; Rumpf, T.; Welke, P.; Dehne, H.W.; Plumer, L.; Steiner, U.; Oerke, E.C. Development of spectral indices for detecting and identifying plant diseases. *Remote Sens. Environ.* **2013**, *128*, 21–30. [CrossRef]
26. Li, H.-D.; Zeng, M.; Tan, B.-B.; Liang, Y.-Z.; Xu, Q.-S.; Cao, D.-S. Recipe for revealing informative metabolites based on model population analysis. *Metabolomics* **2010**, *6*, 353–361. [CrossRef]

27. Mo, X.X.; Sun, T.; Liu, J.; Wu, Y.Q.; Liu, M.H. Qualitative detection of haloxyfop- P- methyl residue in edible oil by near infrared spectroscopy combined with variable selection method. *Chin. J. Anal. Lab.* **2018**, *37*, 125–130.

28. Sun, T.; Wu, Y.Q.; Li, X.Z.; Xu, P.; Liu, M. Discrimination of camellia oil adulteration by NIR spectra and subwindow permutation analysis. *Acta Opt. Sin.* **2015**, *35*, 342–349.

29. Wang, W.Y. *Monitoring Powdery Mildew with Hyperspectral Reflectance in Wheat*; Nanjing Agricultural University: Nanjing, China, 2016.

30. Graeff, S.; Link, J.; Claupein, W. Identification of powdery mildew (*Erysiphe graminis* sp. tritici) and take-all disease (*Gaeumannomyces graminis* sp. tritici) in wheat (*Triticum aestivum* L.) by means of leaf reflectance measurements. *Cent. Eur. J. Biol.* **2006**, *1*, 275–288. [CrossRef]

31. Green, A.A.; Berman, M.; Switzer, P.; Craig, M.D. A transformation for ordering multispectral data in terms of image quality with implications for noise removal. *IEEE Trans. Geosci. Remote Sens.* **1988**, *26*, 65–74. [CrossRef]

32. Haralick, R.M.; Shanmugam, K.; Dinstein, I. Textural Features for Image Classification. *IEEE Trans. Syst. Man Cybern.* **1973**, *SMC-3*, 610–621. [CrossRef]

33. Zheng, H.; Cheng, T.; Zhou, M.; Li, D.; Yao, X.; Tian, Y.; Cao, W.; Zhu, Y. Improved estimation of rice aboveground biomass combining textural and spectral analysis of UAV imagery. *Precis. Agric.* **2019**, *20*, 611–629. [CrossRef]

34. Huang, W.; Guan, Q.; Luo, J.; Zhang, J.; Zhao, J.; Liang, D.; Huang, L.; Zhang, D. New optimized spectral indices for identifying and monitoring winter wheat diseases. *IEEE J. Sel. Top. Appl. Earth Obs. Remote Sens.* **2014**, *7*, 2516–2524. [CrossRef]

35. Chen, B.; Wang, K.; Li, S.; Wang, J.; Bai, J.; Xiao, C.; Lai, J. Spectrum characteristics of cotton canopy infected with verticillium wilt and inversion of severity level. *Comput. Comput. Technol. Agric.* **2008**, *2*, 1169.

36. Gamon, J.A.; Penuelas, J.; Field, C.B. A narrow-waveband spectral index that tracks diurnal changes in photosynthetic efficiency. *Remote Sens. Environ.* **1992**, *41*, 35–44. [CrossRef]

37. Daughtry, C.S.T.; Walthall, C.L.; Kim, M.S.; de Colstoun, E.B.; McMurtrey, J.E. Estimating corn leaf chlorophyll concentration from leaf and canopy reflectance. *Remote Sens. Environ.* **2000**, *74*, 229–239. [CrossRef]

38. Lewis, H.G.; Brown, M. A generalized confusion matrix for assessing area estimates from remotely sensed data. *Int. J. Remote Sens.* **2001**, *22*, 3223–3235. [CrossRef]

39. Penuelas, J.; Baret, F.; Filella, I. Semiempirical indexes to assess carotenoids chlorophyll-a ratio from leaf spectral reflectance. *Photosynthetica* **1995**, *31*, 221–230.

40. Merton, R.; Huntington, J. Early simulation results of the ARIES-1 satellite sensor for multi-temporal vegetation research derived from AVIRIS. In Proceedings of the Eighth Annual JPL Airborne Earth Science Workshop, Pasadena, CA, USA, 8–11 February 1999; pp. 9–11.

41. Thenkabail, P.S.; Smith, R.B.; Pauw, E. Hyperspectral vegetation indices and their relationships with agricultural crop characteristics. *Remote. Sens. Environ.* **2000**, *71*, 158–182. [CrossRef]

42. Filella, I.; Serrano, L.; Serra, J.; Peñuelas, J. Evaluating wheat nitrogen status with canopy reflectance indices and discriminant analysis. *Crop. Sci.* **1995**, *35*, 1400–1405. [CrossRef]

43. Broge, N.H.; Mortensen, J.V. Deriving green crop area index and canopy chlorophyll density of winter wheat from spectral reflectance data. *Remote Sens. Environ.* **2001**, *81*, 45–57. [CrossRef]

44. Haboudane, D.; Miller, J.R.; Pattey, E.; Zarco-Tejada, P.J.; Strachan, I.B. Hyperspectral vegetation indices and novel algorithms for predicting green LAI of crop canopies: Modeling and validation in the context of precision agriculture. *Remote Sens. Environ.* **2004**, *90*, 337–352. [CrossRef]

45. Merzlyak, M.N.; Gitelson, A.A.; Chivkunova, O.B.; Rakitin, V.Y. Non-destructive optical detection of pigment changes during leaf senescence and fruit ripening. *Physiol. Plant.* **1999**, *106*, 135–141. [CrossRef]

46. Mirik, M.; Michels, G.J.; Kassymzhanova-Mirik, S.; Elliott, N.C.; Catana, V.; Jones, D.; Bowling, R. Using digital image analysis and spectral reflectance data to quantify damage by greenbug (Hemitera: Aphididae) in winter wheat. *Comput. Electron. Agric.* **2006**, *51*, 86–98. [CrossRef]

47. Maimaitiyiming, M.; Ghulam, A.; Bozzolo, A.; Wilkins, J.L.; Kwasniewski, M.T. Early detection of plant physiological responses to different levels of water stress using reflectance spectroscopy. *Remote Sens.* **2017**, *9*, 745. [CrossRef]

48. Xie, C.; Shao, Y.; Li, X.; He, Y. Detection of early blight and late blight diseases on tomato leaves using hyperspectral imaging. *Sci. Rep.* **2015**, *5*, 16564. [CrossRef]

49. Tang, X.; Cao, X.; Xu, X.; Jiang, Y.; Luo, Y.; Ma, Z.; Fan, J.; Zhou, Y. Effects of climate change on epidemics of powdery mildew in winter wheat in China. *Plant Dis.* **2017**, *101*, 1753–1760. [CrossRef] [PubMed]

50. Zhang, D.Y.; Zhang, J.C.; Zhu, D.Z.; Wang, J.H.; Luo, J.H.; Zhao, J.L.; Huang, W.-J. Investigation of the hyperspectral image characteristics of wheat leaves under different stress. *Spectrosc. Spectr. Anal.* **2011**, *31*, 1101–1105.

51. Yao, S.; Huo, Z.; Dong, Z.; Li, M.; Chen, X. Indices and modeling of wheat powdery mildew epidemic based on hourly air temperature and humidity data. *Shengtaixue Zazhi* **2013**, *32*, 1364–1370.

52. Sankaran, S.; Mishra, A.; Ehsani, R.; Davis, C. A review of advanced techniques for detecting plant diseases. *Comput. Electron. Agric.* **2010**, *72*, 1–13. [CrossRef]

53. Yuan, L. *Identification and Differentiation of Wheat Disease and Insects with Multi-Source and Multi-Scale Remote Sensing Data*; Zhejiang University: Hangzhou, China, 2015.

remote sensing

Article

Research on Polarized Multi-Spectral System and Fusion Algorithm for Remote Sensing of Vegetation Status at Night

Siyuan Li, Jiannan Jiao and Chi Wang *

Department of Precision Mechanical Engineering, Shanghai University, Shanghai 200444, China; lsyyj@shu.edu.cn (S.L.); m160018@e.ntu.edu.sg (J.J.)
* Correspondence: wangchi@shu.edu.cn; Tel.: +86-021-6613-0801

Abstract: The monitoring of vegetation via remote sensing has been widely applied in various fields, such as crop diseases and pests, forest coverage and vegetation growth status, but such monitoring activities were mainly carried out in the daytime, resulting in limitations in sensing the status of vegetation at night. In this article, with the aim of monitoring the health status of outdoor plants at night by remote sensing, a polarized multispectral low-illumination-level imaging system (PMSIS) was established, and a fusion algorithm was proposed to detect vegetation by sensing the spectrum and polarization characteristics of the diffuse and specular reflection of vegetation. The normalized vegetation index (NDVI), degree of linear polarization (DoLP) and angle of polarization (AOP) are all calculated in the fusion algorithm to better detect the health status of plants in the night environment. Based on NDVI, DoLP and AOP fusion images (NDAI), a new index of night plant state detection (NPSDI) was proposed. A correlation analysis was made for the chlorophyll content (SPAD), nitrogen content (NC), NDVI and NPSDI to understand their capabilities to detect plants under stress. The scatter plot of NPSDI shows a good distinction between vegetation with different health levels, which can be seen from the high specificity and sensitivity values. It can be seen that NPSDI has a good correlation with NDVI (coefficient of determination $R^2 = 0.968$), PSAD ($R^2 = 0.882$) and NC ($R^2 = 0.916$), which highlights the potential of NPSDI in the identification of plant health status. The results clearly show that the proposed fusion algorithm can enhance the contrast effect and the generated fusion image will carry richer vegetation information, thereby monitoring the health status of plants at night more effectively. This algorithm has a great potential in using remote sensing platform to monitor the health of vegetation and crops.

Keywords: vegetation health monitoring; remote sensing; NDVI; polarization; image fusion

Citation: Li, S.; Jiao, J.; Wang, C. Research on Polarized Multi-Spectral System and Fusion Algorithm for Remote Sensing of Vegetation Status at Night. *Remote Sens.* **2021**, *13*, 3510. https://doi.org/10.3390/rs13173510

Academic Editor: Xanthoula Eirini Pantazi

Received: 2 August 2021
Accepted: 1 September 2021
Published: 4 September 2021

Publisher's Note: MDPI stays neutral with regard to jurisdictional claims in published maps and institutional affiliations.

1. Introduction

Vegetation plays an important role in the global ecosystem [1], but natural disasters such as drought, pests and diseases affect their growth of greatly, even leading to their death. In the past 30 years, monitoring the health status of vegetation via remote sensing has been applied in many areas successfully, such as the analysis of crop diseases and pests, monitoring forest coverage and vegetation growth status, etc. However, such remote sensing-based activities are mainly carried out during the daytime, with limitations in sensing the vegetation status at night. Crop diseases and pests are important factors for analyzing crop yields. For example, *Spodoptera frugiperda*, as one kind of nocturnal animal, spreads extremely fast and seriously threatens human food security [2]. Thus, it is necessary to monitor them efficiently with state-of-the-art techniques. Therefore, it is essential and important to develop an effective and accurate method to monitor the health status of vegetation in the night environment.

With the development of remote sensing technologies based on multiple platforms (space, air and ground), various remote sensing-based methods are used to monitor plant health [3–11]. For the optical remote sensing methods, the visible and near-infrared wave bands are mainly used for analysis of plant diseases and insect pests [12–21]. As the vegetation of different healthy status have different absorption and reflection characteristics at different wavelengths, various vegetation indices (VIS) with visible and near-infrared wave bands have been developed to monitor vegetation growth and health [22], such as the normalized vegetation index (NDVI, Rouse et al. 1973) [23], soil-adjusted vegetation index (SAVI, Huete 1988) [24], modified soil-adjusted vegetation index (MSAVI, Qi et al. 1994) [25] and normalized difference water index (NDWI, Gao 1996) [26]. More and more indices are being developed to study and analyze the crop damage caused by pests with remote sensing.

For the ground-based remote sensing technologies, typically, handheld instruments are widely used monitoring crop diseases [27], and tower-based platforms [28–30] and other huge near-ground platforms are used to obtain crop spectral information under different disease states at the leaf and canopy scales. For example, Graeff et al. (2006) [31] used a digital imager (Leica S1 Pro, Leica, Wetzlar, Germany) to analyze the spectrum of wheat leaves infected with powdery mildew and take-all disease, finding this disease resulted in strong spectral response at 490 nm, 510 nm, 516 nm, 540 nm, 780 nm and 1300 nm. Yang et al. (2009) [32] used a hand-held Cropscan radiometer and found that it was feasible to use the ratio vegetation index (400/450 nm and 950 nm/450 nm) to identify and distinguish plants infested by green aphids and Russian wheat aphids. Liu et al. (2010) [33] used a portable spectrophotometer (Field Spec Full Range, ASD Inc., Boulder, CO, USA) to analyze the spectral characteristics of Rice Panicles and found that there was a correlation between the reflectance change in the 450–850-nm band and rice glume blight disease. In addition, Prabhakar et al. (2011) [34] found that the spectral reflectance of healthy plants was significantly different from the plants infested by leafhoppers in the visible and near-infrared bands with a Hi-Res spectroradiometer (ASD Inc., Boulder, CO, USA; spectral range: 350–2500 nm), and the new leaf hopper index (LHI) was applied to monitor the severity of leafhopper. Prabhakar et al. (2013) [35] used a Hi-Res spectroradiometer (ASD Inc., Boulder, CO, USA; spectral range: 350–2500 nm) for the experiment and found that the most sensitive bands to cotton mealybug infestation were concentrated at 492 nm, 550 nm, 674 nm, 768 nm and 1454 nm, and a new cotton mealybug Stress Index (MSI) was developed.

Besides ground-based remote sensing technologies, various air and space-based remote sensing technologies have been developed and widely applied to the monitoring of plant diseases and insect pests, with the advantages of good temporal, spatial and spectral resolutions. In general, the airborne platform used for monitoring crop diseases can be integrated with different sensors, such as an imaging camera, multispectral/hyperspectral spectrometer, infrared camera, lidar and other detection systems [36,37]. For example, Yang et al. (2010) [38] used high-resolution multispectral and hyperspectral aerial image data to extract the occurrence range of cotton root rot and indicated that multispectral data was promising for large-scale disease monitoring. Calder ó n et al. (2013) [39] used UAV with the multispectral camera and thermal infrared camera to diagnose the verticillium wilt of olive trees. Sanches et al. (2014) [40] calculated a new Plant Stress Detection Index (PSDI) from the chlorophyll characteristic center (680 nm) and green edge (560 nm and 575 nm) with an airborne imaging spectrometer (ProSpecTIR-VS) and the continuum division method, showing this index could be used for an analysis of the canopy status successfully. Lehmann et al. (2015) [41] analyzed the UAV multi-spectral images with object-based image processing methods to monitor the pests of oak trees. For the cases of space-based remote sensing applications, Yuan et al. (2016) [42] proposed a monitoring method with the help of SPOT-6 images to analyze the occurrence of Wheat Powdery Mildew in the Guanzhong area of Shaanxi Province, with an accuracy of 78%. Chen et al. (2018) [43] applied high-resolution multi-spectral satellite images to monitor wheat rust, with a classification accuracy of 90%. Zheng et al. (2018) [44] used the Sentinel-2 Multispectral Instrument (MSI) to distinguish the severity of yellow rust infection in the winter and proposed the Red Edge Disease Stress Index (REDSI) to detect yellow rust infections of different severities. Meiforth et al. (2020) [45] used the WorldView-2 (WV2) satellite and

LiDAR data to detect the stress of kauri canopies in New Zealand. The results showed that this method can be used to monitor kauri canopies economically and efficiently in a large area. Li et al. (2021) [46] developed a machine learning model to analyze the vegetation growth by retrieving NDVI from the satellite sensor.

It is clear that vegetation indices based on different remote sensing platforms play an important role in monitoring the status of crops. When the polarization information is supplemented, more complex and accurate indices and models can be developed, and the polarization shows directional characteristics when light interacts with the vegetation [47,48]. Polarization has attracted many researchers to study remote sensing together with polarization information. For example, Vanderbilt et al. pointed out that polarization was affected by vegetation canopy morphology and leaf surface characteristics [49,50]. Since vegetation canopy morphology and leaf surface characteristics are affected by stress, activity and the growth stage, polarization-based parameters, such as the degree of linear polarization (DoLP) of vegetation, can reflect such information well [51]. Compared to traditional remote sensing methods without polarization information, polarization-based remote sensing can provide vegetation canopy information, such as the structure of leaf layers [52] and the emergence of panicles above canopies [53]. Theoretical and practical research indicates the possibility of detecting the geometry of vegetation canopies with polarization [54,55]. Besides, many researchers have measured various leaves in an extensive way [56–60]. Hu et al. studied polarization imaging in low-light environments [61]. For the above-mentioned research, it is worth noting that the researchers are still in the phase of ground experiments in studying remote sensing technologies.

It is shown that the health of vegetation can be monitored successfully with different remote sensing technologies, and various methods have been carried out during the daytime. However, limited efforts have been done in monitoring the health of vegetation at night, which requires high-sensitivity imaging equipment but is quite important and necessary. In this paper, a ground-based remote sensing system is developed and used for monitoring the health of vegetation at night. Since polarization information can supplement traditional spectral imaging to develop more complex and accurate models for vegetation monitoring, in this paper, we developed a polarized multispectral low-illumination-level imaging system, which is used to record the spectral images of vegetation at 680 nm and 760 nm during the night, as well as the polarization images at $0°$, $60°$ and $120°$. Moreover, a fusion algorithm is proposed to improve the contrast of vegetation detection by fusing polarization images and spectral images of vegetation. The vegetation index (NDVI), degree of linear polarization (DoLP) and angle of polarization (AOP) are all calculated in the fusion algorithm for better detection of the health status of plants at night. In addition, a new index, the night plant status detection index (NPSDI), was calculated based on the fusion images of NDVI, DoLP and AOP. The index was compared with physiological parameters of plants, such as nitrogen content (NC) and SPAD, to assess the applicability of the fusion algorithm for the remote sensing of vegetation. Moreover, based on our novel ground-based remote sensing research, it is promising to transform the platform into air and space-based ones for large-scale remote sensing applications in the future.

2. Materials and Methods

2.1. Plant Materials and Experimental Design

The spectral and polarization data analyzed in the study are from the leaves of plants on the campus of Shanghai University: spotted laurel (*Aucuba japonica*), money plant (*Epipremnum aureum*) and Malabar chestnut (*Pachira aquatica*). Figure 1a is a potted money plant in the laboratory, and Figure 1a1,a2 are the regions of interest for collecting vegetation index and polarization data, respectively. Figure 1b shows the leaves of spotted laurel, money plant and Malabar chestnut, and Figure 1b1–b5 shows the regions of interest collected by time series experiment at 3-day intervals. Figure 1c is a spotted laurel on the campus of Shanghai University, and Figure 1c1–c4 is defined as the healthy leaf region, level-2 stress leaf region, level-1 stress leaf region and withered leaf region, respectively.

Figure 1. Plant materials. (**a**) Money plant: (**a1,a2**) are the regions of interest for the NDVI and polarization data collection, respectively. (**b**) The leaves of a spotted laurel, money plant and Malabar chestnut: (**b1–b5**) are the images of the changes of a money plant leaf every 3 days. (**c**) A spotted laurel at Shanghai University: (**c1–c4**) are the healthy leaf region, the level-2 stress leaf region, the level-1 stress leaf region and the withered leaf region, respectively.

Three experiments were designed in this paper, including: (1) an experiment was carried out under different illuminances of 0.01, 0.1, 0.5, 1 and 5 lux on a pot of money plant in a dark laboratory room, and the LED light group and LED controller were used to control the illumination of the dark room. The illumination was measured with a high-precision illuminometer (model: tes1330A, Taishi). (2) In the dark laboratory room, a time series experiment was carried out to collect data on the leaves of spotted laurel, money plant and Malabar chestnut every three days. Vegetation leaves were pasted on a display board printed at Shanghai University and placed in a dark room. The LED light group and LED controller were used to control the illumination of the dark room, and a high-precision illumination meter (model: tes1330A, Taishi) was used to measure the illumination. (3) An outdoor experiment was conducted on a spotted laurel on campus at night.

2.2. Polarized Multispectral for Low-Illumination-Level Imaging System

The polarized multispectral for low-illumination-level imaging system (PMSIS) developed for outdoor plant imaging at night consists of a scientific CMOS (SCMOS) monochrome camera with 2048 pixels × 2048 pixels (model: PCO. edge 4.2, PCO, GER), coupled to a Nikon MF 50-mm (1:1.8) fixed focus camera lens, narrow band interference filters fitted on a rotating filter mount and a linear polarizer fitted on a rotating polarizer mount (Figure 2). Blue (FF02-482/18-25) and green (FF02-520/28-25) band filters of Semrock and red (FB680-10) and near-infrared band (FB760-10) filters of Thorlabs were used. The camera was connected to a desktop computer, and the control and data acquisition were carried out through a Cameralink port.

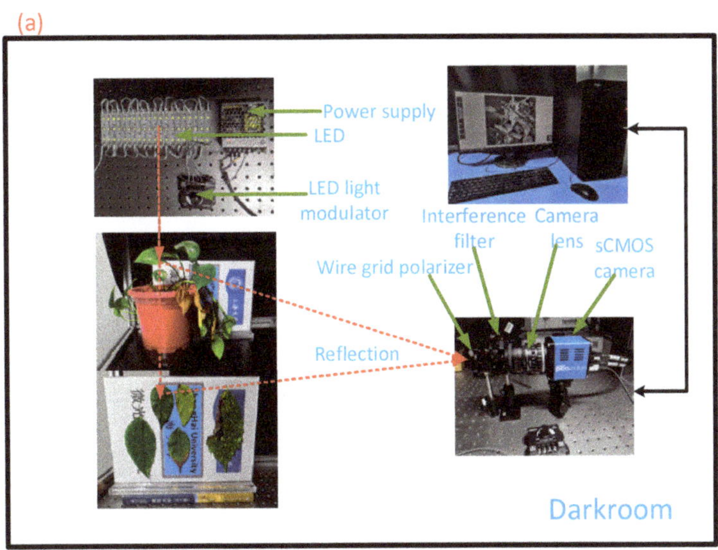

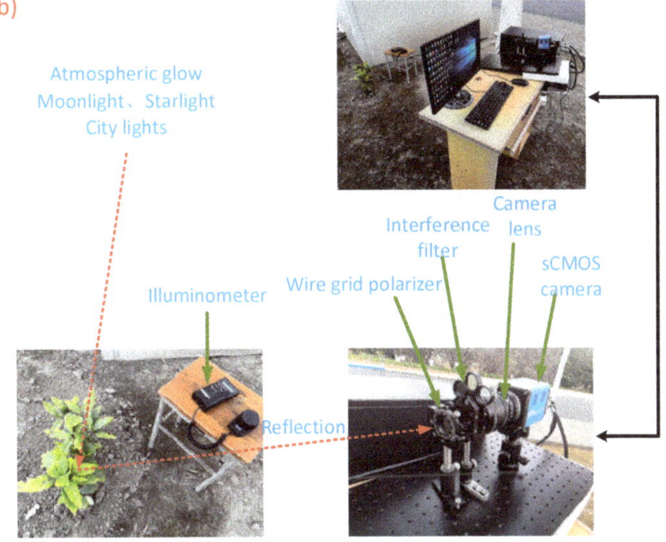

Figure 2. (**a**) Indoor polarized multispectral for a low-illumination-level imaging system. (**b**) Outdoor polarized multispectral for a low-illumination-level imaging system.

As shown in Figure 2a, an indoor polarized multispectral for low-illumination-level imaging system was established. In order to simulate a night environment, the system was placed in a dark room made of optical shading material, and the shading rate of the dark room was greater than 99%. During the experiment, the illuminance of the dark room was controlled at 0.01, 0.1, 0.5, 1 and 5 lux, respectively (Note: dark night: 0.001–0.02 lux, moonlit night: 0.02–0.3 lux and cloudy indoor: 5–50 lux). The illuminance of the dark room was controlled by the LED lamp set and LED controller to carry out the experiment of monitoring the vegetation health status with different illuminances at night.

As shown in Figure 2b, an outdoor polarized multispectral for a low-illumination-level imaging system was built to conduct an outdoor night monitoring experiment on a spotted laurel. The time selected was from 8:00 at night. The ambient light included an atmospheric glow, starlight, moonlight and city light. The outdoor experiment location was on the campus of Shanghai University, Baoshan District, Shanghai (Season: Winter (Longitude: 121°24′1.95″, Latitude: 31°19′10.57″)), and the illuminance of 0.22 lux was actually measured with an illuminance meter.

2.3. Measurement of Chlorophyll and Nitrogen Content

The chlorophyll and nitrogen contents of plant leaves were measured with a handheld chlorophyll analyzer (Medium Kelvin, Model: TYS-4N). The chlorophyll analyzer can measure the relative chlorophyll (SPAD) and nitrogen contents (mg/g), leaf moisture (RH%) and temperature of plants instantly. In the test, two LED light sources emit light in two wavebands, one with a red light (center wavelength: 650 nm), the other with an infrared light (center wavelength: 940 nm). After penetrating the blade, the light is received and processed by the receiver, and finally, the SPAD value is calculated and displayed on the screen.

2.4. Image Acquisition and Analysis

Highly sensitive sCMOS cameras and filters with central bands of 482, 520, 680 and 760 nm were used to obtain the spectral data images of vegetation. Linear polarizers were used to obtain the polarization data images of vegetation with 0, 60 and 120 degrees. In this study, the spectral analysis focused on the absorption characteristics of chlorophyll in the central band near 680 nm and the reflection characteristics of chlorophyll in the central band near 760 nm, and the NDVI was used to retrieve the vegetation health status. The DoLP and AOP of vegetation were calculated using 0, 60 and 120-degree polarization images. Finally, the NDVI, DoLP and AOP were fused with the proposed fusion algorithm of night plant detection.

2.4.1. The Normalized Vegetation Index

The analysis based on the vegetation index has become a main way for scholars to study and practice the remote sensing detection of pests and diseases. So far, many different types of vegetation indices have been proposed one after another. With certain biological or physical and chemical significance, these vegetation indices are an important application form of the plant spectrum. At present, many studies have tried to establish the relationship between remote sensing information and the occurrence and degree of diseases and pests through various vegetation indices. In this study, the spectral image data of 680 nm and 760 nm in the central band were used to calculate the NDVI, of which the formula is as follows:

$$NDVI = \frac{R(760\,nm) - R(680\,nm)}{R(760\,nm) + R(680\,nm)} \tag{1}$$

2.4.2. Polarization of Vegetation

When light is reflected from the leaf surface, its spectral characteristics will be affected by the element composition of the leaf surface, and its polarization characteristics will be affected by the characteristics of the leaf surface [62]. Polarization [63] and multispectral imaging can provide complementary information for vegetation detection. However, few people currently propose to fuse the information of polarization and spectral imaging to obtain better vegetation detection results. Therefore, this paper introduces polarization into vegetation detection and integrates it with the vegetation index to improve the detection of the vegetation health status.

Stokes formula:

$$F = \begin{bmatrix} I \\ Q \\ U \\ V \end{bmatrix} = \begin{bmatrix} \langle A_x{}^2 + A_y{}^2 \rangle \\ \langle A_x{}^2 - A_y{}^2 \rangle \\ \langle 2 A_x A_y \cos \gamma \rangle \\ \langle 2 A_x A_y \sin \gamma \rangle \end{bmatrix} \approx \begin{bmatrix} S_0 \\ S_1 \\ S_2 \\ 0 \end{bmatrix} \tag{2}$$

Intensity of radiation: S_0 is obtained by passing light waves through linear polarizers oriented at 0 degrees, S_1 is obtained by passing light waves through linear polarizers oriented at 60 degrees, and S_2 is obtained by passing light waves through linear polarizers oriented at 120 degrees.

The method for measuring linear Stokes parameters is shown in Figure 3 [64]. It should be noted that, in remote sensing measurements, the degree of circular polarization is usually very small, so we only describe the linear polarization state of the beam.

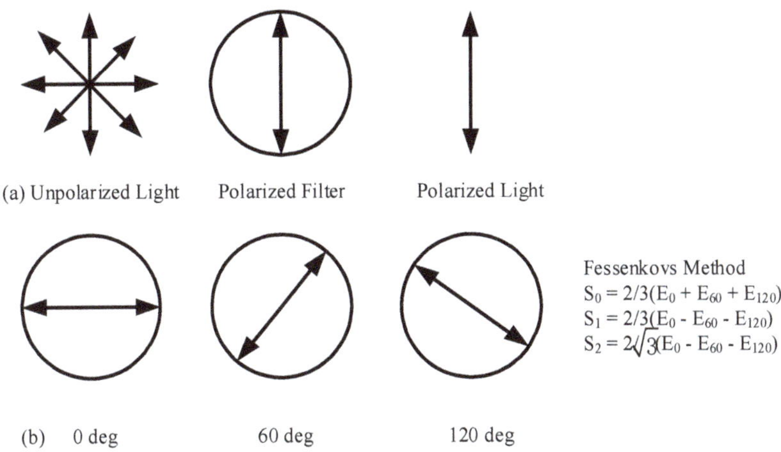

Figure 3. (**a**) Generation of polarized light with a polarized filter. (**b**) Fessenkovs Method to characterize Stokes vectors.

Fessenkovs method was used to measure the Stokes parameters in the paper. Degree of Linear Polarization (DoLP):

$$\text{DoLP} = \frac{\sqrt{S_1{}^2 + S_2{}^2}}{S_0} \tag{3}$$

Angle of Polarization (AOP):

$$\text{AOP} = \frac{1}{2} \tan^{-1}\left(\frac{S_2}{S_1}\right) \tag{4}$$

2.4.3. Fusion Algorithm for Nighttime Plant Detection

In view of the different polarization and spectral characteristics of healthy and stressed vegetation, in this paper, we propose a fusion algorithm to detect the vegetation with the spectral and polarization characteristics of diffuse and specular reflections of vegetation. The NDVI, DoLP and AOP are all calculated in the fusion algorithm to better detect the health status of plants in the night environment.

The schematic diagram of our method is shown in Figure 4. This algorithm can be summarized as follows: (1) Spectral and polarized images are preprocessed (normalization and filtering) and then co-registered together. (2) The NDVI was calculated with a F680-nm red spectral image and F760-nm infrared spectral image (Equation (1)). (3) The DoLP (Equation (3)) and AOP (Equation (4)) were calculated with 0, 60 and 120° polarized images. (4) The NDVI, DoLP and AOP images were fused, and the fused images were converted into HSV with RGB color mapping.

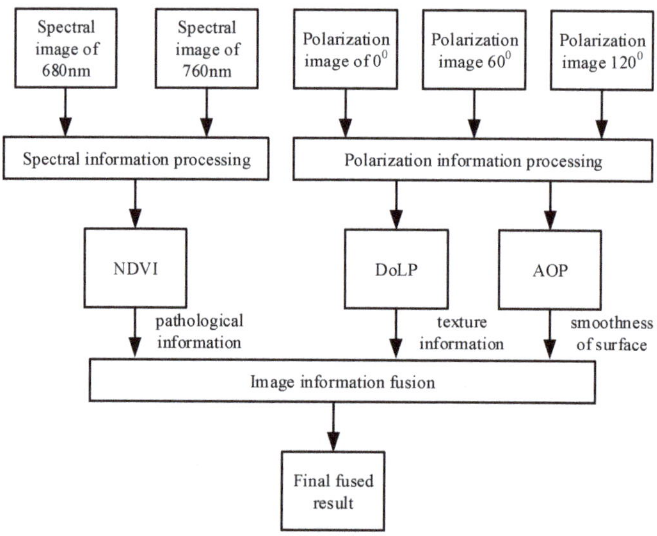

Figure 4. Schematic diagram of the proposed method.

The images are combined following the conversion from HSV to RGB color maps:

$$\sum HSV \rightarrow RGB\{NDVI, DoLP, AOP\} \tag{5}$$

Based on the fusion image (NDAI) of the NDVI, DoLP and AOP, a new index, the night plant state detection index (NPSDI), was proposed.

The NPSDI formula (Equation (1)) is as follows:

$$NPSDI = \left(\frac{R_{(AOP)} + G_{(NDAI)} + B_{(DoLP)}}{3 \times 255} \right) \tag{6}$$

$R_{(AOP)}$, $G_{(NDAI)}$ and $B_{(DoLP)}$ represent the intensity values of the R, G and B channels of the fused image, respectively.

3. Results

3.1. Experiments on Different Illumination of Vegetation

Five groups of illuminance experiments were carried out: 0.01, 0.1, 0.5, 1 and 5 lux, and spectral data were collected to calculate the NDVI. Figure 5 shows the NDVI under different illuminances. The corresponding illuminances are 0.01 (Figure 5a), 0.1 (Figure 5b), 0.5 (Figure 5c), 1.0 (Figure 5d) and 5.0 lux (Figure 5e).

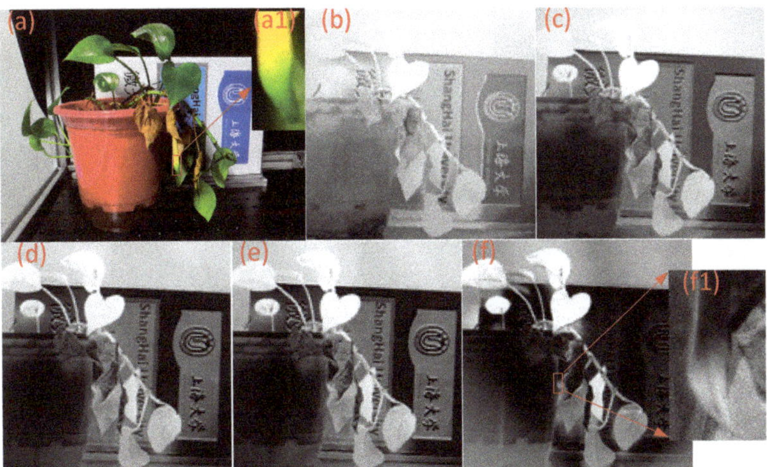

Figure 5. The NDVI under different illuminations. (**a,a1**) Color image of a money plant and illumination of the NDVI: (**b**) 0.01 lux, (**c**) 0.1 lux, (**d**) 0.5 lux, (**e**) 1.0 lux and (**f**) 5.0 lux. (**f1**) The region of interest for the NDVI.

As shown in Figure 5f1, some regions of interest in Figure 5b–f were selected. Fifty-one pixels were selected from the black and white regions, respectively, averaged and then divided. Then, as shown in Figure 6, the NDVI changes with the illumination. It can be clearly seen from Figure 6 that the NDVI changes significantly with the increase of the illumination. Therefore, the NDVI is greatly affected by the illuminance of the experimental environment, so it is not reliable to use the NDVI to monitor the vegetation state at night, and it should be supplemented and corrected with other information.

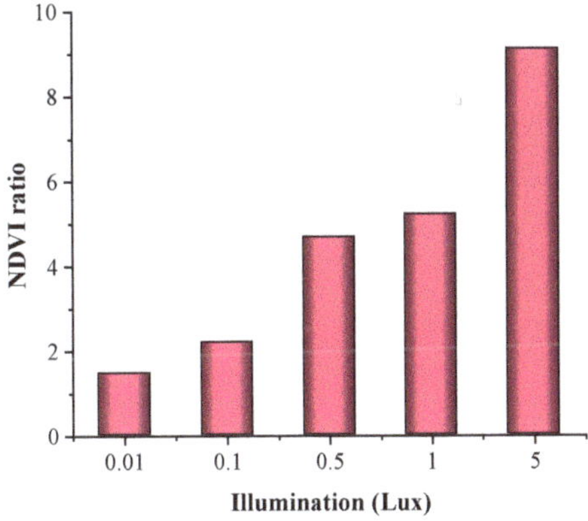

Figure 6. Histogram of the NDVI with different illuminations.

Figure 7a–e is the calculated linear polarization (DoLP) images under illuminations of 0.01, 0.1, 0.5, 1.0 and 5.0 lux, respectively. Some regions of interest in Figure 7a–e are selected, as shown in Figure 7e1, and 51 pixels are selected for the black and white regions in Figure 7e1, averaged and then divided. Then, as shown in Figure 8, the image of the DoLP changes with the illumination. It can be clearly seen from Figure 8 that the DoLP of vegetation does not change significantly with the increase of illumination. Thus, the DoLP was almost insensitive to ambient illumination. It can be seen from Figure 7 that the polarization imaging shows detailed information of the foliage, so the use of the DoLP for monitoring the nighttime vegetation state can be used to supplement the information for NDVI monitoring.

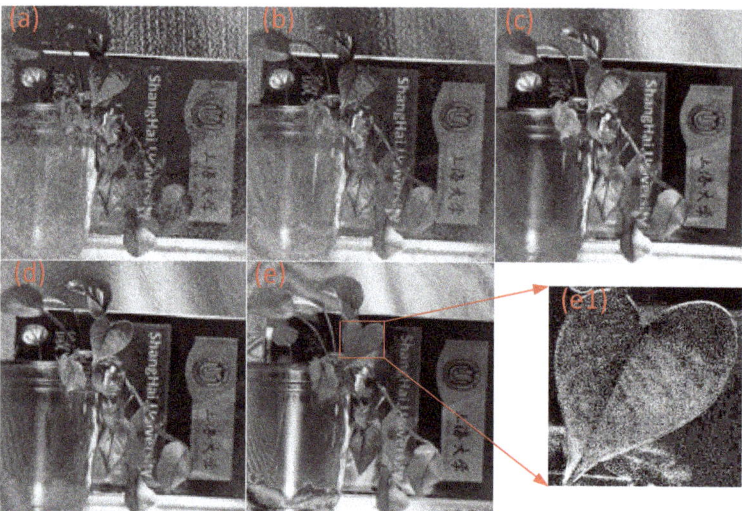

Figure 7. The DoLP with different illuminations. The illumination of the DoLP: (**a**) 0.01 lux, (**b**) 0.1 lux, (**c**) 0.5 lux, (**d**) 1.0 lux and (**e**) 5.0 lux. (**e1**) The region of interest for the DoLP.

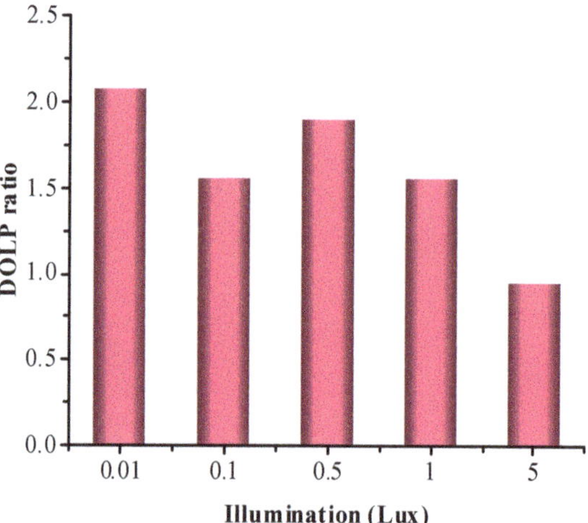

Figure 8. Histogram of the DoLP with different illuminations.

Figure 9a–e is the fusion images (NDAI) calculated by the NDVI, DoLP and AOP, and the illuminance is 0.01, 0.1, 0.5, 1.0 and 5.0 lux, respectively. The results show that healthy and nonhealthy vegetation can be well-distinguished from each other. The fusion images (NDAI) highlight more detailed information of the leaf surface than the NDVI image, and the contrast by target monitoring is enhanced.

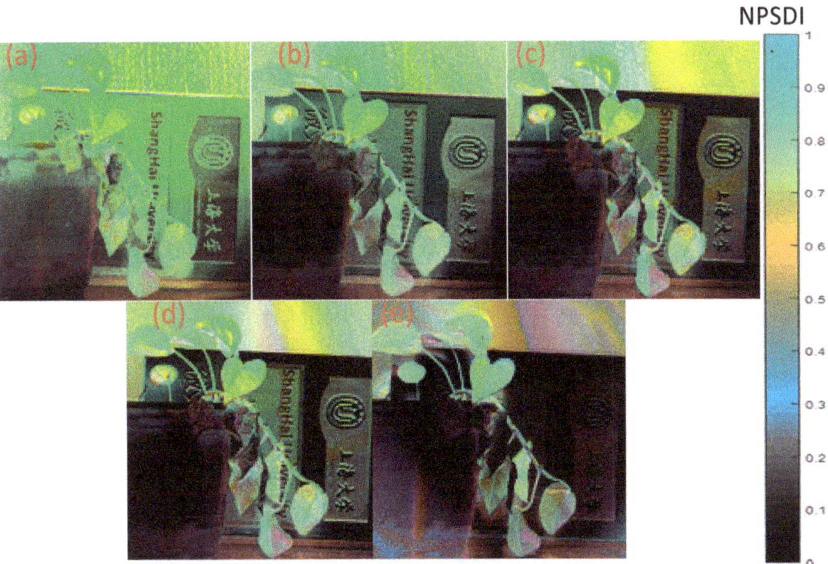

Figure 9. The fusion image (NDAI) with different illuminations. Illumination of the NDAI: (**a**) 0.01 lux, (**b**) 0.1 lux, (**c**) 0.5 lux, (**d**) 1.0 lux and (**e**) 5.0 lux.

3.2. Time Series Experiment of Vegetation

During the experiments, measurements were made once every 3 days, and five groups of experiments were carried out, respectively. Spectral data of the leaves were collected to calculate the NDVI. Figure 10 shows the NDVI time series picture, in which the red arrow refers to the corresponding color picture, and the illumination is 0.5 lux.

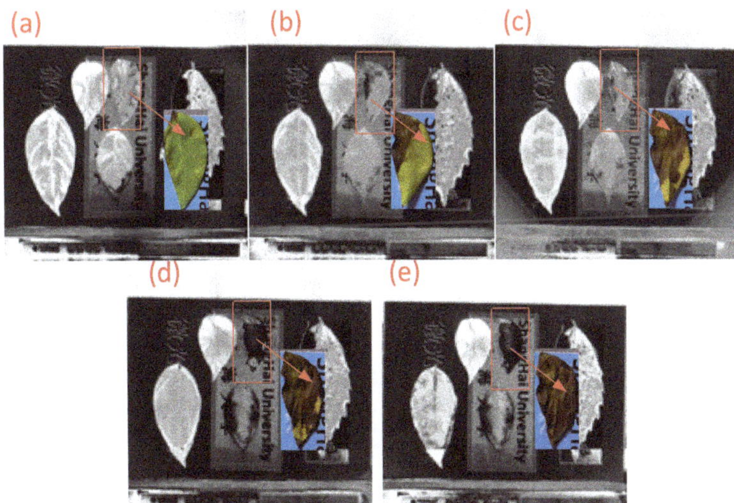

Figure 10. Vegetation index (NDVI) in a time series. Illumination of the NDVI: (**a**) 0.01 lux, (**b**) 0.1 lux, (**c**) 0.5 lux, (**d**) 1.0 lux and (**e**) 5.0 lux.

Figure 11a–e is a fusion image (NDAI) calculated from the NDVI, DoLP and AOP. It can be seen that the fusion image can track the information of the changes in the leaf health status.

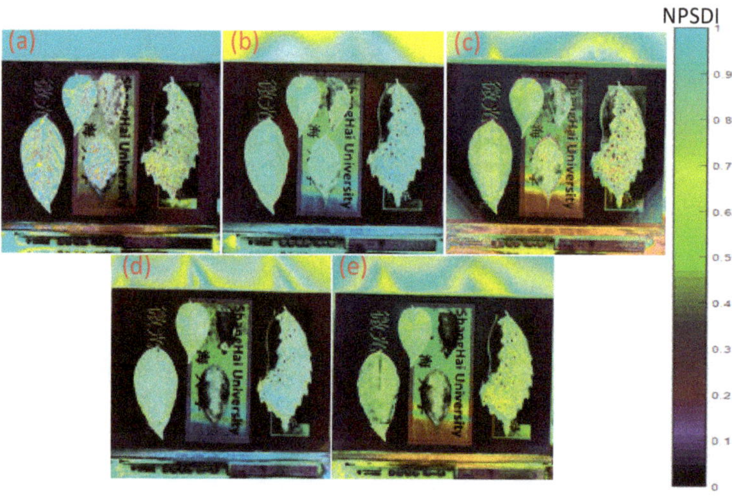

Figure 11. The fusion image in a time series. Illumination of the NDAI: (**a**) 0.01 lux, (**b**) 0.1 lux, (**c**) 0.5 lux, (**d**) 1.0 lux and (**e**) 5.0 lux.

3.3. Outdoor Experiment at Night

The developed polarized multispectral for a low-illumination-level imaging system was used in outdoor experiments, and as measured by the illuminator, the ambient illumination was 0.22 lux. As shown in the figure below, Figure 12a is the NDVI image calculated by Formula (1), Figure 12b the AOP image by Formula (4), Figure 12c the DoLP image by Formula (3), Figure 12d the NDAI image by Formula (5) and Figure 12e the gray-scale image. Figure 12e1–e4 are four different health state regions of plants, respectively.

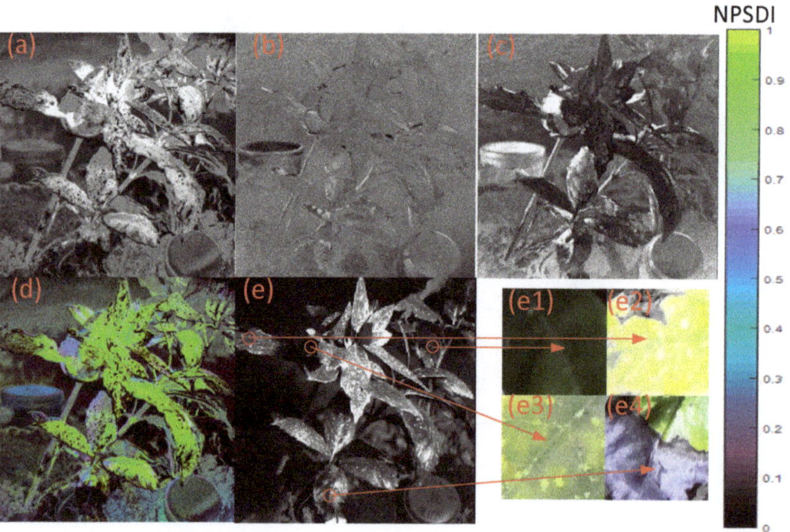

Figure 12. (**a**) NDVI image, (**b**) AOP image, (**c**) DoLP image, (**d**) fusion image (NDAI), (**e**) gray-scale image, (**e1**) healthy leaf region, (**e2**) level-2 stress leaf region, (**e3**) level-1 stress leaf region and (**e4**) withered leaf region.

The spectral image, polarization image and fusion image (NDAI) were obtained through PMSIS. Four regions: healthy leaf region (Figure 12e1), level-2 stress leaf region (Figure 12e2), level-1 stress leaf region (Figure 12e3) and withered leaf region (Figure 12e4) were selected to draw the scatter graph of the discriminant function.

Figure 13a shows the 3D scatter plot of the NDVI image obtained from the F680 nm and F760 nm bands of the selected location points in a specific region, in which the X-axis stands for the health status level of vegetation, the Y-axis the number of groups and the Z-axis the NDVI value. The change trend of vegetation health can be seen intuitively through the change of color and trend of the curved surface. The curved surface is projected on the XY plane and the YZ plane. The cut-off lines are drawn from the arithmetic mean value of the status grades of adjacent vegetation on the YZ plane. For example, a cut-off line with a value of 0.196 distinguished a withered leaf from a level-2 stressed leaf, one with a value of 0.494 distinguished a level-1 stressed leaf from a level-2 stressed leaf and one with a value of 0.773 of distinguished a healthy leaf from a level-1 stressed leaf. According to the misclassification rate of the adjacent plant states, the sensitivity and specificity of the classification of different plant health states were determined, and the positive predictive value (PPV) and negative predictive value (NPV) of the classification are shown in Table 1.

Table 1. Discrimination accuracy between different grades of the plant health states obtained with the NDVI, NPSDI, nitrogen content (NC) and PSAD.

| | Level of Vegetation Health Status | | | | | | | | | | | |
| | Withered Leaf-Level-2 Stress Leaf | | | | Level-2 Stress Leaf-Level-1 Stress Leaf | | | | Level-1 Stress Leaf-Healthy Leaf | | | |
	Se	Sp	PPV	NPV	Se	Sp	PPV	NPV	Se	Sp	PPV	NPV
NPSDI	1	0.91	0.93	1	0.91	0.87	0.88	0.91	0.89	0.92	0.91	0.90
NDVI	1	0.96	0.96	1	0.9	0.88	0.88	0.9	0.98	1	1	0.98
NC	1	1	1	1	0.98	0.96	0.96	0.98	0.96	1	1	0.96
SPAD	1	0.98	0.98	1	0.98	0.96	0.96	0.98	0.96	1	1	0.96

Notes: Se (Sensitivity) = true positive/(true positive + false negative), Sp (Specificity) = true negative/(true negative + false positive), PPV (positive predictive value) = true positive/(true positive + false positive) and NPV (negative predictive value) = true negative/(true negative + false negative).

According to the NDVI, DoLP and AOP image data, the image fusion was carried out, and the 3D scatter plot of the NPSDI was made in the selected four feature regions; the X-axis stands for the health status level of vegetation, the Y-axis the number of groups and the Z-axis the NPSDI value, as shown in Figure 13b. In the YZ plane, the cut-off line between a withered leaf and level-2 stressed leaf laid at 0.153 and that between a level-1 stressed leaf and level-2 stressed leaf and that between a healthy leaf and level-1 stressed leaf laid at 0.399 and 0.574, respectively. Table 1 also contains the sensitivity and specificity for classifying different levels of the plant health status according to the scatter plot shown in Figure 13b, as well as the corresponding PPV and NPV.

In the four plant health state regions, 51 pixels were selected and averaged to draw the NDVI, NPSDI and DoLP histogram of the four regions. As shown in Figure 14, the NDVI values show an increasing trend with the improvement of the plant health status, and the NPSDI values show the same trend. This also indicates that there is a correlation between the NDVI and NPSDI.

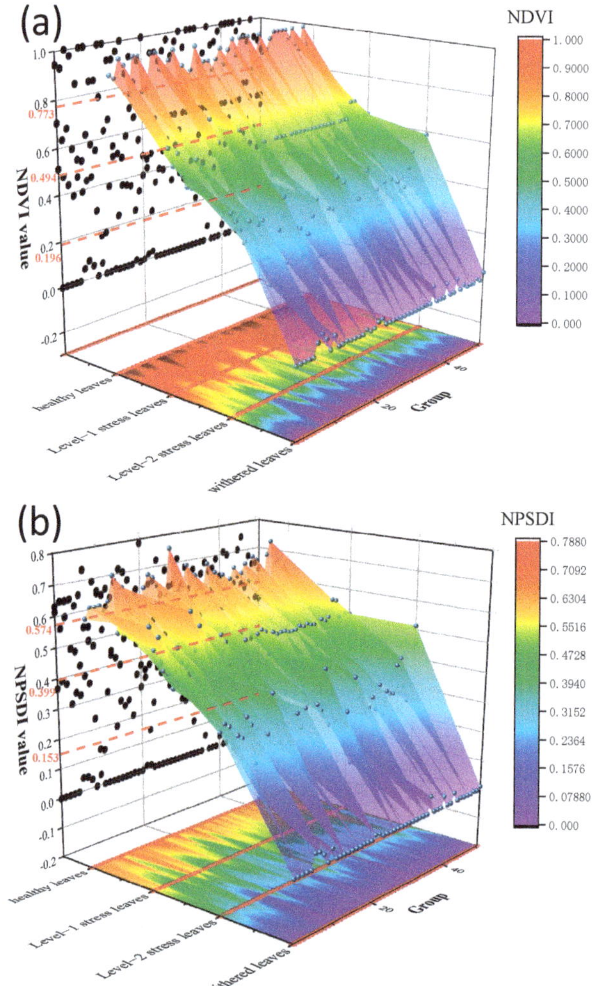

Figure 13. (**a**) 3D scatter plot of the NDVI. (**b**) 3D scatter plot of the NPSDI.

The reason for the higher value of the DoLP in the level-1 stress leaf region may lie in that it is also the leaf shadow region. Since the target was inversely proportional to the DoLP and surface reflectance [65], the DoLP of the shadow area is higher, which is helpful for monitoring the shadow area. The value of the DoLP in the withered leaf region was higher than that in the healthy leaf region. The reason may lie in that the completely withered leaves with more diffuse reflections lead to a higher DoLP value. Under low night illumination, the healthy leaves are smooth, but the mirror reflection is low, so the DoLP value is lower.

The NC and SPAD were measured for spotted laurel with leaves at varying levels of health status. It was observed that the NC of healthy leaves was above 16.8 mg/g, and it decreased to a value between 10.61 and 16.8 mg/g for level-1 stress leaves. The NC of level-2 stress leaves was in the range of 3.53–10.61 mg/g and that of withered leaves was below 3.53 mg/g. In comparison, the SPAD content of healthy leaves was above 44.71, that of level-1 stress leaves in the range of 25.22–44.71, that of level-2 stress leaves in the range of 7.01–25.22 and that of withered leaves below 7.01. The three-dimensional scatter plots of the NC and SPAD contents in the leaves of different health states are shown in Figure 15a,b. The sensitivity and specificity of using the NC and SPAD to distinguish different levels of the plant health status, as well as the respective PPV and NPV values, are shown in Table 1.

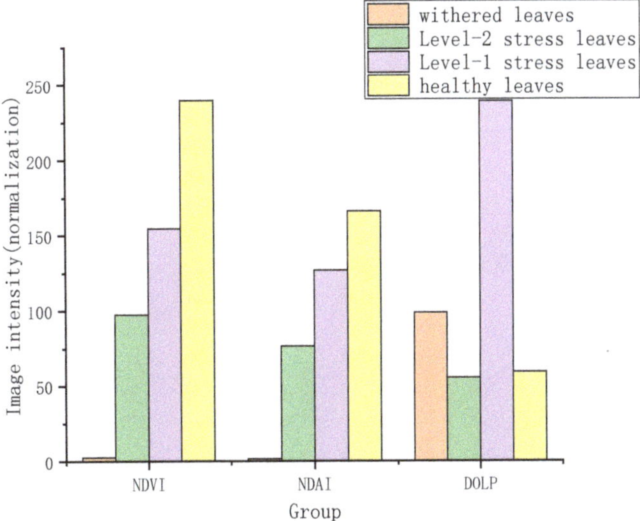

Figure 14. The NDVI, NPSDI and DoLP histograms of the plant in the four health status regions.

The NDVI image obtained through PMSIS system imaging and the calculated fusion image (NDAI) have a significant positive linear correlation with the NC and SPAD, and so do the NDVI and NPSDI (Figures 16a,b and 17a,b). The results of the different health states showed that the correlation coefficient between the NC and NDVI was the highest ($R^2 = 0.931$) and that of the SPAD under the same NDVI was 0.911. In comparison, the R^2 values of the NPSDI with the SPAD and NC of the plant leaves were 0.882 and 0.916, respectively.

The relationship between the NDVI and NPSDI obtained using PMSIS was studied under different health states of plants, separately. As seen from Figure 18, there is a positive linear relationship between the NDVI and NPSDI ($R^2 = 0.968$), which can be used to evaluate stress-induced changes in plants.

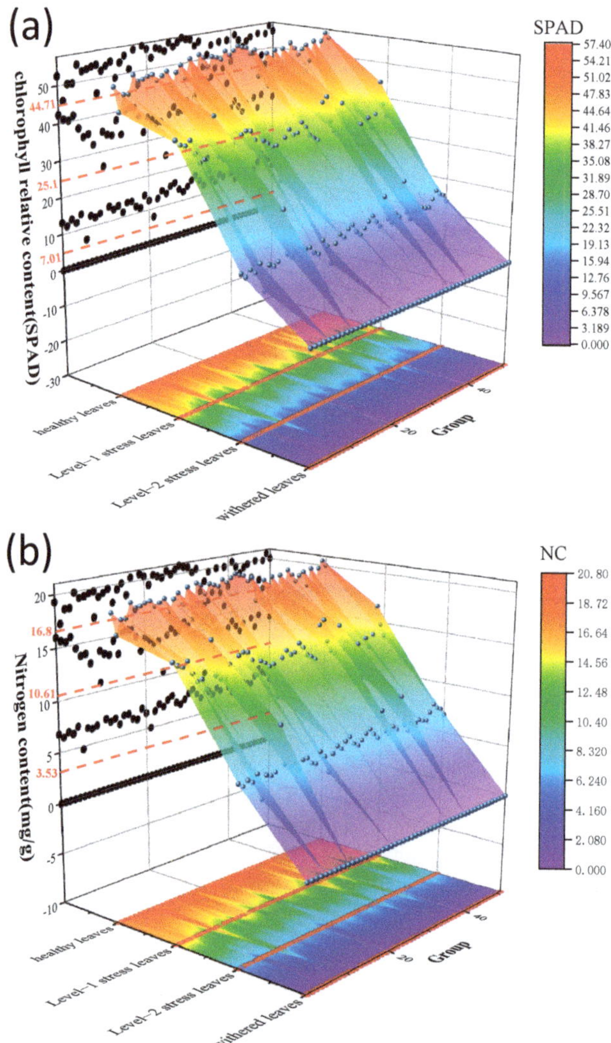

Figure 15. (**a**) 3D scatter plot of the chlorophyll content. (**b**) 3D scatter plot of the nitrogen content.

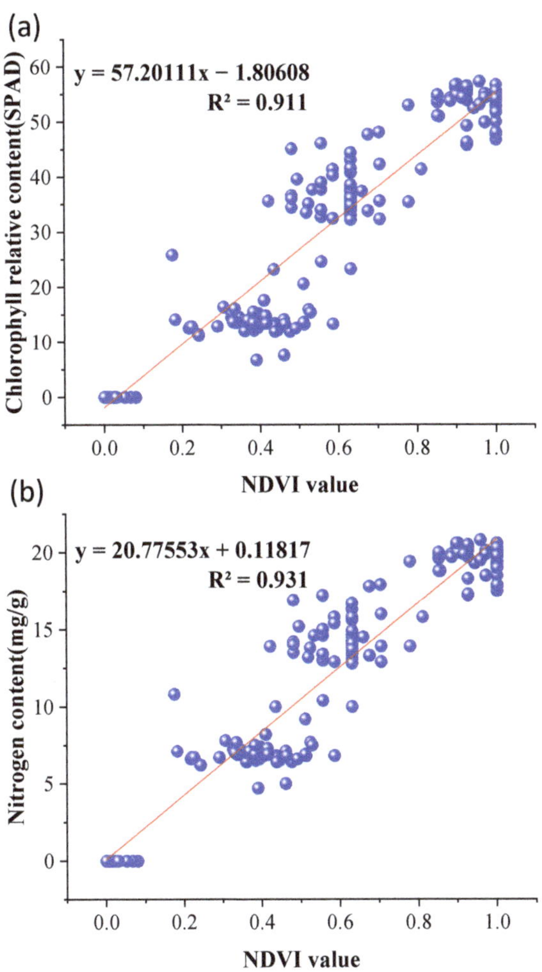

Figure 16. (**a**) Correlation between the NDVI and SPAD. (**b**) Correlation between the NDVI and NC.

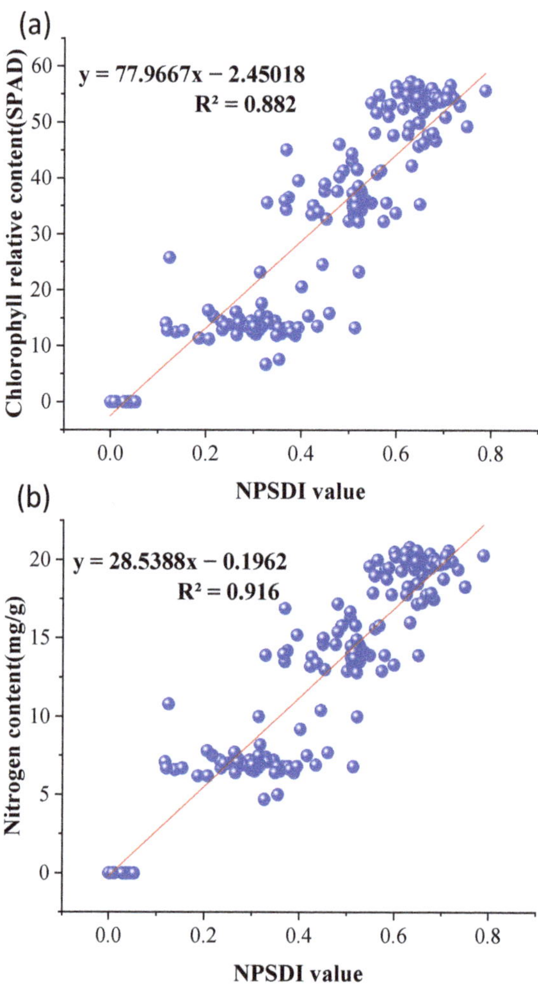

Figure 17. (**a**) Correlation between the NDVI and SPAD. (**b**) Correlation between the NDVI and NC.

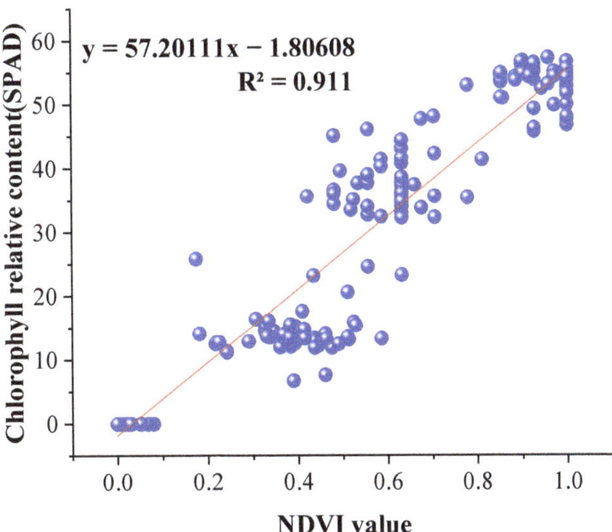

Figure 18. Correlation between the NDVI and NPSDI.

4. Discussion

A polarized multispectral for low-illumination-level imaging system was developed to monitor the plant status at night and evaluate stress-induced plant changes. The reflectance images of leaves on 680-nm and 760-nm narrow spectral lines were recorded by spectral technology, and the 0°, 60° and 120° polarization images of vegetation were recorded by polarization technology. The NDVI, DoLP and AOP images of vegetation were calculated by Formulas (1), (3) and (4), respectively. Then, the NDVI, DoLP and AOP images were fused by Formula (5). For example, a DOLP image can reflect the smoothness and rich texture information of the leaf surface, an AOP image can reflect the smoothness of the leaf surface and a NDVI image contains the pathological information of vegetation. Therefore, if the NDVI, DOLP and AOP images are fused, the generated images will carry more abundant vegetation information.

The research on remote sensing of the vegetation health status based on the vegetation index has become a main research approach for scholars. However, for the study of the health status of vegetation at night, the vegetation index is greatly affected by the illumination [66–68]. As shown in Figure 6, the NDVI value decreases with the decrease of the environmental illumination. Therefore, it is not reliable to only use the NDVI as an indicator of the plant stress degree in a low illumination environment at night.

It can be seen clearly in Figure 8 that, with the increase of illumination, there is no significant change in the DoLP of vegetation. Therefore, the DoLP is hardly affected by the illumination of the experimental environment. In addition, polarized images have the potential to provide additional information about the target's shape, shadow, roughness and surface characteristics [69,70]. There are relatively few studies on polarization technology providing contrast enhancement information for the detection of vegetation health. Therefore, in this paper, we proposed the fusion of the vegetation index (NDVI), DoLP and AOP to better detect the physiological state of plants in a nighttime environment. As shown in Figures 9, 11 and 12d, different health states of plants correspond to different colors, so the proposed enhancement algorithm for nighttime plant detection has the effect of enhancing the contrast.

Under stress, the decrease of the NDVI ratio was related to the decrease of the SPAD content, and this proportional relationship could be explained by the optical characteristics of leaves. Compared with healthy leaves, those under stress had lower chlorophyll contents [71], and the reduced chlorophyll content led to a decreased chlorophyll absorption and enhanced reflectance at F680 nm. Moreover, the reflectance at F760 nm was reduced. According to Formula (1), the NDVI ratio was bound to decrease. In the NDVI images, the NDVI ratio of the stressed leaves was low (Figure 12a). Due to the fusion of polarization, the contrast of the fused images was enhanced, and leaf lesions could be identified more clearly (Figure 12d). It can be pointed out that the increase of the stress degree will lead to a lower vegetation index and leaf chlorosis [72,73].

Three-dimensional scatter plots (Figure 13a,b) were drawn using NDVI and NPSDI data obtained from the polarized multispectral low-light-level imaging monitoring system to classify different health status levels of plants. The classification discrimination accuracy is shown in Table 1. It can be seen from Table 1 that the sensitivity and specificity of the fusion image for the classification of different plant health conditions are generally close to the NDVI image data. It is important that the fusion image shows a color contrast, and the fusion image carries richer object information, so it is excellent to use the proposed fusion image algorithm to monitor the health of plants. In addition, the sensitivity and specificity of the fusion image (NDAI) to classify healthy leaves from stress level-1 were 89% and 92%, which was significant, as they demonstrate the potential of the fusion algorithm for detecting early symptoms of stressed plants at night.

The decrease of the chlorophyll and nitrogen contents under stress was also consistent with previous studies. We believe that the decrease of the chlorophyll content in stressed leaves may be caused by the degradation of chloroplasts in stressed leaves [74,75]. The scatter plots of the chlorophyll and nitrogen contents (Figure 15a,b) were similar to those of the NDVI and NPSDI and tended to decrease with the increase of the leaf stress degree. By drawing the distinguishing lines of different stress levels, the diagnostic accuracy of different stress levels can be determined (Table 1). It can be seen that the NDVI and NPSDI have similar diagnostic accuracies in distinguishing different levels of stress, while the SPAD has a higher correlation with stress and higher diagnostic accuracy. Furthermore, the NDVI determined through a spectra analysis follows a linear relationship with the SPAD and NC (Figure 16a,b), and the NPSDI obtained through a fusion algorithm for nighttime plant detection follows a linear relationship with the SPAD and NC (Figure 17a,b). There was also a linear correlation between the NDVI and NPSDI (Figure 18), which shows the potential of applying the fusion image determined by PMSIS in understanding the health status of outdoor plants at night from proximal sensing platforms.

In contrast to point monitoring devices such as spectroradiometers and hand-held chlorophyll analyzers, imaging devices provide spatial distribution information on the properties of the objects being detected. The advantage of PMSIS imaging is that it is a passive technology. It is a relatively inexpensive and compact instrument that can be easily carried anywhere and deployed on platforms (such as cars and drones) to feel the effects of various types of stress on farmlands and forests at night.

5. Conclusions

We built a polarized multispectral for a low-illumination-level imaging system and a polarized multispectral low-light imaging system (PMSIS), incorporating a SCOMS camera, rotating filter wheel and linear polarizer, for monitoring the health status of plants outdoors at night. With a compact size, the developed system is portable, consisting of a minimum number of moving parts, which makes it ideal for field measurements. We have done some research on the potential of polarization technology to provide contrast enhancement information for vegetation assessments. Polarization imaging and multispectral imaging can provide complementary discrimination information for target monitoring and other applications. However, few people have proposed to fuse the information of polarized images and multispectral images to obtain better target monitoring results. Therefore, a fusion algorithm based on the spectral and polarization characteristics of the diffuse and specular reflection of vegetation is proposed to detect vegetation at night. The NDVI, DoLP and AOP were all computed within a fusion framework to better monitor the health status of plants in a nighttime environment.

In order to evaluate the remote sensing power of the fusion algorithm in monitoring the plant health status at night, the leaf regions of spotted laurel in different health states were measured. The changes of the SPAD, NC, NDVI and NPSDI in different health levels of plants were determined, and the health degrees of spotted laurel were classified. The value of the NPSDI decreased with the decrease of the plant health status, and it was in a positive linear correlation with the physiological parameters such as the SPAD and NC. The correlation of the NPSDI with NC was 0.916, while, with the SPAD, it was 0.882. In applying the fusion algorithm in classifying healthy leaves from level-1 stress leaves, the sensitivity obtained was 89% and the specificity 92%, which suggests the effectiveness of the method for the early detection of the health status in field-grown plants.

In this paper, the fusion of the NDVI, DoLP and AOP was proposed for the first time. The proposed fusion algorithm enhanced the contrast effect, and the resultant image carried more abundant information of the object; thus, monitoring the health status of plants at night becomes more effective. The fusion algorithm makes up for the unreliability of the NDVI in a low-illumination environment. Based on the polarimetric imaging system that we developed, polarization information improved the monitoring accuracy at night with traditional NDVI information. In the study of monitoring the plant health status at night, we found that the NPSDI was excellent in tracking stress-induced plant changes. The NPSDI scatter plots were used to study for classifying different grades of the health state in spotted laurel, and the sensitivity and specificity were good, which indicated that this technique was suitable for the classification of the outdoor plant health status and various stresses. This study provided the basis for the promotion and use of PMSIS from various remote sensing platforms (such as cars and drones) to assess the crop health status by polarization spectral imaging at night. In the future, the system can be improved (a three-camera real-time imaging system), so that it can be installed on vehicle and airborne devices, enabling its application in a large area for monitoring the health of vegetation at night. The system has the following applications: (1) the detection of the vegetation status in low-illumination environments, (2) the detection of the vegetation status during the day, (3) the detection of the polarization of vegetation leaves and (4) the detection of artificial targets in vegetation environments.

We supplement a table at the end of paper. Acronyms are shown in Table 2.

Table 2. List of acronyms with their meanings.

	Acronyms	English Full Name
1	PMSIS	Polarized multispectral low-illumination-level imaging system
2	NDVI	Normalized vegetation index
3	DoLP	Degree of linear polarization
4	AOP	Angle of polarization
5	NDAI	NDVI, DoLP and AOP fusion image
6	NPSDI	Night plant state detection index
7	SPAD	Chlorophyll content
8	NC	Nitrogen content

Author Contributions: J.J. and C.W. conceived and designed the experiments; S.L. performed the experiments and analyzed and interpreted the results. All authors have read and agreed to the published version of the manuscript.

Funding: This research was funded by the National Natural Science Foundation of China (No. 61773249) and Shanghai Science and Technology Innovation Action Plan (No. 20142200100) for project funding.

Acknowledgments: Thanks to the colleagues and students who participated in the outdoor data collection.

Conflicts of Interest: The authors declare no conflict of interest.

References

1. Food and Agriculture Organization of the United Nations. *The State of the World's Forests 2018—Forest Pathways to Sustainable Development*; Food and Agriculture Organization of the United Nations: Rome, Italy, 2018.
2. Zhou, Y.; Wu, L.; Zhang, H.; Wu, K. Spread of invasive migratory pest spodoptera frugiperda and management practices throughout china. *J. Integr. Agric.* **2021**, *20*, 637–645. [CrossRef]
3. Tong, A.; He, Y. Estimating and mapping chlorophyll content for a heterogeneous grassland: Comparing prediction power of a suite of vegetation indices across scales between years. *ISPRS J. Photogramm. Remote Sens.* **2017**, *126*, 146–167. [CrossRef]
4. Sharma, R.C.; Kajiwara, K.; Honda, Y. Estimation of forest canopy structural parameters using kernel-driven bi-directional reflectance model based multi-angular vegetation indices. *ISPRS J. Photogramm. Remote Sens.* **2013**, *78*, 50–57. [CrossRef]
5. Yue, J.; Yang, G.; Tian, Q.; Feng, H.; Xu, K.; Zhou, C. Estimate of winter-wheat above-ground biomass based on UAV ultrahigh-ground-resolution image textures and vegetation indices. *ISPRS J. Photogramm. Remote Sens.* **2019**, *150*, 226–244. [CrossRef]
6. Gao, L.; Wang, X.; Johnson, B.A.; Tian, Q.; Wang, Y.; Verrelst, J.; Mu, X.; Gu, X. Remote sensing algorithms for estimation of fractional vegetation cover using pure vegetation index values: A review. *ISPRS J. Photogramm. Remote Sens.* **2020**, *159*, 364–377. [CrossRef]
7. Zhou, X.; Zheng, H.; Xu, X.; He, J.; Ge, X.; Yao, X.; Cheng, T.; Zhu, Y.; Cao, W.; Tian, Y. Predicting grain yield in rice using multi-temporal vegetation indices from UAV-based multispectral and digital imagery. *ISPRS J. Photogramm. Remote Sens.* **2017**, *130*, 246–255. [CrossRef]
8. Bajgain, R.; Xiao, X.; Wagle, P.; Basara, J.; Zhou, Y. Sensitivity analysis of vegetation indices to drought over two tallgrass prairie sites. *ISPRS J. Photogramm. Remote Sens.* **2015**, *108*, 151–160. [CrossRef]
9. Barbosa, H.A.; Kumar, T.L.; Paredes, F.; Elliott, S.; Ayuga, J.G. Assessment of Caatinga response to drought using Meteosat-SEVIRI Normalized Difference Vegetation Index (2008–2016). *ISPRS J. Photogramm. Remote Sens.* **2019**, *148*, 235–252. [CrossRef]
10. Sakamoto, T.; Shibayama, M.; Kimura, A.; Takada, E. Assessment of digital camera-derived vegetation indices in quantitative monitoring of seasonal rice growth. *ISPRS J. Photogramm. Remote Sens.* **2011**, *66*, 872–882. [CrossRef]
11. Rahimzadeh-Bajgiran, P.; Omasa, K.; Shimizu, Y. Comparative evaluation of the Vegetation Dryness Index (VDI), the Temperature Vegetation Dryness Index (TVDI) and the improved TVDI (iTVDI) for water stress detection in semi-arid regions of Iran. *ISPRS J. Photogramm. Remote Sens.* **2012**, *68*, 1–12. [CrossRef]
12. Adam, E.; Deng, H.; Odindi, J.; Abdel-Rahman, E.M.; Mutanga, O. Detecting the early stage of phaeosphaeria leaf spot infestations in maize crop using in situ hyperspectral data and guided regularized random forest algorithm. *J. Spectrosc.* **2017**, *8*, 6961387. [CrossRef]
13. Apan, A.; Datt, B.; Kelly, R. Detection of pests and diseases in vegetable crops using hyperspectral sensing: A comparison of reflectance data for different sets of symptoms. In Proceedings of the 2005 Spatial Sciences Institute Biennial Conference 2005: Spatial Intelligence, Innovation and Praxis (SSC2005), Melbourne, Australia, 12–16 September 2005; pp. 10–18.
14. Dhau, I.; Adam, E.; Mutanga, O.; Ayisi, K.K. Detecting the severity of maize streak virus infestations in maize crop using in situ hyperspectral data. *Trans. R. Soc. S. Afr.* **2018**, *73*, 1–8. [CrossRef]
15. Iordache, M.; Mantas, V.; Baltazar, E.; Pauly, K.; Lewyckyj, N. A machine learning approach to detecting pine wilt disease using airborne spectral imagery. *Remote Sens.* **2020**, *12*, 2280. [CrossRef]
16. Jo, M.H.; Kim, J.B.; Oh, J.S.; Lee, K.J. Extraction method of damaged area by pinetree pest (Bursaphelenchus Xylophilus) using high resolution IKONOS image. *J. Korean Assoc. Geogr. Inform. Stud.* **2001**, *4*, 72–78.
17. Mota, M.M.; Vieira, P. Pine wilt disease: A worldwide threat to forest ecosystems. *Nematology* **2009**, *11*, 5–14.
18. Nguyen, V.T.; Park, Y.S.; Jeoung, C.S.; Choi, W.I.; Kim, Y.K.; Jung, I.H.; Shigesada, N.; Kawasaki, K.; Takasu, F.; Chon, T.S. Spatially explicit model applied to pine wilt disease dispersal based on host plant infestation. *Ecol. Model.* **2017**, *353*, 54–62. [CrossRef]
19. Selvaraj, M.G.; Vergara, A.; Montenegro, F.; Ruiz, H.A.; Safari, N.; Raymaekers, D.; Ocimati, W.; Ntamwira, J.; Tits, L.; Bonaventure, A.; et al. Detection of banana plants and their major diseases through aerial images and machine learning methods: A case study in DR Congo and Republic of Benin. *ISPRS J. Photogramm. Remote Sens.* **2020**, *169*, 110–124. [CrossRef]
20. Shi, Y.; Huang, W.; Luo, J.; Huang, L.; Zhou, X. Detection and discrimination of pests and diseases in winter wheat based on spectral indices and kernel discriminant analysis. *Comput. Electron. Agric.* **2017**, *141*, 171–180. [CrossRef]
21. Son, M.H.; Lee, W.K.; Lee, S.H.; Cho, H.K.; Lee, J.H. Natural spread pattern of damaged area by pine wilt disease using geostatistical analysis. *J. Korean Soc. Forest Sci.* **2006**, *95*, 240–249.
22. Baranoski, G.V.G.; Rokne, J.G. A practical approach for estimating the red edge position of plant leaf reflectance. *Int. J. Remote Sens.* **2005**, *26*, 503–521. [CrossRef]
23. Rouse, J.W.; Hass, R.H.; Schell, J.A.; Deering, D.W. Monitoring vegetation systems in the great plains with ERTS. In Proceedings of the Third Earth Resources Technology Satellite-1 Symposium, Washington, DC, USA, 10–14 December 1973; Volume 1, pp. 309–317.
24. Huete, A.R. A soil-adjusted vegetation index (SAVI). *Remote Sens. Environ.* **1988**, *25*, 295–309. [CrossRef]
25. Qi, J.; Chehbouni, A.; Huete, A.R.; Kerr, Y.H.; Sorooshian, S. A modified adjusted vegetation index (MSAVI). *Remote Sens. Environ.* **1994**, *48*, 119–126. [CrossRef]
26. Gao, B. NDWI—A normalized difference water index for remote sensing of vegetation liquid water from space. *Remote Sens. Environ.* **1996**, *58*, 257–266. [CrossRef]

27. Steddom, K.; Bredehoeft, M.W.; Khan, M.; Rush, C.M. Comparison of visual and multispectral radiometric disease evaluations of cercospora leaf spot of sugar beet. *Plant Dis.* **2005**, *89*, 153–158. [CrossRef]

28. Arens, N.; Backhaus, A.; Döll, S.; Fischer, S.; Seiffert, U.; Mock, H.P. Non-invasive presymptomatic detection of cercospora beticola infection and identification of early metabolic responses in sugar beet. *Front. Plant Sci.* **2016**, *7*, 1377. [CrossRef] [PubMed]

29. Deery, D.; Jimenez-Berni, J.; Jones, H.; Sirault, X.; Furbank, R. Proximal remote sensing buggies and potential applications for field-based phenotyping. *Agronomy* **2014**, *4*, 349–379. [CrossRef]

30. Vigneau, N.; Ecarnot, M.; Rabatel, G.; Roumet, P. Potential of field hyperspectral imaging as a non-destructive method to assess leaf nitrogen content in wheat. *Field Crop. Res.* **2011**, *122*, 25–31. [CrossRef]

31. Graeff, S.; Link, J.; Claupein, W. Identification of powdery mildew (*Erysiphe graminis* sp. tritici) and take-all disease (*Gaeumanno-myces graminis* sp. tritici) in wheat (*Triticum aestivum* L.) by means of leaf reflectance measurements. *Open Life Sci.* **2006**, *1*, 275–288. [CrossRef]

32. Yang, Z.; Rao, M.N.; Elliott, N.C.; Kindler, S.D.; Popham, T.W. Differentiating stress induced by greenbugs and Russian wheat aphids in wheat using remote sensing. *Comput. Electron. Agric.* **2009**, *67*, 64–70. [CrossRef]

33. Liu, Z.Y.; Wu, H.F.; Huang, J.F. Application of neural networks to discriminate fungal infection levels in rice panicles using hyperspectral reflectance and principal components analysis. *Comput. Electron. Agric.* **2010**, *72*, 99–106. [CrossRef]

34. Prabhakar, M.; Prasad, Y.; Thirupathi, M.; Sreedevi, G.; Dharajothi, B.; Venkateswarlu, B. Use of ground based hyperspectral remote sensing for detection of stress in cotton caused by leafhopper (Hemiptera: Cicadellidae). *Comput. Electron. Agric.* **2011**, *79*, 189–198. [CrossRef]

35. Prabhakar, M.; Prasad, Y.G.; Vennila, S.; Thirupathi, M.; Sreedevi, G.; Ramachandra Rao, G.; Venkateswarlu, B. Hyperspectral indices for assessing damage by the solenopsis mealybug (Hemiptera: Pseudococcidae) in cotton. *Comput. Electron. Agric.* **2013**, *97*, 61–70. [CrossRef]

36. Nebiker, S.; Lack, N.; Abächerli, M.; Läderach, S. Light-weight multispectral UAV sensors and their capabilities for predicting grain yield and detecting plant diseases. ISPRS—International Archives of the Photogrammetry. *Remote Sens. Spat. Inf. Sci.* **2016**, *XLI-B1*, 963–970.

37. Dash, J.P.; Watt, M.S.; Pearse, G.D.; Heaphy, M.; Dungey, H.S. Assessing very high resolution UAV imagery for monitoring forest health during a simulated disease outbreak. *ISPRS J. Photogramm. Remote Sens.* **2017**, *131*, 1–14. [CrossRef]

38. Yang, C.H.; Everitt, J.H.; Fernandez, C.J. Comparison of airborne multispectral and hyperspectral imagery for mapping cotton root rot. *Biosyst. Eng.* **2010**, *107*, 131–139. [CrossRef]

39. Calderón, R.; Navas-Cortés, J.A.; Lucena, C.; Zarco-Tejada, P.J. High-resolution airborne hyperspectral and thermal imagery for early detection of Verticillium wilt of olive using fluorescence, temperature and narrow-band spectral indices. *Remote Sens. Environ.* **2013**, *139*, 231–245. [CrossRef]

40. Sanches, I.; Filho, C.R.S.; Kokaly, R.F. Spectroscopic remote sensing of plant stress at leaf and canopy levels using the chlorophyll 680nm absorption feature with continuum removal. *ISPRS J. Photogramm. Remote Sens.* **2014**, *97*, 111–122. [CrossRef]

41. Lehmann, J.; Nieberding, F.; Prinz, T.; Knoth, C. Analysis of Unmanned Aerial System-Based CIR Images in Forestry—A New Perspective to Monitor Pest Infestation Levels. *Forests* **2015**, *6*, 594–612. [CrossRef]

42. Yuan, L.; Pu, R.; Zhang, J.; Wang, J.; Yang, H. Using high spatial resolution satellite imagery for mapping powdery mildew at a regional scale. *Precis. Agric.* **2016**, *17*, 332–348. [CrossRef]

43. Chen, D.; Shi, Y.; Huang, W.; Zhang, J.; Wu, K. Mapping wheat rust based on high spatial resolution satellite imagery. *Comput. Electron. Agric.* **2018**, *152*, 109–116. [CrossRef]

44. Zheng, Q.; Huang, W.; Cui, X.; Shi, Y.; Liu, L. New Spectral Index for Detecting Wheat Yellow Rust Using Sentinel-2 Multispectral Imagery. *Sensors* **2018**, *18*, 868. [CrossRef]

45. Meiforth, J.J.; Buddenbaum, H.; Hill, J.; Shepherd, J.D.; Dymond, J.R. Stress Detection in New Zealand Kauri Canopies with WorldView-2 Satellite and LiDAR Data. *Remote Sens.* **2020**, *12*, 1906. [CrossRef]

46. Li, X.; Yuan, W.; Dong, W. A Machine Learning Method for Predicting Vegetation Indices in China. *Remote Sens.* **2021**, *13*, 1147. [CrossRef]

47. Rvachev, V.P.; Guminetskii, S.G. The structure of light beams reflected by plant leaves. *J. Appl. Spectrosc.* **1966**, *4*, 303–307. [CrossRef]

48. Woolley, J.T. Reflectance and transmission of light by leaves. *Plant Physiol.* **1971**, *47*, 656–662. [CrossRef] [PubMed]

49. Vanderbilt, V.C.; Grant, L.; Daughtry, C.S.T. Polarization of light scattered by vegetation. *Proc. IEEE* **1985**, *73*, 1012–1024. [CrossRef]

50. Vanderbilt, V.C.; Grant, L.; Biehl, L.L.; Robinson, B.F. Specular, diffuse, and polarized light scattered by two wheat canopies. *Appl. Opt.* **1985**, *24*, 2408–2418. [CrossRef] [PubMed]

51. Duggin, M.J.; Kinn, G.J.; Schrader, M. Enhancement of vegetation mapping using Stokes parameter images. In *Proceedings of SPIE—The International Society for Optical Engineering*; Society of Photo-Optical Instrumentation Engineers: Bellingham, WA, USA, 1997; Volume 3121, pp. 307–313.

52. Rondeaux, G.; Herman, M. Polarization of light reflected by crop canopies. *Remote Sens. Environ.* **1991**, *38*, 63–75. [CrossRef]

53. Fitch, B.W.; Walraven, R.L.; Bradley, D.E. Polarization of light reflected from grain crops during the heading growth stage. *Remote Sens. Environ.* **1984**, *15*, 263–268. [CrossRef]

54. Perry, G.; Stearn, J.; Vanderbilt, V.C.; Ustin, S.L.; Diaz Barrios, M.C.; Morrissey, L.A.; Livingston, G.P.; Breon, F.-M.; Bouffies, S.; Leroy, M. Remote sensing of high-latitude wetlands using polarized wide-angle imagery. In *Proceedings of SPIE—The International Society for Optical Engineering*; Society of Photo-Optical Instrumentation Engineers: Bellingham, WA, USA, 1997; Volume 3121, pp. 370–381.

55. Shibayama, M. Prediction of the ratio of legumes in a mixed seeding pasture by polarized reflected light. *Jpn. J. Grassl. Sci.* **2003**, *49*, 229–237.

56. Vanderbilt, V.C.; Grant, L. Polarization photometer to measure bidirectional reflectance factor R (55°, 0°; 55°, 180°) of leaves. *Opt. Eng.* **1986**, *25*, 566–571. [CrossRef]

57. Vanderbilt, V.C.; Grant, L.; Ustin, S.L. *Polarization of Light by Vegetation*; Springer: Berlin/Heidelberg, Germany, 1991; Volume 7, pp. 191–228.

58. Grant, L.; Daughtry, C.S.T.; Vanderbilt, V. Polarized and specular reflectance variation with leaf surface-features. *Physiol. Plant.* **1993**, *88*, 1–9. [CrossRef]

59. Raven, P.; Jordan, D.; Smith, C. Polarized directional reflectance from laurel and mullein leaves. *Opt. Eng.* **2002**, *41*, 1002–1012. [CrossRef]

60. Georgiev, G.; Thome, K.; Ranson, K.; King, M.; Butler, J. The effect of incident light polarization on vegetation bidirectional reflectance factor. In Proceedings of the 2010 IEEE International Geoscience and Remote Sensing Symposium (IGARSS), Honolulu, HI, USA, 25–30 July 2010.

61. Hu, H.; Lin, Y.; Li, X.; Qi, P.; Liu, T. IPLNet: A neural network for intensity-polarization imaging in low light. *Opt. Lett.* **2020**, *45*, 6162–6165. [CrossRef] [PubMed]

62. Zhao, Y.; Zhang, L.; Zhang, D.; Pan, Q. Object separation by polarimetric and spectral imagery fusion. *Comput. Vis. Image Underst.* **2009**, *113*, 855–866. [CrossRef]

63. Shibata, S.; Hagen, N.; Otani, Y. Robust full Stokes imaging polarimeter with dynamic calibration. *Opt. Lett.* **2019**, *44*, 891–894. [CrossRef] [PubMed]

64. Schott, J.R. *Fundamentals of Polarimetric Remote Sensing*; SPIE Press: Bellingham, WA, USA, 2009; Volume 81.

65. Goldstein, D.H. Polarimetric characterization of federal standard paints. In *Proceedings of SPIE—The International Society for Optical Engineering*; Society of Photo-Optical Instrumentation Engineers: Bellingham, WA, USA, 2000; Volume 4133, pp. 112–123.

66. Bajwa, S.G.; Tian, L. Multispectral cir image calibration for cloud shadow and soil background influence using intensity normalization. *Appl. Eng. Agric.* **2002**, *18*, 627–635. [CrossRef]

67. Rahman, M.M.; Lamb, D.W.; Samborski, S.M. Reducing the influence of solar illumination angle when using active optical sensor derived NDVIAOS to infer fAPAR for spring wheat (*Triticum aestivum* L.). *Comput. Electron. Agric.* **2019**, *156*, 1–9. [CrossRef]

68. Martín-Ortega, P.; García-Montero, L.G.; Sibelet, N. Temporal Patterns in Illumination Conditions and Its Effect on Vegetation Indices Using Landsat on Google Earth Engine. *Remote Sens.* **2020**, *12*, 211. [CrossRef]

69. Lavigne, D.A.; Breton, M.; Fournier, G.R.; Pichette, M.; Rivet, V. Development of performance metrics to characterize the degree of polarization of man-made objects using passive polarimetric images. In *Proceedings of SPIE—The International Society for Optical Engineering*; Society of Photo-Optical Instrumentation Engineers: Bellingham, WA, USA, 2009; Volume 7336, p. 73361A.

70. Lavigne, D.A.; Breton, M.; Fournier, G.; Charette, J.-F.; Pichette, M.; Rivet, V.; Bernier, A.-P. Target discrimination of man-made objects using passive polarimetric signatures acquired in the visible and infrared spectral bands. In *Proceedings of SPIE—The International Society for Optical Engineering*; Society of Photo-Optical Instrumentation Engineers: Bellingham, WA, USA, 2011; Volume 8160, p. 2063.

71. Raji, S.N.; Subhash, N.; Ravi, V.; Saravanan, R.; Mohanan, C.; Nita, S.; Kumar, T.M. Detection of mosaic virus disease in cassava plants by sunlight-induced fluorescence imaging: A pilot study for proximal sensing. *Remote Sens.* **2015**, *36*, 2880–2897. [CrossRef]

72. Lichtenthaler, H.K. Chlorophyll Fluorescence Signatures of Leaves during the Autumnal Chlorophyll Breakdown*). *J. Plant Physiol.* **1987**, *131*, 101–110. [CrossRef]

73. Saito, Y.; Kurihara, K.J.; Takahashi, H.; Kobayashi, F.; Kawahara, T.; Nomura, A.; Takeda, S. Remote Estimation of the Chlorophyll Concentration of Living Trees Using Laser-induced Fluorescence Imaging Lidar. *Opt. Rev.* **2002**, *9*, 37–39. [CrossRef]

74. Esau, K. An anatomist's view of virus diseases. *Am. J. Bot.* **1956**, *43*, 739–748. [CrossRef]

75. Fang, Z.; Bouwkamp, J.C.; Theophanes, S. Chlorophyllase activities and chlorophyll degradation during leaf senescence in non-yellowing mutant and wild type of *Phaseolus vulgaris* L. *J. Exp. Bot.* **1998**, *1998*, 503–510.

remote sensing

MDPI

Article

Integrating Spectral Information and Meteorological Data to Monitor Wheat Yellow Rust at a Regional Scale: A Case Study

Qiong Zheng [1], Huichun Ye [2,3,*], Wenjiang Huang [2], Yingying Dong [2], Hao Jiang [1], Chongyang Wang [1], Dan Li [1], Li Wang [1] and Shuisen Chen [1]

[1]   Key Lab of Guangdong for Utilization of Remote Sensing and Geographical Information System, Guangdong Open Laboratory of Geospatial Information Technology and Application, Research Center of Guangdong Province for Engineering Technology Application of Remote Sensing Big Data, Guangzhou Institute of Geography, Guangdong Academy of Sciences, Guangzhou 510070, China; zhengqiong@gdas.ac.cn (Q.Z.); jianghao@gdas.ac.cn (H.J.); wangchongyang@gdas.ac.cn (C.W.); lidan@gdas.ac.cn (D.L.); wangli1990@nwsuaf.edu.cn (L.W.); css@gdas.ac.cn (S.C.)

[2]   Key Laboratory of Digital Earth Science, Aerospace Information Research Institute, Chinese Academy of Sciences, Beijing 100094, China; huangwj@aircas.ac.cn (W.H.); dongyy@aircas.ac.cn (Y.D.)

[3]   Key Laboratory for Earth Observation of Hainan Province, Sanya 572029, China

*   Correspondence: yehc@aircas.ac.cn

Abstract: Wheat yellow rust has a severe impact on wheat production and threatens food security in China; as such, an effective monitoring method is necessary at the regional scale. We propose a model for yellow rust monitoring based on Sentinel-2 multispectral images and a series of two-stage vegetation indices and meteorological data. Sensitive spectral vegetation indices (single- and two-stage indices) and meteorological features for wheat yellow rust discrimination were selected using the random forest method. Wheat yellow rust monitoring models were established using three different classification methods: linear discriminant analysis (LDA), support vector machine (SVM), and artificial neural network (ANN). The results show that models based on two-stage indices (i.e., those calculated using images from two different days) significantly outperform single-stage index models (i.e., those calculated using an image from a single day), the overall accuracy improved from 63.2% to 78.9%. The classification accuracies of models combining a vegetation index with meteorological feature are higher than those of pure vegetation index models. Among them, the model based on two-stage vegetation indices and meteorological features performs best, with a classification accuracy exceeding 73.7%. The SVM algorithm performed best for wheat yellow rust monitoring among the three algorithms; its classification accuracy (84.2%) was ~10.5% and 5.3% greater than those of LDA and ANN, respectively. Combined with crop growth and environmental information, our model has great potential for monitoring wheat yellow rust at a regional scale. Future work will focus on regional-scale monitoring and forecasting of crop disease.

Keywords: wheat yellow rust; vegetation indices; meteorological information; food security; regional remote sensing

Citation: Zheng, Q.; Ye, H.; Huang, W.; Dong, Y.; Jiang, H.; Wang, C.; Li, D.; Wang, L.; Chen, S. Integrating Spectral Information and Meteorological Data to Monitor Wheat Yellow Rust at a Regional Scale: A Case Study. *Remote Sens.* **2021**, *13*, 278. https://doi.org/10.3390/rs13020278

Received: 11 December 2020
Accepted: 12 January 2021
Published: 14 January 2021

Publisher's Note: MDPI stays neutral with regard to jurisdictional claims in published maps and institutional affiliations.

1. Introduction

Wheat is the main grain crop for mankind [1]. Yellow rust (*Puccinia striiformis f.* sp. *tritici Erikss*) is a devastating disease in wheat planting that affects wheat growth, thus seriously affecting the quality and yield of wheat in China [1,2]. The average annual area of wheat yellow rust is 4 million hm², resulting in a reduction in wheat production of more than 1 billion kg per year [3]. Traditional methods of wheat yellow rust involve manual surveys that are time-consuming, laborious, and inefficient [4]. In recent decades, remote sensing technology has been proved to be an effective tool for monitoring of crop disease and pest, with advantages of large-scale and real time simultaneous monitoring [4,5]. Therefore, timely, effective, and accurate monitoring of wheat yellow rust based on remote sensing technology and multi-source data of disease occurrence is essential to ensure food security and agricultural sustainability in China.

It is well-known that the phenology of vegetation is closely related to the seasonality of meteorological variables [6]. In addition, the integration of meteorological data with satellite hydrological models can improve irrigation scheduling management and help farmers to properly make rational irrigation plans [7]. For pathogens, the propagation, spread, and infection of pathogen spores require suitable environmental support; thus, the monitoring of crop pests and diseases is related to environmental conditions [4,8]. Meteorological data, such as temperature, humidity, sunshine, and rainfall are the key factors to determining the occurrence, development, and prevalence of crop diseases [4]. Zhang et al. showed that integrating meteorological data (precipitation, temperature, humidity, and solar radiation) and remote sensing features has significant potential to forecast the occurrence probability of wheat powdery mildew [9]. Papastamati et al. stated that the inoculation concentrations, temperature, and rainfall time were related to light leaf spot epidemics in winter oilseed rape disease and proposed a new model for predicting the disease [10]. Habitat monitoring has great potential for evaluating the occurrence and distribution of plant diseases on a regional scale [8].

In addition, biophysical parameters of the host have a certain effect on the infection of diseases [11]. In different infestation stages, plants exhibit specific host–pathogen interactions, such as a reduction in leaf area index, pigment content changes, and canopy structure morphology destruction [12]. These changes respond in the visible to near-infrared regions of the spectrum and can even be captured through multi-temporal observations [11–13]. Zhang et al. used single- and multi-temporal images from HJ satellite to monitor wheat powdery mildew, and proved that multi-stage images were superior to single-temporal images for wheat disease monitoring, with monitoring accuracy reaching 78% [14]. Ma et al. developed a multi-temporal vegetation index that indicated crop growth status and habitat characteristics to monitor wheat powdery mildew based on the k-nearest neighbor approach [15]. Therefore, the integration of habitat information and the host growth status related to disease and pest occurrence have great potential for crop stress monitoring.

Classifiers can learn the characteristics of target classes from training samples and apply this information to unclassified data. Methods, such as decision tree (DT), k-nearest neighbors (kNN) method, support vector machine (SVM), and artificial neural network (ANN), have been developed and applied in objects recognition and species classification in the remote sensing field [16–18]. Among these, the SVM and ANN methods are popular for crop classification and disease remote sensing monitoring [4,16].

Satellites with different spatial resolutions have been used in agricultural disease identification and monitoring [4]. Razz et al. and Yue et al. effectively utilized the high-resolution of PlantScope imagery (3 m) to detect soybean sudden death syndrome and rice diseases, respectively [19,20]. Calderón et al. used high spatial resolution and hyperspectral imagery for Verticillium wilt of olive for early detection [21]. Yuan et al. used high spatial resolution (Worldview 2) and medium-spatial resolution (Landsat 8) satellite images to monitor the distribution of wheat disease and pest, achieving a monitoring accuracy of more than 71.0% [8]. Therefore, satellite imagery has proven to be an efficient tool for monitoring crop disease and pest in farms on large regional scales. The Sentinel-2 multispectral satellite launched in 2015, has a 290 km swath width and a 10 days revisit cycle, although it can reach up to 5 days if two satellites work simultaneously [22]. Sentinel-2 multispectral images have 13 spectral bands with a spectral range that includes visible, near-infrared, and shortwave infrared regions; their spatial resolutions are 10, 20, and 60 m, respectively (Figure 1). The most innovative feature of the Sentinel-2 satellite is that it contains rich red-edge bands information with center wavelengths of 705, 740, and 783 nm, which provides abundant information for vegetation biophysical status monitoring and estimation [23,24]. In summary, Sentinel-2 has an unpreceded spatial, temporal resolution and revisit cycle, which is suitable for monitoring crop growth processes such as crop diseases or pest stress [25].

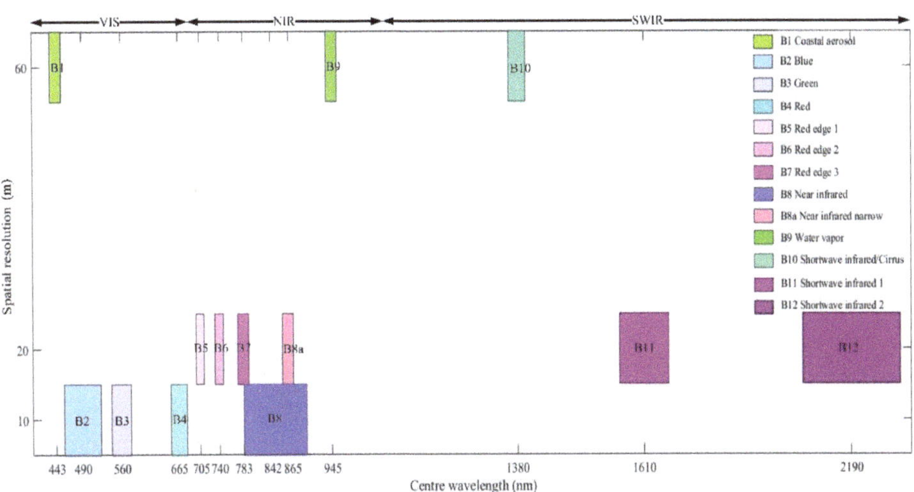

Figure 1. Parameters and information of Sentinel-2 multispectral satellite.

Yellow rust is an air-dispersed pandemic disease [2]; its occurrence is strongly related to habitat conditions, such as humidity, sunshine, and temperature. However, most of the existing remote sensing identification methods for wheat yellow rust depend on spectral data; few studies have considered the habitat characteristics of yellow rust disease [8]. In this study, we considered the characteristics of spectral changes and meteorological factors in the occurrence period of wheat yellow rust; and developed a large-scale and high-precision monitoring method for wheat yellow rust on a regional scale that integrates environment conditions and growth status. The primary aims of this research are: (1) To present a series of two-temporal vegetation indices for monitoring wheat yellow rust using two-stages satellite remote sensing images on a regional scale; (2) to explore the feasibility of combining remote sensing and meteorological information for yellow rust monitoring; and (3) to develop an optimal classification method for monitoring wheat yellow rust utilizing both spectral information and meteorological data on a regional scale.

2. Materials and Methods

2.1. Field Survery and Data Collection

2.1.1. Field Survey Area

Field surveys of wheat yellow rust were conducted in Ningqiang county (118°35′19.5″ E, 37°35′51.75″ N), Shaanxi province, China from 11 May 2018 to 14 May 2018 (i.e., the filling stage, Figure 2). In Ningqiang County, wheat is a major crop, and yellow rust is the dominant wheat disease; a severe infestation occurred in 2018. The average annual temperature and precipitation of this area are 13 °C, and 1812.2 mm, respectively; low temperature and high humidity environment conditions are conductive to the occurrence, spread, and epidemic of wheat yellow rust [23]. Shaanxi Province is considered to be an important spring epidemic area and winter breeding area of wheat yellow rust in China [26]. Therefore, there is an urgent demand to monitor wheat yellow rust disease in this region.

Three data types, including field survey data of yellow rust disease, multispectral satellite images, and meteorological information, were collected to develop a wheat yellow rust monitoring model on a regional scale.

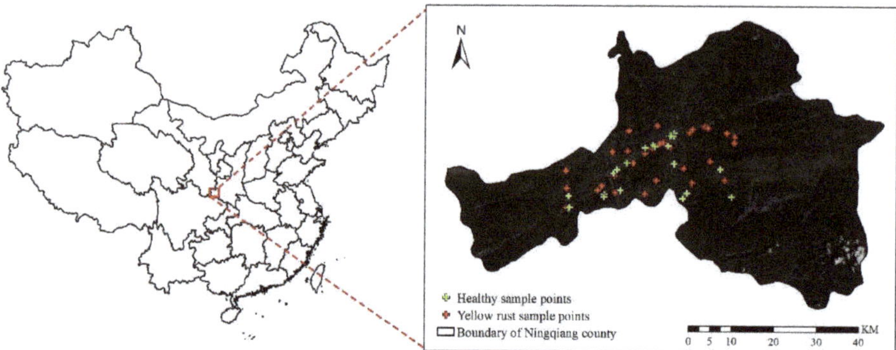

Figure 2. Map of Ningqiang county, Shaanxi province in China. Green crosses denote field survey locations of healthy wheat, red denote field survey locations of yellow rust-infested wheat.

2.1.2. Disease Survey Data Collection

In this field survey, 58 samples (22 healthy, 36 infected) were investigated from 11 May 2018 to 14 May 2018. Considering the pixels size of remote sensing images, uniformly growing wheat samples were randomly selected with a continuous area of 10×10 m, and the severity of disease was surveyed. We selected five representative 1×1 m plots (located at the four corners and centers of the 10×10 m plots), and the average severity of yellow rust for the five plots was used to represent the disease degree for one sample [8]. The center coordinate of each sample was recorded by a differential global positioning system (GPS) sensor (Trimble GeoXH). The field severity survey and disease index calculation of wheat yellow rust referenced the rule of the National Rules for the Investigation and Forecasting of Crop Diseases (GB/T 15795-1995) [12]. The information of wheat growth, disease incidence, and location are recorded in Appendix A Table A1.

2.1.3. Remote Sensing Data Collection and Wheat Planting Area Extraction

Sentinel-2A remote sensing images (processing level 1C) on 2 April 2018 (early onset of disease) and 12 May 2018 (disease outbreak stage) were downloaded from the European Space Agency Sentinels Scientific Date Hub (https://scihub.copernicus.eu/) for the study region [23]. The preprocessing of Sentinel-2A images included atmospheric correction, and clipping. Atmospheric correction was performed using the Sen2cor module (version 2.2.1) within the Sentinel-2 Toolbox, and image mosaic and cropping were implemented in the Sentinel Application Platform software (SNAP, 4.0.2) [23]. In addition, the Sentinel-2A multispectral data carried 13 bands that include three different spatial resolution (Figure 1). For subsequent analysis, the spatial resolutions of the 13 bands were resampled to 10 m using the resampling tool in the software. The large-scale crop disease monitoring was based on the extraction of wheat planting area; therefore, we used the decision tree and multi-temporal phenological information methods, as proposed by Zhang et al. and Xu et al., to extract the planting area of wheat [14,27]. Field survey points were used to verify the accuracy of the extracted wheat area, which reached 94%. This result meets the demand of subsequent remote sensing monitoring of crop diseases.

2.1.4. Meteorological Data

Meteorological data are the basis for analyzing and describing climate characteristics and their laws of change [28]. The occurrence and prevalence of wheat yellow rust depend on the interaction among wheat varieties, amount of yellow rust disease, and environmental conditions [1,4,8]. When both pathogen and host have the potential for an epidemic, environmental conditions, specifically meteorological conditions, become the dominant factor in a wheat yellow rust epidemic [4,29].

Considering the influence of climate conditions on the wheat infection of yellow rust pathogens, five types of meteorological data were collected for March to May 2018 from the National Meteorological Information Center in 37 sampling sites around Ningqiang county, including average temperature (TEM), precipitation (PRE), sunshine hours (SSD), wind speed (WIN), and relative humidity (RHU). From each of these, we calculated a monthly mean value for March, April, and May. Therefore, a total of 15 meteorological features were calculated in this study.

2.2. Vegetation Indices for Plant Diseases Discrimination

Crop under the stress of pests and diseases often undergo changes that impact their spectral properties, including pigmentation, moisture, and biomass. The sensitive spectral bands were combined to construct vegetation indices (VIs) in the relevant mathematical forms. VIs related to plant growth status, vegetation coverage, and pigmentation content were used to capture the physiological and biochemical changes caused by wheat yellow rust infection (Table 1).

Table 1. Multispectral vegetation indices for wheat yellow rust discrimination.

Vegetation Indices	Name	Formula	Reference
NDVI	Normalized differece vegetation index	$(R_{NIR} - R_R)/(R_{NIR} + R_R)$	[30]
GNDVI	Green normalized difference vegetation index	$(R_{NIR} - R_G)/(R_{NIR} + R_G)$	[31]
SAVI	Soil adjusted vegtation index	$((R_{NIR} - R_R) * 1.5)/(R_{NIR} + R_R + 0.5)$	[32]
SIPI	Structural independent pigment index	$(R_{NIR} - R_B)/(R_{NIR} - R_R)$	[33]
EVI	Enhanced vegetation index	$2.5(R_{NIR} - R_R)/(R_{NIR} + 6R_R - 0.5R_B + 1)$	[34]
RDVI	Re—noramalied difference vegetation index	$(R_{NIR} - R_R)/\sqrt{(R_{NIR} + R_R)}$	[35]
RGR	Ration of red and green	R_R/R_G	[36]
VARIgreen	Visible atmospherically resistant index	$(R_G - R_R)/(R_G + R_R)$	[37]
NDVIre1	Normalied difference vegetation index red—edge1	$(R_{NIR} - R_{Re1})/(R_{NIR} + R_{Re1})$	[30]
NREDI1	Normalied red—edge1 index	$(R_{Re2} - R_{Re1})/(R_{Re2} + R_{Re1})$	[38]
NREDI2	Normalied red—edge2 index	$(R_{Re3} - R_{Re1})/(R_{Re3} + R_{Re1})$	[38]
NREDI3	Normalied red—edge3 index	$(R_{Re3} - R_{Re2})/(R_{Re3} + R_{Re2})$	[38]
PSRI1	Plant senescence reflectance index	$(R_R - R_G)/R_{Re1}$	[24]
REDSI	Red—edge disease stress index	$((705 - 665)(R_{Re3} - R_R) - (783 - 665)(R_{Re1} - R_R))/(2R_R)$	[23]

2.3. Two-Stage Vegetation Index for Wheat Yellow Rust Monitoring

The study area belongs to the wheat region of southwest China. Generally, the initial stage of wheat yellow rust disease in this region is from the end of March to the beginning of April, the yellow rust outbreak occurred in mid-May in 2018. Accordingly, we selected Sentinel-2 images acquired on 2 April and 12 May 2018. Based on the commonly used vegetation indices listed in Table 1, we calculated the change in magnitude from 2 April to 12 May using the normalization quantification formula:

$$nVIs = \frac{VI_{12May} - VI_{2April}}{VI_{12May} + VI_{2April}} \tag{1}$$

where nVIs represents the change of vegetation index features between two-stages; VI_{2April} and VI_{12May} indicate the values of the vegetation index extracted from the images at the time of the first occurrence (2 April 2018) of yellow rust and at the large outbreak of yellow rust (12 May 2018), respectively.

2.4. Spectral VIs and Meteorological Features Importance Ranking

There are many features including vegetation indices and meteorological parameters that are potentially relevant to crop diseases monitoring, however the sensitivity of these features varies substantially. It is necessary to describe the degree that how much a feature will impact on the model predictions. In this study, random forest (RF) was applied for classification and feature importance analysis and was first described by Breimen et al. [39]. It is an ensemble approach for building decision trees for predictions. The feature importance in RF is computed as the average contribution of each feature on each tree in the RF [39]. We used the out-of-bag (OOB) data to calculate the error ($errOOB_t$) for each tree in the RF algorithm. Subsequently, we compared the difference in the OOB error of each feature before ($errOOB_t$) and after adding noise ($errOOB_t^i$) to calculate the importance of the feature (X), where Ntree denotes the number of trees in RF [22]. Finally, the importance of feature X^i was defined as:

$$V(X^i) = \frac{1}{N}\sum\nolimits_t \left(errOOB_t - errOOB_t^i\right) \tag{2}$$

In addition, we used the analysis of variance methods to test the significance of the selected features [40]. The statistical significance expressed by the ρ value reflects the suitability of the feature [9,40]. Finally, we selected the features that were highly important and significant as the optimal features for yellow rust detection.

2.5. Monitoring Methods

The main purpose of this research is to explore the feasibility of remote sensing monitoring of crop diseases by meteorological data information. In addition, because of the small sample size in this experiment, three commonly used methods of liner discriminant analysis (LDA), SVM, and ANN and regular parameter settings were selected to construct wheat yellow rust monitoring models.

LDA is a dimensionality reduction method based on the best classification effect [41], usually by finding a set of linear feature combinations to classify two or more targets. The primary idea is to find a linear combination of variables to maximize in-between variance and minimize within-class variance [41]. The LDA model was implemented using the Statistical Package for the Social Sciences (SPSS 20.0). The parameters were set as default value.

In the SVM classification algorithm, the primary idea is to determine an optimal decision boundary and maximize the distance of the closest samples in two categories as much possible across the boundary [42]. Using the radial basis function (RBF) as the kernel function for SVM classification exhibited superior performance in the case of inseparable linearity [17]. The key parameters of SVM are shown in Table 2. The model is trained and tested in the Matlab R2016 software.

ANNs can be described as parallel and complex computing systems composed of large numbers of interconnected simple processors (neurons, also called nodes) [43]. As an important data mining tool, ANNs have comprehensive mathematical mechanisms and have been applied in various fields of remote sensing, such as ground objects identification and change detection [18,43]. In this study, the vegetation indices and meteorological data were the ANN input parameters. The transfer functions of logarithm sigmoid (logsig) transfer and linear (purelin) were used to activate the hidden layers and weighted output layers, respectively [44,45]. The learning rule takes the approach of a gradient descent backpropagation (traingd) training function. The key parameters of ANN are shown in Table 2. We used the MATLAB R2016 software to run the ANN models.

Table 2. Key parameters used for support vector machine (SVM) and artificial neural network (ANN) classification.

Classifier	Parameter Name	Description	Optimized Parameter
SVM	s	Type of SVM	0 (SVC)
	t	Type of kernel function	2 (RBF)
	c	Regularization parameter	1
	γ	Kernel coefficient for "RBF"	0.5
ANN	Number of hidden neurons	-	[2,10]
	net.trainParam.lr	Learning rate	0.01
	net.trainParam.epochs	Learning number	13
	net.trainParam.goal	Target error	0.01

Once sensitive spectral features and meteorological features were identified, the three classification algorithms (LDA, SVM, and ANN) were used to establish a classification model for wheat yellow rust. According to the three methods, wheat yellow rust classification was conducted using the following datasets: case 1: spectral vegetation indices (containing single temporal VIs and two-temporal nVIs); case 2: a combination of spectral vegetation indices and meteorological information.

2.6. Classification Accuracy Assessment

Considering the number of samples (n = 58, of which healthy samples = 22, and disease samples = 36), the samples were divided into three parts, two of which (n = 39) were used as training samples for model training; with the remaining part (n = 19) used as a validation dataset to verify the accuracy of the model. Accuracy evaluation was determined by the overall classification accuracy (OA), kappa coefficient, user's accuracy (U.a), and producer's accuracy (P.a) from confusion matrices [46].

3. Results

3.1. Vegetation Index Response for Yellow Rust Disease

The responses of the 14 VIs and corresponding two-stage vegetation indices (nVIs) are shown in Figure 3, where their mean and difference values are compared for healthy and yellow rust-infected wheat. The values of the re-normalized difference vegetation index (RDVI), and red-edge disease stress index (REDSI) were reduced by 100 times to maintain the same magnitude as that of the other indices. On the single-stage indices, according to the difference information of each vegetation index in healthy and diseased wheat, RDVI, REDSI, enhance vegetation index (EVI), soil adjusted vegetation index (SAVI), and normalized red-edge3 index (NREDI3) were most suitable for discriminating healthy and yellow rust-infected wheat (Figure 3a). Among them, the difference value of RDVI for healthy and disease samples was the largest, reaching 5.5. However, the magnitude of the difference value in the entire single-stage VIs was relatively small, except for RDVI and REDSI.

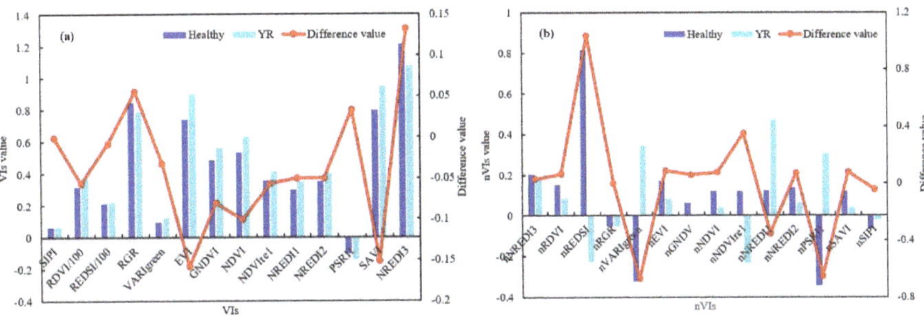

Figure 3. Mean and difference values for healthy wheat and yellow rust infected wheat. (**a**) Single—stage vegetation indices and (**b**) normalized two—stage vegetation indices.

Regarding the normalized two-stage indices, the difference between health samples and yellow rust infection samples was evident, specifically for the normalized REDSI (nREDSI), normalized visible atmospherically resistant index (nVARIgreen), normalized difference vegetation index reg-edge 1 (nNDVIre1), and normalized plant senescence reflectance index (nPSRI1). Among them, the difference between healthy and diseased samples in nREDSI was up to 1.1 (Figure 3b). The two-stage nVIs (using images from both 2 April and 12 May 2018) exhibited a greater difference between healthy and yellow rust-infested wheat compared with the corresponding single-stage VIs (using the images from 12 May). This confirms that nVIs are closely related to the pathological progress of the crop, which can more clearly reflect leaf wilting, leaf tissue death, and canopy structure changes caused by the yellow rust pathogen.

3.2. Meteorological Data Processing and Selection

Each type of meteorological data was averaged by month to obtain its monthly average value. Then, the meteorological factors from March to May 2018 were spatially interpolated using an inverse distance weighted method in the ArcGIS software for subsequent continuous spatial pixel-scale analysis [9]. All meteorological factors were interpolated at a spatial resolution of 10 m, which matches the resolution of the Sentinel-2 satellite imagery. Figure 4 presents the results of meteorological data spatial resolution in May Finally, based on continuous meteorological data, meteorological characteristics were extracted for wheat yellow rust habitat monitoring.

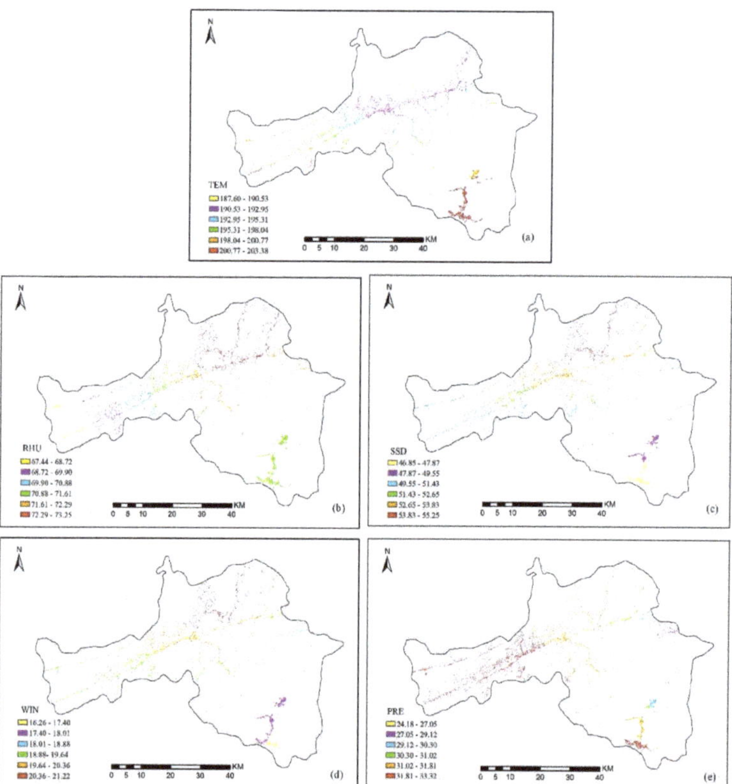

Figure 4. Spatial interpolation of meteorological data in Ningqiang county in May 2018. (**a**) Average temperature (TEM; 0.1 °C), (**b**) average relative humidity (RHU; %), (**c**) average sunshine hours (SSD; 0.1 h), (**d**) average wind speed (WIN; 0.1 m/s), (**e**) average of precipitation (PRE; 0.1 mm).

3.3. VIs and Meteorological Features Sensitivity of Yellow Rust Monitoring

We observed a strong correlation between VIs as well as wheat physiological and, biochemical parameters caused by the development of yellow rust; however, correlations and multiple collinearities among different VIs limit the extraction of sensitive information for wheat yellow rust discrimination. Therefore, we selected the single- and two-stage VIs most sensitive to reflect the state of the crop after being stressed by disease using the important criterion in the RF method. The selected single-stage vegetation index, normalized two-stage vegetation index, and meteorological data were used to determine the important features for yellow rust detection (Figure 5).

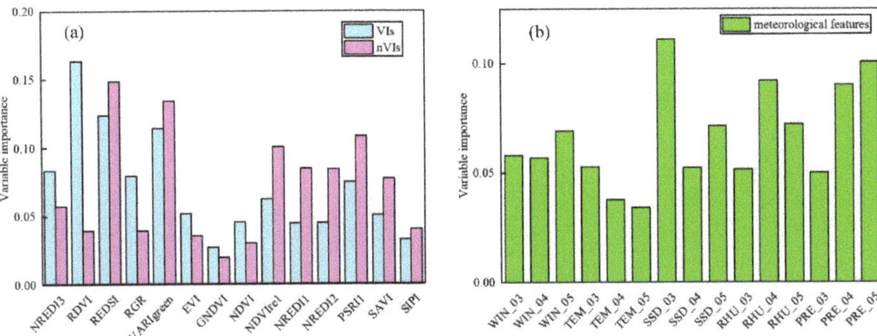

Figure 5. Variable importance of vegetation indices and meteorological factors in identifying yellow rust-infected wheat. (a) Vegetation indices (VIs) and normalized VIs (nVIs) based on a single- and two-stage imagery; (b) meteorological data from March to May 2018.

The relative importance of the three features for wheat yellow rust discrimination were analyzed using the RF method (Figure 5). According to the importance ranking of spectral VIs (Figure 5a), we selected the features with variable importance greater than 0.05 for subsequent analysis. In terms of single-stage VIs, the RDVI, REDSI, VARIgreen, NREDI3, RGR, PSRI1, NDVIre1, SAVI, and EVI were selected; for nVIs, the nREDSI, nVARIgreen, nPSRI1, nNDVIre1, nNREDI1, nNREDI2, nSAVI, and nNREDI3 were selected. For meteorological data, the SSD_03, PRE_05, RHU_04, PRE_04, RHU_05, WIN_03, WIN_04, TEM_03 and SSD_05 were selected. To avoid information redundancy of selected features, we used the analysis of variance (ANOVA) method to optimize important features (Table 3). For VIs, the EVI, NREDIre1, PSRI1, and SAVI showed no significant differences ($p > 0.05$); and for nVIs, the differences of nNREDI3, nNREDI2, and nSAVI with other nVIs were insignificant ($p > 0.05$).

The variable importance values of the vegetation indices based on a single image and two images taken at different times in the discrimination of wheat yellow ruts differed. The five most important vegetation indices were selected for subsequent analysis. For single-stage imagery, the RDVI, REDSI, NREDI3, RGR, and VARIgreen were most sensitive to wheat yellow rust (Figure 5a, indigo histogram); for two-stage imagery, nREDSI, nVARIgreen, nPSRI1, nNREDI1, and nNDVIre1 were most sensitive to wheat yellow rust (Figure 5a, magenta histogram). This is generally consistent with the results shown in Figure 3b and allows us to distinguish between healthy wheat and yellow rust infection. Similarly, the meteorological features of WIN_03, WIN_04, TEM_03, and SSD_05 were excluded using ANOVA.

Figure 5b demonstrated the importance ranking of meteorological features. Due to the strong correlation between the same type of meteorological data, we selected the features above the average value (0.07) of all variable importances for yellow rust identification. Accordingly, five of the most important meteorological features were selected for monitoring wheat yellow rust on a regional scale: average sunshine hours in March (SSD_03), average relative humidity (RHU_04, RHU_05) in April and May, and average precipitation (PRE_04, PRE_05) in April and May.

Table 3. ANOVA for vegetation indices feature sets (VIs and nVIs).

VIs	RDVI	NDVIre1	NREDI3	VARIg	EVI	REDSI	RGR	SAVI	PSRI1
RDVI	-								
NDVIre1	0.0180 [a]	-							
NREDI3	0.0047 [b]	0.31	-						
VARIg	0.0087 [b]	0.066	0.043 [a]	-					
EVI	0.1250	0.058	0.032 [a]	0.04 [a]	-				
REDSI	0.0016 [b]	0.132	0.0007 [c]	0.002 [b]	0.15	-			
RGR	0.0003 [c]	0.027 [a]	0.0035 [b]	0.012 [a]	0.03 [a]	0.015 [a]	-		
SAVI	0.0910	0.132	0.001 [b]	0.019 [a]	0.16	0.001b	0.05	-	
PSRI1	0.0370 [a]	0.224	0.0009 [c]	0.32	0.06	0.129	0.009 [b]	0.078	-

nVIs	nREDSI1	nNREDI3	nVARIg	nNDVIre1	nNREDI1	nPSRI1	nNREDI2	nSAVI
nREDSI1	-							
nNREDI3	0.14	-						
nVARIgr	0.0001 [c]	0.14	-					
nNDVIre1	0.072	0.06	0.019 [a]	-				
nNREDI1	0.002 [b]	0.27	0.0004 [c]	0.0002 [c]	-			
nPSRI1	0.0005 [c]	0.38	0.0027 [b]	0.0004 [c]	0.0002 [c]	-		
nNREDI2	0.35	0.06	0.0003	0.026 [a]	0.0004 [c]	0.003 [b]	-	
nSAVI	0.0004 [c]	0.074	0.15	0.0001 [c]	0.11	0.0009 [c]	0.26	-

Note: "a" indicates the difference is significant at the 0.95 confidence level, "b" indicates the difference is significant at the 0.99 confidence level, and "c" indicates the difference is significant at the 0.999 confidence level; "VARIg" = VARIgreen, "nVARIg" = nVARIgreen," and "mete" = meteorological.

3.4. Wheat Yellow Rust Monitoring Based on Spectral Vegetation Indices

Monitoring models for wheat yellow rust were built using the LDA, SVM, and ANN algorithms. The RDVI, REDSI, NREDI3, RGR, and VARIgreen were selected for use in single-stage monitoring models; nREDSI, nVARIgreen, nPSRI1, nNREDI1, and nNDVIre1 were selected for use in two-stage monitoring models. Table 4 presents the classification results of the three algorithms using the different VIs.

Table 4. Classification accuracies of three classification algorithms using single- and two-stage vegetation indices (VIs and nVIs, respectively; n = 19).

Classifier	VIs					nVIs				
LDA	Healthy	YR	P.a (%)	OA (%)	Kappa	Healthy	YR	P.a (%)	OA (%)	Kappa
Healthy	3	3	42.9	63.2	0.18	4	3	57.1	68.4	0.32
YR	4	9	75.0			3	9	75.0		
U.a (%)	50.0	69.2				57.1	75			
SVM										
Healthy	4	2	57.1	73.7	0.42	5	2	71.4	78.9	0.55
YR	3	10	83.3			2	10	83.3		
U.a (%)	66.7	76.9				71.4	83.3			
ANN										
Healthy	4	4	57.1	63.2	0.23	4	3	57.1	68.4	0.32
YR	3	8	66.7			3	9	75.0		
U.a (%)	50.0	72.7				57.1	75			

Note: P.a = producer's accuracy, U.a = user's accuracy, OA = overall classification accuracy.

For the single-stage vegetation index model, the overall classification accuracy and kappa coefficient were 63.2% and 0.18 for the LDA algorithm, respectively; 73.7% and 0.42 for the SVM algorithm, respectively; and 63.2% and 0.23 for the ANN algorithm, respectively. For the nVIs models, the overall classification accuracy and kappa coefficient were 68.4% and 0.32 for the LDA algorithm, respectively; 78.9% and 0.55 for the SVM algorithm, respectively; and 68.4% and 0.32 for the ANN algorithm, respectively. Based on these results, the classification accuracy of wheat yellow rust monitoring models using nVIS as the input features is better than that of models using VIs; the overall accuracy is improved by 5.2%. Compared with the VIs model, the P.a of healthy wheat identification and yellow rust wheat exceeded 57.1% and 83.3%, respectively, for nVIs. Among the algorithms, SVM performed the best.

3.5. Wheat Yellow Rust Monitoring Based on Meteorological Data and Spectral Information

The classification results of the three algorithms based on both vegetation indices (VIs and nVIs) and meteorological data are shown in Table 5. For the single-stage vegetation index model, the overall classification accuracy and kappa coefficient were 68.4% and 0.32 for the LDA algorithm, respectively; 78.9% and 0.55 for the SVM algorithm, respectively; and 73.7% and 0.45 for the ANN algorithm, respectively. For the two-stage models, the overall classification accuracy and kappa coefficient were 73.7% and 0.42 for the LDA algorithm, respectively; 84.2% and 0.65 for the SVM algorithm, respectively; and 78.9% and 0.55 for the ANN algorithm, respectively.

Table 5. Classification accuracies for three classification algorithms using single- and two-stage vegetation indices (VIs and nVIs, respectively) combined with meteorological data (n = 19).

Classifier	VIs_Meteorological Data					nVIs_Meteorological Data				
LDA	Healthy	YR	P.a (%)	OA (%)	Kappa	Healthy	YR	P.a (%)	OA (%)	Kappa
Healthy	4	3	57.1	68.4	0.32	4	2	57.1	73.7	0.42
YR	3	9	75.0			3	10	83.3		
U.a (%)	57.1	75.0				66.7	76.9			
SVM										
Healthy	5	2	71.4	78.9	0.55	5	1	71.4	84.2	0.65
YR	2	10	83.3			2	11	91.7		
U.a (%)	71.4	83.3				83.3	84.6			
ANN										
Healthy	5	3	71.4	73.7	0.45	5	2	71.4	78.9	0.55
YR	2	9	75.0			2	10	83.3		
U.a (%)	62.5	81.8				71.4	83.3			

According to these results, the accuracies of wheat yellow rust monitoring models with the nVIs and meteorological data as the input features are higher than those based on VIs and meteorological data. This is consistent with the results based on the pure VIs model (see Section 3.4). Among the three algorithms, SVM again had the highest classification accuracy. Moreover, the U.a of healthy and yellow rust wheat identification was 83.3% and 84.6%, respectively; the P.a of yellow rust reached 91.7% in the nVIs and meteorological model (nVIs_meteorological data). The results confirm that the inclusion of meteorological data improves model accuracy and offers the potential for crop disease monitoring on a regional scale.

3.6. Mapping Wheat Yellow Rust Using the Optimal Monitoring Model

Figure 6 shows a map of wheat yellow rust in Ningqiang county, Shaanxi Province during the filling period based on the optimal model (SVM algorithm using two-stage spectral vegetation indices and meteorological data). The wheat yellow rust infected region is highly consistent with the field observation, which verifies the feasibility of the model for crop disease monitoring. This remote sensing method has the potential for effective, rapid (near real-time), and a spatially continuous regional monitoring of crop disease, offering substantial labor-, time-, and cost-savings.

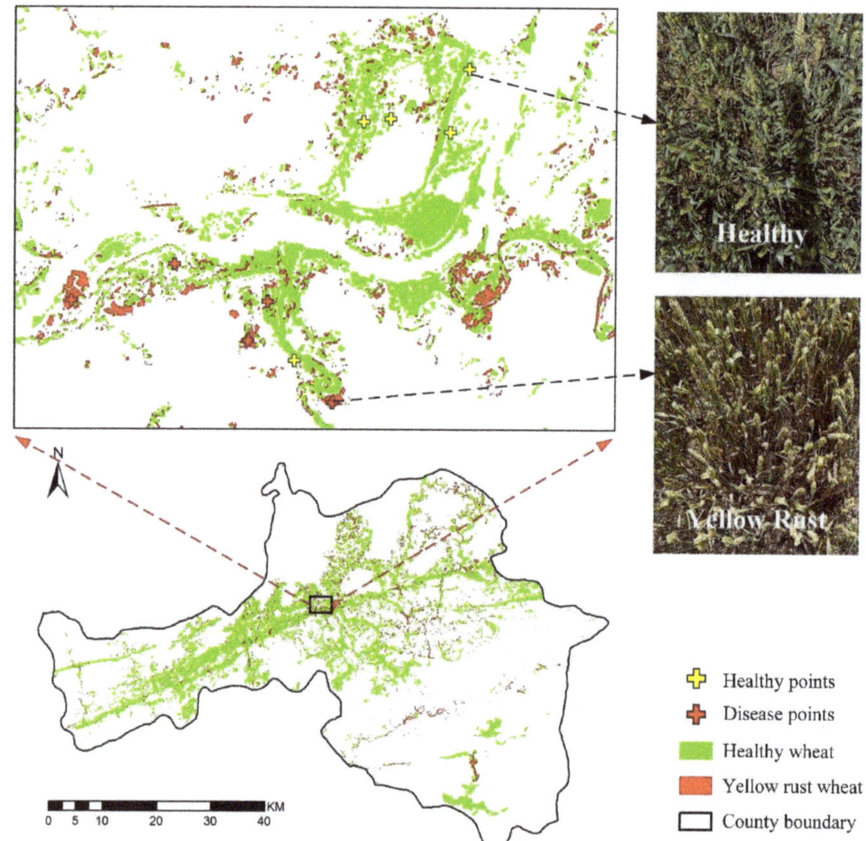

Figure 6. Mapping wheat yellow rust disease in Ningqiang county during the filling period (May 2018) based on the support vector machine (SVM) algorithm using two-stage spectral vegetation indices and meteorological data.

4. Discussion

Remote sensing data has the characteristics of spatial continuity and rich information, which facilitates the acquisition of crop growth and environmental information, and provided a basis for crop pest monitoring [8,47]. This study explored the potential of spectral VIs and meteorological information related to disease occurrence to monitor wheat yellow rust infestation on a regional-scale.

4.1. Performance of Spectral Vegetation Indices in Wheat Yellow Rust Discrimination

VIs can reflect the biophysical and biochemical change of crops, and can be used for detection and identification of plant diseases [37,48]. Yellow rust primarily infects wheat leaves, causing green fading and deformation of leaf tissue, thereby significantly changing the chlorophyll content and biomass [23]. We selected VIs with highly sensitive yellow rust discrimination during the wheat milking stage based on single- and two-stage remote sensing images. The Sentinel-2 satellite has rich red-edge information that are significant for crop growth status and stress monitoring [22,23]. In particular, REDSI consists of red edges and bands and was proposed by Zheng for monitoring wheat yellow rust, particularly during the filling stage [12,23]. In this study, REDSI and VARIgreen were more important for wheat yellow rust discrimination among the single-stage indices; PSRI1, NREDI1, and

NDVIre1 were most sensitive wheat yellow rust among the two-stage indices. This is primarily related to the destruction of the tissue structure of leaf cells and the decrease of leaves chlorophyll content under yellow rust disease stress, resulting in the shift of the spectrum on the red edge [22,49]. PSRI1 can be used to assess the crop pigment content and status [10]. Moreover, the band combination of nPSRI1, nNREDI1, and nNDVIre1 contains red-edge information that can capture changes in physiological and biochemical parameter, and better eliminate the effects of growth factors compared with single-stage vegetation index models (classification accuracy is 5.2% higher, Table 4) [23]. Here, the optimal model (i.e., that using the two-stage vegetation indices) captured changes caused by yellow rust disease with a classification accuracy of 78.9%.

4.2. Performance of Meteorological Data in Wheat Yellow Rust Discrimination

The propagation, spread, and infection of pathogen spores require suitable environmental conditions (such as, precipitation, humidity, and temperature). Wheat yellow rust disease occurs in high humidity and low-temperature environments. Favorable climate conditions such as warm winters and heavy rainfall in early spring are external causes of wheat yellow rust occurrence and epidemics in Shaanxi Province [26]. In this study, the average relative humidity (RHU_04, RHU_05) and average precipitation (PRE_04, PRE_05) were sensitive to yellow rust discrimination (Figure 5b). Moreover, the TEM in Shaanxi Province reaches a suitable range for the incidence of wheat yellow rust in April and May. The study area belongs to the winter breeding region of wheat yellow rust in China [50]. That is, the yellow rust pathogen in this area infects wheat during the winter, making WIN less important in the monitoring of wheat yellow rust than RHU and PRE [26]. However, as wheat yellow rust disease is an air-borne bacterium, WIN can provide important information for forecasting. In summary, our results confirm that meteorological information can provide crop disease monitoring, which is consistent with the conclusions of Yuan et al. [8].

4.3. Performance of Wheat Yellow Rust Monitoring Classification Algorithms

Among LDA, SVM, and ANN, the SVM algorithm exhibited the best performance for distinguishing healthy and yellow rust-infested wheat, with a classification accuracy of 73.7–84.2%. The SVM classifier is based on the threshold discriminant rule and maps the samples to appropriate feature space. Some researchers have also shown that the SVM is superior to LDA in remote sensing classification or extraction in plants [40,41,51]. For example, Yue et al. reported that SVM-based models achieve higher classification accuracies than those using LDA in wheat yellow rust monitoring on leaf scale [40].

In terms of classification accuracy, SVM outperformed the ANN classifiers by 5.2–10.5% in different feature spaces. These results differ from those of Raczko et al. who found that ANN performed better than the SVM model [18]. However, ANNs are more difficult to use and optimize, and require many parameters. The number of samples in this study was limited, and ANN requires a large number of parameters to set the initial value of the network topology, weights, and thresholds, thereby making it difficult to optimize the model [18]. Compared with ANNs, the SVM algorithm can solve classification problems for nonlinear and small sample situations, and avoid the neural network structure selection and local minima problem [52]. Overall, considering that monitoring and positioning crop disease on a regional scale are more complicated and challenging than at the canopy and on leaf scales, the classification accuracy achieved in this study (73.7–84.2% based on SVM classifier) is acceptable.

Although the current classification accuracy is lower than that obtained based on airborne hyperspectral images (for example, Zhang et al. used deep convolutional neural network to identify wheat yellow rust based on airborne hyperspectral images with an accuracy of 85.0% [53]), it meets the practical demands of disease monitoring and management. It is majorly based on airborne hyperspectral images, for which fine spectral resolution enables more abundant spectral information to be extracted and analyzed, which may lead to a certain improvement in the accuracy of disease mapping. Many researchers have used medium- and high-resolution satellite images to monitor crop disease. Chemura et al. used spectral indices to identify coffee leaf rust infection based on Sentinel-2 satellite data, with the discrimination accuracy of 82.5% [22]. Yuan et al. used the crop growth index (GNDVI and VARIred-edge) and environmental characteristics to monitor crop disease and pests based on the Wordview2 and Landsat 8 satellite, and proved that the accuracy (82.0%) of models combining VIs and environmental characteristics are better than those of traditional monitoring models that only rely on spectral information [8]. However, despite the significant potential for crop disease monitoring, we should optimize the parameters of the methods to build more robust and reasonable models under the condition of enough samples, and improve the accuracy of crop disease monitoring for practical applications.

In this study, adding monthly average meteorological data to the remote sensing monitoring of crop diseases, we established an effective remote sensing monitoring model for wheat yellow rust. However, crop disease occurrence is also the result of environmental factors, the amount of pathogen, crop planting landscape patterns, and farmland management [4,29]. Therefore, future work should integrate more multi-source data (remote sensing and non-remote sensing data) with well-characterized mechanisms and high stability to further improve crop disease monitoring and forecasting. Moreover, due to the influence of weather and manpower, the sample size of wheat yellow rust in this study was small. In the future, we will collect wheat yellow rust data from large areas in different years to verify and improve the wheat yellow rust monitoring model. Furthermore, we will attempt to effectively utilize the complementary features of meteorological data (for example, various types of meteorological data and 10-day average meteorological data), terrain features, and remote sensing data to establish a collaborative scheme for forecasting crop disease at an early stage.

The rapid and large-scale monitoring of crop disease and pests relieves huge pressure on plant protection personnel. It is a weapon to prevent and control disease, promote healthy development of agriculture, and achieve the goal of sustainable agricultural development. In addition, this will contribute to eradicating hunger, achieving food security, improving nutrition and promoting sustainable agriculture as outlined in the United Nations Sustainable Development Goals.

5. Conclusions

In this study, multispectral satellite imagery (Sentinel-2A) and meteorological data were used to monitor wheat yellow rust disease based on three classification methods (linear discriminant analysis, a support vector machine, and an artificial neural network) on a regional scale. Five meteorological features (sunshine hours in March (SSD_03), average relative humidity in April and May (RHU_04, RHU_05), and average precipitation in April and May (PRE_04, PRE_05)) combined with two-stage vegetation indices using the SVM algorithm were found to be optimal for wheat yellow rust monitoring. In addition, the model for yellow rust monitoring base of two-stage vegetation indices significantly outperformed single-stage vegetation index models, with the overall classification accuracy increasing from 63.2% to 78.9%. Moreover, the addition of meteorological data, which is closely related to yellow rust occurrence, increased the accuracy of the two-stage index SVM model to 84.2%. The proposed model is suitable for rapid, large-scale monitoring and forecasting of biotic (bacterial and fungal disease) stress in crops and offers an effective approach for reducing the impacts of crop disease, including the implications for global food security. In the future, we will consider information from multiple sources to develop further comprehensive and reliable crop disease forecasting models.

Author Contributions: Q.Z.: field survey, methodology, writing—original draft. H.Y. and W.H.: conceived and designed the experiments; Y.D.: data collection; H.J.: software and processed the data; C.W., D.L., and L.W.: provided advices to improve manuscript; S.C.: modified the structure of the paper and grammar. All authors have read and agreed to the published version of the manuscript.

Funding: This research was funded by GDAS' Project of Science and Technology Development (2020GDASYL-20200103004), Guangdong Province Agricultural Science and Technology Innovation and Promotion Project (No.2020KJ102), Guangzhou Basic Research Project (202002020076), National special support program for high-level personnel recruitment (Wenjiang Huang).

Institutional Review Board Statement: This study not involving humans.

Informed Consent Statement: This study not involving humans.

Data Availability Statement: Data sharing is not application to this article.

Conflicts of Interest: The authors declare no conflict of interest.

Appendix A

Table A1. Field survey record form for wheat yellow rust.

Contents		1	2	...
Survey date				
Location (longitude, latitude)				
Crop growth information	Cultivated varieties			
	Growth stage			
	Plant height			
	Planting density			
Disease information	Number of leaves investigated			
	Number of diseased leaves			
	Disease severity (%)			
	Damage percentage (%)			
Remarks (Altitude, irrigation information, etc.)				

References

1. Wan, A.M.; Chen, X.M.; He, Z.H. Wheat stripe rust in China. *Aust. J. Agric. Res.* **2007**, *58*, 605–619. [CrossRef]
2. Wang, H.; Yang, X.B.; Ma, Z. Long-Distance Spore Transport of Wheat Stripe Rust Pathogen from Sichuan, Yunnan, and Guizhou in Southwestern China. *Plant Dis.* **2010**, *94*, 873–880. [CrossRef] [PubMed]
3. Chen, W.Q.; Kang, Z.S.; Ma, Z.H.; Xu, S.C.; Jiang, Y.Y. Integrated Management of Wheat Stripe Rust Caused by Puccinia striiformis f.sp.tritici in China. *Sci. Agric. Sin.* **2013**, *46*, 4254–4262.
4. Zhang, J.C.; Huang, Y.B.; Pu, R.L.; Gonzalez-Moreno, P.; Yuan, L.; Wu, K.H.; Huang, W.J. Monitoring plant diseases and pests through remote sensing technology: A review. *Comput. Electron. Agric.* **2019**, *165*, 104943. [CrossRef]
5. Huang, W.; Guan, Q.; Luo, J.; Zhang, J.; Zhao, J.; Liang, D.; Huang, L.; Zhang, D. New optimized spectral indices for identifying and monitoring winter wheat diseases. *IEEE J. Sel. Top. Appl. Earth Obs. Remote Sens.* **2014**, *7*, 2516–2524. [CrossRef]
6. Coluzzi, R.; D'Emilio, M.; Imbrenda, V.; Giorgio, G.A.; Lanfredi, M.; Macchiato, M.; Ragosta, M.; Simoniello, T.; Telesca, V. Investigating climate variability and long-term vegetation activity across heterogeneous Basilicata agroecosystems. *Geomat. Nat. Hazards Risk* **2019**, *10*, 168–180. [CrossRef]
7. Corbari, C.; Salerno, R.; Ceppi, A.; Telesca, V.; Mancini, M. Smart irrigation forecast using satellite LANDSAT data and meteo-hydrological modeling. *Agric. Water Manag.* **2019**, *212*, 283–294. [CrossRef]
8. Yuan, L.; Bao, Z.; Zhang, H.; Zhang, Y.; Liang, X. Habitat monitoring to evaluate crop disease and pest distributions based on multi-source satellite remote sensing imagery. *Opt. Int. J. Light Electron. Opt.* **2017**, *145*, 66–73. [CrossRef]
9. Zhang, J.; Pu, R.; Yuan, L.; Huang, W.; Nie, C.; Yang, G. Integrating Remotely Sensed and Meteorological Observations to Forecast Wheat Powdery Mildew at a Regional Scale. *IEEE J. Sel. Top. Appl. Earth Obs. Remote Sens.* **2017**, *7*, 4328–4339. [CrossRef]
10. Papastamati, K.; Bosch, F.v.d.; Welham, S.J.; Fitt, B.D.L.; Evans, N.; Steed, J.M. Modelling the daily progress of light leaf spot epidemics on winter oilseed rape (Brassica napus), in relation to Pyrenopeziza brassicae inoculum concentrations and weather factors. *Ecol. Model.* **2002**, *148*, 169–189. [CrossRef]
11. Cunniffe, N.J.; Koskella, B.E.; Metcalf, C.J.; Parnell, S.; Gottwald, T.R.; Gilligan, C.A. Thirteen challenges in modelling plant diseases. *Epidemics* **2015**, *10*, 6–10. [CrossRef] [PubMed]
12. Zheng, Q.; Huang, W.; Cui, X.; Dong, Y.; Shi, Y.; Ma, H.; Liu, L. Identification of Wheat Yellow Rust Using Optimal Three-Band Spectral Indices in Different Growth Stages. *Sensors* **2018**, *19*, 35. [CrossRef]

13. Sankaran, S.; Mishra, A.; Ehsani, R.; Davis, C. A review of advanced techniques for detecting plant diseases. *Comput. Electron. Agric.* **2010**, *72*, 1–13. [CrossRef]

14. Zhang, J.; Pu, R.; Yuan, L.; Wang, J.; Huang, W.; Yang, G. Monitoring powdery mildew of winter wheat by using moderate resolution multi-temporal satellite imagery. *PLoS ONE* **2014**, *9*, e93107. [CrossRef] [PubMed]

15. Ma, H.; Jing, Y.; Huang, W.; Shi, Y.; Dong, Y.; Zhang, J.; Liu, L. Integrating Early Growth Information to Monitor Winter Wheat Powdery Mildew Using Multi-Temporal Landsat-8 Imagery. *Sensors* **2018**, *18*, 3290. [CrossRef]

16. Hou, J.; Hu, Y.; Hou, L.; Guo, K.; Satake, T. Classification of ripening stages of bananas based on support vector machine. *Int. J. Agric. Biol. Eng.* **2015**, *8*, 99–103.

17. Thanh Noi, P.; Kappas, M. Comparison of Random Forest, k-Nearest Neighbor, and Support Vector Machine Classifiers for Land Cover Classification Using Sentinel-2 Imagery. *Sensors* **2018**, *18*, 18. [CrossRef]

18. Raczko, E.; Zagajewski, B. Comparison of support vector machine, random forest and neural network classifiers for tree species classification on airborne hyperspectral APEX images. *Eur. J. Remote Sens.* **2017**, *50*, 144–154. [CrossRef]

19. Raza, M.M.; Harding, C.; Liebman, M.; Leandro, L.F. Exploring the Potential of High-Resolution Satellite Imagery for the Detection of Soybean Sudden Death Syndrome. *Remote Sens.* **2020**, *12*, 1213. [CrossRef]

20. Yue, S.; Wenjiang, H.; Huichun, Y.; Chao, R.; Naichen, X.; Yun, G.; Yingying, D.; Dailiang, P. Partial Least Square Discriminant Analysis Based on Normalized Two-Stage Vegetation Indices for Mapping Damage from Rice Diseases Using PlanetScope Datasets. *Sensors* **2018**, *18*, 1901. [CrossRef]

21. Calderón, R.; Navas-Cortés, J.A.; Lucena, C.; Zarco-Tejada, P.J. High-resolution airborne hyperspectral and thermal imagery for early detection of Verticillium wilt of olive using fluorescence, temperature and narrow-band spectral indices. *Remote Sens. Environ.* **2013**, *139*, 231–245. [CrossRef]

22. Chemura, A.; Mutanga, O.; Dube, T. Separability of coffee leaf rust infection levels with machine learning methods at Sentinel-2 MSI spectral resolutions. *Precis. Agric.* **2017**, *18*, 859–881. [CrossRef]

23. Zheng, Q.; Huang, W.; Cui, X.; Shi, Y.; Liu, L. New spectral index for detecting wheat yellow rust using Sentinel-2 multispectral imagery. *Sensors* **2018**, *18*, 868. [CrossRef] [PubMed]

24. Fernández-Manso, A.; Fernández-Manso, O.; Quintano, C. Sentinel-2A red-edge spectral indices suitability for discriminating burn severity. *Int. J. Appl. Earth Obs. Geoinf.* **2016**, *50*, 170–175. [CrossRef]

25. Zarco-Tejada, P.J.; Hornero, A.; Beck, P.S.A.; Kattenborn, T.; Kempeneers, P.; Hernández-Clemente, R. Chlorophyll content estimation in an open-canopy conifer forest with Sentinel-2A and hyperspectral imagery in the context of forest decline. *Remote Sens. Environ.* **2019**, *223*, 320–335. [CrossRef]

26. Dengke, L.; Zhao, W.; Feizhou, X. Occurrence regularity and meteorological influencing factors of wheat stripe rust in Shaanxi province. *J. Catastrophol.* **2019**, *34*, 59–65.

27. Xu, Q.; Yang, G.; Long, H.; Wang, C.; Li, X.; Huang, D. Crop information identification based on MODIS NDVI time-series data. *Nongye Gongcheng Xuebao Trans. Chin. Soc. Agric. Eng.* **2014**, *30*, 134–144.

28. Grinn-Gofroń, A.; Nowosad, J.; Bosiacka, B.; Camacho, I.; Pashley, C.; Belmonte, J.; De Linares, C.; Ianovici, N.; Manzano, J.M.M.; Sadyś, M. Airborne Alternaria and Cladosporium fungal spores in Europe: Forecasting possibilities and relationships with meteorological parameters. *Sci. Total Environ.* **2019**, *653*, 938–946. [CrossRef]

29. Newlands, N.K. Model-Based Forecasting of Agricultural Crop Disease Risk at the Regional Scale, Integrating Airborne Inoculum, Environmental, and Satellite-Based Monitoring Data. *Front. Environ. Sci.* **2018**, *6*, 16. [CrossRef]

30. Rouse, J.W., Jr.; Haas, R.H.; Schell, J.A.; Deering, D.W. Monitoring vegetation systems in the great plains with Erts. *NASA Spec. Publ.* **1973**, *351*, 309.

31. Gitelson, A.A.; Kaufman, Y.J.; Merzlyak, M.N. Use of a green channel in remote sensing of global vegetation from EOS-MODIS. *Remote Sens. Environ.* **1996**, *58*, 289–298. [CrossRef]

32. Huete, A.R. A soil-adjusted vegetation index (SAVI). *Remote Sens. Environ.* **1988**, *25*, 295–309. [CrossRef]

33. Penuelas, J.; Filella, I.; Lloret, P.; Munoz, F.; Vilajeliu, M. Reflectance assessment of mite effects on apple trees. *Int. J. Remote Sens.* **1995**, *16*, 2727–2733. [CrossRef]

34. Huete, A.; Didan, K.; Miura, T.; Rodriguez, E.P.; Gao, X.; Ferreira, L.G. Overview of the radiometric and biophysical performance of the MODIS vegetation indices. *Remote Sens. Environ.* **2002**, *83*, 195–213. [CrossRef]

35. Roujean, J.L.; Breon, F.M. Estimating PAR absorbed by vegetation from bidirectional reflectance measurements. *Remote Sens. Environ.* **1995**, *51*, 375–384. [CrossRef]

36. Gamon, J.A.; Surfus, J.S. Assessing leaf pigment content and activity with a reflectometer. *New Phytol.* **1999**, *143*, 105–117. [CrossRef]

37. Gitelson, A.A.; Kaufman, Y.J.; Stark, R.; Rundquist, D. Novel algorithms for remote estimation of vegetation fraction. *Remote Sens. Environ.* **2002**, *80*, 76–87. [CrossRef]

38. Gitelson, A.; Merzlyak, M.N. Quantitative estimation of chlorophyll-a using reflectance spectra: Experiments with autumn chestnut and maple leaves. *J. Photochem. Photobiol. B Biol.* **1994**, *22*, 247–252. [CrossRef]

39. Breiman, L. Random Forests. *Mach. Learn.* **2001**, *45*, 5–32. [CrossRef]

40. Shi, Y.; Huang, W.; González-Moreno, P.; Luke, B.; Dong, Y.; Zheng, Q.; Ma, H.; Liu, L. Wavelet-Based Rust Spectral Feature Set (WRSFs): A Novel Spectral Feature Set Based on Continuous Wavelet Transformation for Tracking Progressive Host–Pathogen Interaction of Yellow Rust on Wheat. *Remote Sens.* **2018**, *10*, 525. [CrossRef]

41. Zheng, Q.; Huang, W.; Ye, H.; Dong, Y.; Shi, Y.; Chen, S. Using continous wavelet analysis for monitoring wheat yellow rust in different infestation stages based on unmanned aerial vehicle hyperspectral images. *Appl. Opt.* **2020**, *59*, 8003–8013. [CrossRef] [PubMed]
42. Vapnik, V.N. *The Nature of Statistical Learning Theory*; Springer: Berlin, Germany, 1995.
43. Ye, H.; Huang, W.; Huang, S.; Cui, B.; Jin, Y. Identification of banana fusarium wilt using supervised classification algorithms with UAV-based multi-spectral imagery. *Int. J. Agric. Biol. Eng.* **2020**, *13*, 136–142.
44. Martí, P.; Shiri, J.; Duran-Ros, M.; Arbat, G.; de Cartagena, F.R.; Puig-Bargués, J. Artificial neural networks vs. Gene Expression Programming for estimating outlet dissolved oxygen in micro-irrigation sand filters fed with effluents. *Comput. Electron. Agric.* **2013**, *99*, 176–185. [CrossRef]
45. Azarmdel, H.; Jahanbakhshi, A.; Mohtasebi, S.S.; Munoz, A.R. Evaluation of image processing technique as an expert system in mulberry fruit grading based on ripeness level using artificial neural networks (ANNs) and support vector machine (SVM). *Postharvest Biol. Technol.* **2020**, *166*, 12. [CrossRef]
46. Congalton, R.G.; Mead, R.A. A quantitative method to test for consistency and correctness in photointerpretation. *Photogramm. Eng. Remote Sens.* **1983**, *49*, 69–74.
47. Das, P.K.; Laxman, B.; Rao, S.V.C.K.; Seshasai, M.V.R.; Dadhwal, V.K. Monitoring of bacterial leaf blight in rice using ground-based hyperspectral and LISS IV satellite data in Kurnool, Andhra Pradesh, India. *PANS Pest Artic. News Summ.* **2015**, *61*, 359–368. [CrossRef]
48. Chemura, A.; Mutanga, O.; Sibanda, M.; Chidoko, P. Machine learning prediction of coffee rust severity on leaves using spectroradiometer data. *Trop. Plant Pathol.* **2017**, *43*, 117–127. [CrossRef]
49. Devadas, R.; Lamb, D.W.; Backhouse, D.; Simpfendorfer, S. Sequential application of hyperspectral indices for delineation of stripe rust infection and nitrogen deficiency in wheat. *Precis. Agric.* **2015**, *16*, 477–491. [CrossRef]
50. Shi, S.; Ma, Z.; Wang, H.; Zhao, Z.; Jiang, Y. Climate-based regional classification for overwintering of Puccinia striiformis in China with GIS and geostatistics. *J. Northwest Sci-Tech Univ. Agric. For.* **2005**, *32*, 29–32.
51. Bandos, T.V.; Bruzzone, L.; Camps-Valls, G. Classification of Hyperspectral Images with Regularized Linear Discriminant Analysis. *IEEE Trans. Geosci. Remote Sens.* **2009**, *47*, 862–873. [CrossRef]
52. Pal, M.; Mather, P.M. Support vector machines for classification in remote sensing. *Int. J. Remote Sens.* **2005**, *26*, 1007–1011. [CrossRef]
53. Zhang, X.; Han, L.X.; Dong, Y.Y.; Shi, Y.; Huang, W.J.; Han, L.H.; Gonzalez-Moreno, P.; Ma, H.Q.; Ye, H.C.; Sobeih, T. A Deep Learning-Based Approach for Automated Yellow Rust Disease Detection from High-Resolution Hyperspectral UAV Images. *Remote Sens.* **2019**, *11*, 1554. [CrossRef]

Article

A Modified Geometrical Optical Model of Row Crops Considering Multiple Scattering Frame

Xu Ma and Yong Liu *

College of Earth and Environmental Sciences, Lanzhou University, Lanzhou 730000, China; max15@lzu.edu.cn
* Correspondence: liuy@lzu.edu.cn; Tel.: +86-139-1940-7135

Received: 21 September 2020; Accepted: 27 October 2020; Published: 2 November 2020

Abstract: The canopy reflectance model is the physical basis of remote sensing inversion. In canopy reflectance modeling, the geometric optical (GO) approach is the most commonly used. However, it ignores the description of a multiple-scattering contribution, which causes an underestimation of the reflectance. Although researchers have tried to add a multiple-scattering contribution to the GO approach for forest modeling, different from forests, row crops have unique geometric characteristics. Therefore, the modeling approach originally applied to forests cannot be directly applied to row crops. In this study, we introduced the adding method and mathematical solution of integral radiative transfer equation into row modeling, and on the basis of improving the overlapping relationship of the gap probabilities involved in the single-scattering contribution, we derived multiple-scattering equations suitable for the GO approach. Based on these modifications, we established a row model that can accurately describe the single-scattering and multiple-scattering contributions in row crops. We validated the row model using computer simulations and in situ measurements and found that it can be used to simulate crop canopy reflectance at different growth stages. Moreover, the row model can be successfully used to simulate the distribution of reflectances (RMSEs < 0.0404). During computer validation, the row model also maintained high accuracy (RMSEs < 0.0062). Our results demonstrate that considering multiple scattering in GO-approach-based modeling can successfully address the underestimation of reflectance in the row crops.

Keywords: canopy model of row crops; multiple scattering for geometric optical approach; the gap probabilities of row crops; overlapping relationship; hotspot

---

1. Introduction

In recent canopy modeling studies, row modeling has been extensively studied. The canopy modeling of crops based on the row model can more accurately estimate the seasonal changes of biophysical canopy parameters in remote sensing [1]. In the inversion of remote sensing, physical modeling is key, and its approach can be separated into three categories based on different physical mechanisms, name the computer simulation (CS) approach, radiative transfer (RT) approach, and geometric optical (GO) approach [2]. The GO approach can describe the geometric characteristics of an individual canopy. It is most suitable for heterogeneous canopy modeling [3]. The CS approach has the highest accuracy and has been applied primarily for understanding radiation regimes. It has been used as a "truth value" ("surrogate truth" or benchmark in Radiation Transfer Model Intercomparison (RAMI)) [4–6] to validate GO and RT approaches. However, compared to the GO approach, the computational time of the CS approach is too slow; hence, it is rarely used in remote sensing inversion. The RT approach is based on volume scattering in the radiative transfer equation. It is very suitable for describing the scattering issue in canopies [7]. However, compared to the GO approach, the RT approach lacks a description of the geometric characteristics of an individual canopy and is mostly used for high-density homogeneous (or continuous) canopy modeling [2]. The row crop is a heterogeneous canopy. It has the typical geometric characteristics of an individual canopy, such as row structure (e.g., height, row width, between-row distance, etc.) [8–11]. Therefore, considering the advantages of accurately describing the row structure, the GO approach is widely used in the canopy modeling of row crops [9,10,12].

As canopy reflectance models, GO models have to deal with the interactions of light occurring within and between individual plant canopies and calculate the reflectance at the top of the canopy [13]. These interactions are divided into two basic physical processes. The first physical process is surface reflectance in the GO approach, also known as the single-scattering contribution. The single-scattering contribution represents the one-order interaction between light and medium, which is calculated by the four-component area fraction (sunlit and shaded canopy and sunlit and shaded soil [14]) and corresponding representative reflectance for each component in the GO approach [3,15]. For calculating the four-component area fraction, gap probabilities that reflect the transmission of light in the canopy are key [16,17]. An initial approach [18,19] for calculating gap probabilities only considered the overlapping relationship between leaves and ignored the overlapping relationship between plant canopies, which caused computational deviations in reflectance near the hotspot (a reflectance peak around a viewing direction that is exactly opposite the solar illumination direction [2]) [20]. The reflectance peak (sum of the reflectance) near the hotspot is caused by a single-scattering contribution [2]. To improve computational accuracy for the single-scattering contribution near hotspots, Li et al. [21] and Chen et al. [22] attempted to establish a new approach considering the overlapping relationship between leaves and discrete crowns (individual tree crowns) to calculate gap probabilities. With the improvement of the mathematical description of the gap probabilities, the accuracy of the single scattering near the hotspot is further improved.

The second physical process is a multiple-scattering contribution. When a single-scattering contribution interacts with the medium again, two-order and high-order interactions between light and medium are formed. The cumulative sum of these interactions is called the multiple-scattering contribution [23]. From the models presented in the GO approach, previous studies largely ignored multiple-scattering contributions [3,15,24,25]. GO models are generally accurate in the visible part of the solar spectrum but less accurate in the near-infrared (NIR) part where multiple scattering in plant canopies is the strongest [26]. In general, this issue can be tackled by two different methods to establish a multiple-scattering equation in the GO approach. In the first method, similar to the single-scattering principle, the multiple-scattering equation continues to be established using the GO approach. The second method uses the radiative transfer (RT) approach to establish the multiple-scattering equation, combines it with the first scattering equation constructed by the GO approach, and calculates the reflectance at the top of the canopy. In the first method, Chen et al. [26] attempted to introduce view factors between the various sunlit and shaded components within a canopy to perfect the multiple scattering framework in the GO approach. However, Chen et al. [23,26] suggested that some treatments remain to be improved in the multiple-scattering equations they proposed: (1) because calculating the view factors is quite complicated, the study used an a priori numerical experiment to simplify the complex integral involved in the view factor, which caused some situations to be inapplicable; (2) because the geometric relationship for two-order and high-order scattering is unclear, the two-order and high-order scattering angles are difficult to accurately determine. Regarding the second method, Li et al. [27] developed a GORT (an Analytical Hybrid geometric optical and radiative transfer approaches) model. They used the GO approach to describe the surface reflectance (single-scattering contribution) and adopted the RT approach (numerical method of successive order) to estimate the multiple-scattering contribution. Subsequently, Ni et al. [13] reported that the GORT model was overestimated in the multiple scattering of a sunlit canopy. Therefore, they established a simplified analysis model considering the path scattering effect on a sunlit canopy, thereby reducing the overestimation of the multiple scattering of the sunlit canopy. However, the results shown by Ni et al. have a computational deviation at a small solar zenith angle, and this phenomenon is especially obvious in the principal plane of the NIR part. Although the established multiple-scattering equation of the RT approach was coupled with the GO approach to address some issues of improvement in reflectance accuracy, it still has a limitation. The major difficulty is that the geometric characteristics of independent canopies are not considered in the multiple-scattering contribution [26].

The above studies for the single- and multiple-scattering contributions all aimed at heterogeneous forest modeling but rarely row modeling. Therefore, it is necessary to consider these issues of the single- and multiple-scattering contributions in a row model based on the GO approach. Similar to the heterogeneous forest, row crops are also heterogeneous canopies in agriculture [15], and the tree crown mentioned in the forest is similar to the canopy closure (vegetation material area, also called a row). Therefore, row crops also have an overlapping relationship between individual plant canopies, which implies that the calculation of gap probabilities in the row crops should also consider the overlapping relationship between leaves and canopy closures. Similar to the heterogeneous forest canopy, gap probabilities in the row crops are accurately calculated. The single-scattering contribution near the hotspot can be accurately calculated further.

In addition to the issue of the single-scattering contribution in row modeling, the multiple-scattering contribution is more important and has often been ignored in previous studies [14,15,28,29]. The multiple scattering of row crops includes the multiple scattering of the between-row area (the between-row area is the area of bare soil between the canopy closures) in addition to the multiple scattering of the canopy closure [30,31]. The calculation area involved in the multiple scattering of the canopy closure is an area with vegetation material, and its multiple scattering characteristics are similar to those of continuous crops. Therefore, its calculation is uncomplicated. However, in the between-row area, the multiple scattering of the between-row area involves multiple-scattering contributions between the soil and the adjacent canopy closures (two mediums: soil particles and vegetation leaves) and needs to consider the row structure in the calculation [30–32]. Therefore, the calculation for multiple scattering in the between-row area is more complicated. Although an integral form of the radiative transfer equation was proposed to address the multiple scattering of the between-row area, its study remains in the RT field [30]. The multiple scattering of the between-row area used in the GO approach in row crops is an issue requiring further study.

This study focuses on the limitations of the GO approach currently used in row modeling, especially in describing the multiple scattering framework. To achieve this goal, there are two subproblems to be dealt with for the single- and multiple-scattering contributions: (1) How can we address the issue of ignoring the geometric characteristics of independent canopies in the establishment of multiple-scattering equations using the RT approach? (2) How can we establish a gap probability approach that considers the overlapping relationship between leaves and canopy closures and addresses computational deviations in the reflectance near the hotspot in a single-scattering contribution? Specifically, we introduced the adding method and the mathematical solution of integral radiative transfer equation into row modeling. On the basis of improving the overlapping relationship of the gap probabilities involved in the single-scattering contribution, we derived multiple-scattering equations suitable for the GO approach. Finally, we established a row model to accurately calculate the canopy reflectance of row crops.

2. Materials and Methods

2.1. Description of the Row Model

2.1.1. General Form of Row Crops Based on a Geometric Optics Approach

In previous studies [17,30,31], the row canopy has been assumed to comprise periodic box-shaped plant materials with bare soil between the box-shaped scene (Figure 1a). Our study uses this assumption of the row canopy. The box-shaped plant materials are isotropic along the row (on the y-axis in Figure 1b). The reflectance at the top of the canopy of row crops (r) is of two proportions (i.e., $A_1/(A_1 + A_2)$ for the canopy closure and $A_2/(A_1 + A_2)$ between-row area) as well as the corresponding representative reflectance for each component, i.e., the reflectance at the top of the canopy closure ($r_{canopy\_closure}$) and the reflectance at the top of the between-row area ($r_{between\_row}$). The equation of reflectance at the top of the canopy of row crops is

$$r = \frac{A_1}{A_1+A_2} r_{canopy\_closure} + \frac{A_2}{A_1+A_2} r_{between\_row}$$
$$= \frac{A_1 r_{canopy\_closure} + A_2 r_{between\_row}}{A_1+A_2} \tag{1}$$

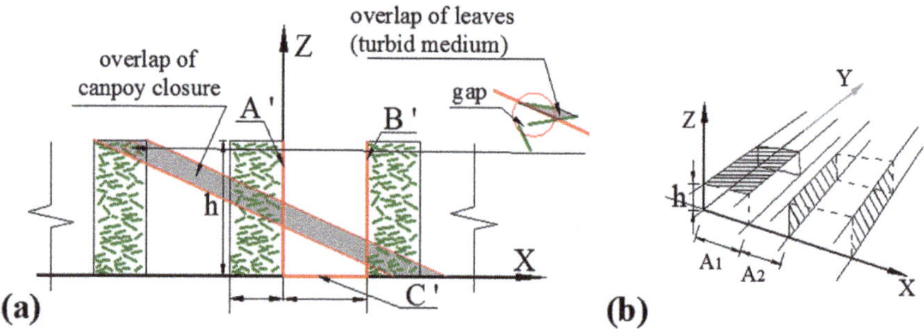

Figure 1. Sketch of the scene of row crops. (**a**) Overlapping relationship between leaves and canopy closures involved in the calculation of gap probabilities; (**b**) three-dimensional map of row crops. Here, A_1 is the row width, A_2 is the between-row distance, and h is the canopy height.

Here, A_1 is the row width and A_2 is the between-row distance. To facilitate a full understanding, r, $r_{canopy\_closure}$, and $r_{between\_row}$ are, respectively, called the sum of the reflectance of row crops, the reflectance of the canopy closure, and reflectance of the between-row area for short. In Equation (1), there are differences in the radiation mechanism of the canopy closure and between-row area. Therefore, the sum of the reflectance of row crops (r) is separated into the reflectance of the canopy closure ($r_{canopy\_closure}$) and between-row area ($r_{between\_row}$) for consideration. According to Equation (1), the width of the row (A_1) and between-row distance (A_2) can be measured as input parameters. Therefore, $r_{canopy\_closure}$ and $r_{between\_row}$ are key in row modeling. When $r_{canopy\_closure}$ and $r_{between\_row}$ are calculated, it is implied that r can be calculated. As a result, we can establish a row model. Moreover, the sum of the reflectance of row crops (r) is composed of single- and multiple-scattering contributions [30,31]. Therefore, more detailed modeling examples for the single- and multiple-scattering contributions in the two areas are presented in the next two sections.

2.1.2. Reflectance of the Canopy Closure

The reflectance of the canopy closure is the sum of the single- and multiple-scattering contributions inside the canopy [30,31]. We therefore have the following equation

$$r_{canopy\_closure} = r_{cc\_1} + r_{cc\_m} \tag{2}$$

Here, r_{cc_1} is the single-scattering contribution of the canopy closure, and r_{cc_m} is the multiple-scattering contribution of the canopy closure. In the next two sections, we explain the modeling for r_{cc_1} and r_{cc_m}.

1. Single-scattering contribution of the canopy closure

In the GO approach, the single scattering of the canopy closure is a linear combination of the four-component (illuminated leaf, shaded leaf, illuminated soil, and shaded soil [16]) area fraction and the corresponding representative reflectance for each component in the viewing direction [3,24,25]. Therefore

$$r_{cc\_1} = S_c r_c + S_i r_i + S_z r_z + S_g r_g \tag{3}$$

Here, r_c is the reflectance of the illuminated leaf, r_i is the reflectance of the shaded leaf, r_z is the reflectance of the illuminated soil, and r_g is the reflectance of the shaded soil. S_c is the area fraction of the illuminated vegetation in the canopy closure, S_i is the area fraction of the shaded vegetation in the canopy closure, S_z is the area fraction of the illuminated soil in the canopy closure, S_g is the area fraction of the shaded soil in the canopy closure.

We modified the area fraction equations of the four-component area fraction proposed by Verheof [33] in continuous crops and derived an expression suitable for the four-component area fraction in the canopy closure. The specific derivation can be seen in Supplementary Materials A, and the equations are

$$S_z = P_{so}(\theta_s, \theta_o, x, h) \tag{4}$$

$$S_g = P_o(\theta_o, x, h) - P_{so}(\theta_s, \theta_o, x, h) \tag{5}$$

$$S_c = k \int_0^h P_{so}(\theta_s, \theta_o, x, z)dz \tag{6}$$

$$S_i = 1 - P_o(\theta_o, x, h) - k \int_0^h P_{so}(\theta_s, \theta_o, x, z)dz \tag{7}$$

Here, $P_o(\theta_o,x,h)$ is the gap probability of the canopy closure in viewing direction, $P_{so}(\theta_s,\theta_o,x,h)$ is the bidirectional gap probability of the canopy closure, and $\int_0^h P_{so}(\theta_s, \theta_o, x, z)dz$ is the bidirectional vegetation probability of the canopy closure. k is the extinction coefficient of the canopy closure in the viewing direction

$$k = \frac{2}{\pi}L_{row}\left\{\left[ar\cos\left(\frac{-1}{\tan\theta_o\tan\theta_l}\right) - \frac{\pi}{2}\right]\cos\theta_l + \sin\left[ar\cos\left(\frac{-1}{\tan\theta_o\tan\theta_l}\right)\right]\tan\theta_o\sin\theta_l\right\} \tag{8}$$

in which

$$L_{row} = (A_1 + A_2)Lf(\theta_l)d\theta_l/A_1h \tag{9}$$

Here, h is the canopy height, L is the leaf area index, and $f(\theta_l)$ is the leaf inclination distribution function (LADF). This study used an elliptic distribution function [34,35]. Combining Equations (4)–(7), $P_o(\theta_o,x,h)$, $P_{so}(\theta_s,\theta_o,x,h)$ and $\int_0^h P_{so}(\theta_s, \theta_o, x, z)dz$ are the most important parameters for the calculation of area fraction of the canopy closure (S_c, S_i, S_z, and S_g).

In the calculation of gap and vegetation probabilities, we proposed a new approach to calculate gap probabilities. In this new approach, we used a penetration function [21] to calculate gap probabilities in each path length. In this step, we considered the overlapping relationship between the leaves to calculate an average value of gap probabilities of the canopy closure (Supplementary Materials B-2). Furthermore, the average value of gap probabilities of the canopy closure has been used to represent the whole geometric characteristics of the canopy closure and was substituted into the calculation to analyze the overlapping relationship between the average value of gap probabilities of the canopy closure in the solar direction or the viewing direction. Therefore, the overlapping relationship between individual canopy closures could be considered in the calculation of gap probabilities (Supplementary Materials B-3). According to the above calculation ideas, we can consider both the overlapping relationship between leaves and individual canopy closures. In order to describe the bidirectional gap probability and vegetation probabilities, we modified the hotspot kernel function [25] originally used in forests to be suitable for row crops (Equation (B-25) in Supplementary Materials B-3), which can control the peak value near the hotspot. Based on the above, we attempted to address the computational deviations in the single-scattering contribution near the hotspot. For a detailed mathematical derivation, please refer to Supplementary Materials B.

2.   Multiple-scattering contribution of the canopy closure

To calculate the multiple-scattering contribution, we assumed that the canopy closure is an isotropic scattering layer. The principle of adding in the RT approach (adding method) [36–38] is introduced (Figure 2). The derived reflectance of the canopy closure is

$$r_{canopy\_closure} = r_{cc\_1} + \tau_s r_{cc\_1} \tau_0 + \tau_s r_s r_{cc\_1} \tau_0 + \cdots$$
$$= r_{cc\_1} + \frac{\tau_s \tau_0 r_s}{1 - r_{cc\_1} r_s} \tag{10}$$

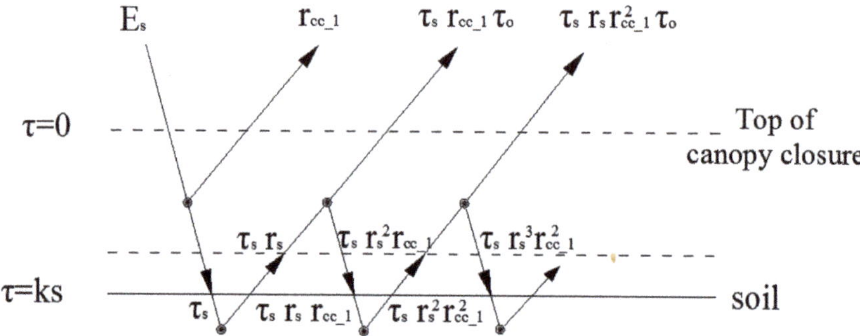

Figure 2. Sketch of radiative transfer inside an isotropic scattering layer and the soil. E_s is the downward irradiance on the horizontal plane, r_{cc_1} is the single scattering contribution of the canopy closure, τ_s is the transmittance of the canopy closure in the solar direction, and τ_0 is the transmittance of the canopy closure in the solar direction. τ is the optical thickness, k is the extinction coefficient, and s is the path length.

Here, r_s is the reflectance of soil and $r_s = (S_z r_z + S_g r_g)/(S_z + S_g)$. τ_s is the transmittance of the canopy closure in the solar direction, and τ_0 is the transmittance of the canopy closure in the view direction. Equation (10) reflects the interaction (scattering) between light, vegetation, and soil, including the single-scattering contribution of the canopy closure and the multiple-scattering contribution of the canopy closure.

We removed the single-scattering contribution (r_{cc_1}) in Equation (10), hence the multiple-scattering contribution of the canopy closure is

$$r_{cc\_m} = \frac{\tau_s \tau_0 \frac{S_z r_z + S_g r_g}{S_z + S_g}}{1 - r_{cc\_1} \frac{S_z r_z + S_g r_g}{S_z + S_g}} = \frac{\tau_s \tau_0 \left(S_z r_z + S_g r_g\right)}{S_z + S_g - r_{cc\_1}\left(S_z r_z + S_g r_g\right)} \tag{11}$$

To address the lack of leaf transmittance in the optical input parameters of the GO approach, we introduce the study by Lang [39]. In this study, the transmittance of the canopy is approximately the same as the gap probabilities and replaces τ_s and τ_0 in Equation (11). Therefore, the multiple scattering of the canopy closure is

$$r_{cc\_m} = \frac{P_0(\theta_o, h) P_s(\theta_s, h)\left(S_z r_z + S_g r_g\right)}{S_z + S_g - r_{cc\_1}\left(S_z r_z + S_g r_g\right)} \tag{12}$$

2.1.3. Reflectance of the Between-Row Area

Previous studies have pointed out that there is a radiation energy exchange between vegetation and soil in the between-row area, which is further affected by multiple scattering between the soil and adjacent canopy closure between the rows [30–32]. The between-row area is an area where both the soil (C′ in Figure 3a) and the vegetation (A′ and B′ in Figure 3a) exist, hence the multiple-scattering equation is very complicated. To establish the multiple-scattering equation of the between-row area, we introduced the mathematical solution of integral radiative transfer equation derived by Ma et al. [30]. We parameterized this equation according to the requirements of the GO approach. Then, an equation of reflectance in the between-row suitable for the GO approach was derived.

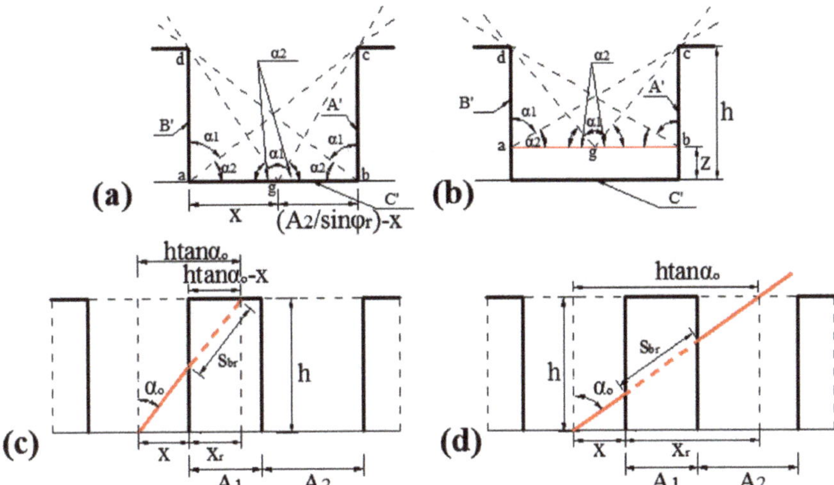

Figure 3. Sketch of the between-row area. (**a**) Angle relationship between the escape surface and the bottom of the between-row area. (**b**) Angle relationship between the escape surface and the *z* position of the between-row area. (**c**) Geometric relationship for the viewing probability of the between-row soil when $x_r \geq A_1$. (**d**) Geometric relationship for the viewing probability of the between-row soil when $x_r < A_1$. Here, α_o is the inclined angle projected in the perpendicular plane in the viewing direction, α_1 is the openness angle of the between-row area, and α_2 is the nonopenness angle of the between-row area. s_{br} is the path length of the light of the canopy closure for the between-row soil between being observed, φ_r is the row azimuth angle, A_1 is the row width, A_2 is the between-row distance, and *h* is the canopy height.

The reflectance of the between-row area ($r_{between\_row}$) is divided into two components: one is the single scattering of the between-row area (r_{br_1}) and the other is the multiple scattering of the between-row area (r_{br_m}), and the expression is

$$r_{between\_row} = r_{br\_1} + r_{br\_m} \tag{13}$$

1. Single-scattering contribution of the between-row area

The single-scattering contribution of the between-row area is the reflectance of the soil that can be observed. According to the projection relationship between canopy closures in the GO approach, the single scattering of the between-row area is

$$r^*_{br\_1} = \begin{cases} \frac{l}{A_2} r_g + \frac{(A_2 - l)}{A_2} r_s & l < A_2 \\ r_g & l \geq A_2 \end{cases} \tag{14}$$

Here, l is the horizontal projected length of the row height on the ground in the solar or viewing directions (for its expression, please refer to Supplementary Materials B-2). When the viewing direction is considered (Figure 3b,c), Equation (14) becomes

$$r_{br\_1} = \overline{P_{o\_br}(\theta_o, x, h)}\, r^*_{br\_1} \tag{15}$$

where $\overline{P_{o\_br}(\theta_o, x, h)}$ is the average viewing probability of the between-row soil, and

$$\overline{P_{o\_br}(\theta_o, x, h)} = \frac{1}{A_2}\int_0^{A_2} e^{-ks_{br}}\, dx \tag{16}$$

Here, $s_{br}(\theta_o, x, h)$ is the path length of the light in the canopy closure in the between-row soil between being observed (Figure 3b,c), and

$$s_{br}(\theta_o, x, h) = \begin{cases} (N_{br}+1)\dfrac{A_1}{\sin\alpha_o \sin\beta_o} & x_r \geq A_1 \\[2mm] (N_{br}+1)\dfrac{z\tan\alpha - x}{\sin\alpha_o \sin\beta_o} & x_r < A_1 \end{cases} \tag{17}$$

where α_o is the inclined angle projected in the perpendicular plane in the viewing direction (its specific description can be found in Supplementary Materials B-1), β_o is the azimuth of the inclined angle in the viewing direction, and $\sin\beta_o = \sin\varphi_{ro}\sin|\theta_o|/\sin\alpha_o$. x_r is a remainder on the x-axis, and $x_r = (h\tan\alpha - x)\mathrm{mod}(A_1 + A_2)$. N_{br} is the number of row cycles $(A_1 + A_2)$ involved in $s_{br}(\theta_o, x, h)$, and $N_{br} = \lfloor (h\tan\alpha - x)/(A_1 + A_2)\rfloor$. Here, mod is the mathematical symbol of modulus, and $\lfloor\cdot\rfloor$ is the mathematical notation for rounding down. Their sketches are shown in Figure 3b,c. Similarly, the average viewing probability at z of the between-row area $(\overline{P_{o\_br}(\theta_o, x, z)})$ only needs to replace h with z.

2. Multiple-scattering contribution of the between-row area

Multiple scattering of the between-row area (r_{br_m}) occurs between the soil and adjacent canopy closure between the rows. Ma et al. gave a multiple-scattering equation of the between-row area based on the operation of the differentia integral operator [40], and its solution is

$$\begin{aligned} r^*_{br\_m} &= K_{br}r^*_{br\_1} + K^2_{br}r^*_{br\_1} + K^3_{br}r^*_{br\_1} + \cdots \\ &= \frac{r^*_{br\_1}K_{br}}{1 - K_{br}} \end{aligned} \tag{18}$$

Here, K_{br} is the transfer probability of the collision, and $K_{br} = k_{br}P_{br}$. Here, k_{br} is the extinction coefficient of the between-row area, and P_{br} is the visual probability of each surface in the between-row area. The between-row area consists of four surfaces (vegetation surface (A′ in Figure 3a), vegetation surface (B′ in Figure 3a), soil surface (C′ in Figure 3a), and escape surface). To calculate P_{br}, we defined the openness angle of the between-row area (α_1) and nonopenness angle of the between-row area (α_2) (Figure 3a,b). Then, we can calculate the visual probability of the escape surface (P_{open}) as $\alpha_1/(\alpha_1 + \alpha_2)$, e.g., $\angle dac/(\pi/2)$, $\angle dgc/\pi$, and $\angle dbc/(\pi/2)$, shown in Figure 3a,b. The simplified equation is

$$P_{open}(x, \theta) = \begin{cases} \dfrac{2}{\pi}\arctan\left(\dfrac{A_2}{\sin\varphi_r h}\right) & (x = A_2) \wedge (x = 0) \\[3mm] \dfrac{1}{\pi}\left[\pi - \arctan\left(\dfrac{h\sin\varphi_r}{A_2 - x\sin\varphi_r}\right) - \arctan\left(\dfrac{h}{x}\right)\right] & x \in (0, A_2) \end{cases} \tag{19}$$

Here, φ_r is the row azimuth angle. Equation (19) is the radiation escape probability at the bottom of the between-row area (Figure 3a). For the escape probability at z of the between-row area, h in Equation (19) needs to be replaced by z (Figure 3b). Since different positions (different coordinate points (x,y) in Figure 3b) in the between-row area have a corresponding radiation escape probability, there are many radiation escape probabilities in the between-row area. In terms of modeling, we focus on the average radiation escape probability value in the between-row area, which is

$$\overline{P_{open}} = \frac{1}{h}\int_0^h \frac{1}{A_2}\int_0^{A_2} P_{open}(x,z,\theta)dxdz \approx \frac{1}{h}\sum_{z=0}^{h}\frac{1}{A_2}\sum_{x=0}^{A_2} P_{open}(x,z,\theta)dxdz$$

$$P_{open}(x,z,\theta) = \begin{cases} \frac{2}{\pi}\arctan\left(\frac{A_2}{\sin\varphi_r h}\right) & [(x=A_2)\wedge(x=0)]\wedge[z\in[0,h]] \\ \frac{1}{\pi}\left[\pi - \arctan\left(\frac{z\sin\varphi_r}{A_2-x\sin\varphi_r}\right) - \arctan\left(\frac{z}{x}\right)\right] & [x\in(0,A_2)]\wedge[z\in[0,h]] \\ 0 & z=h \end{cases} \quad (20)$$

When the viewing direction is considered, Equation (18) can become

$$r_{br\_m} = r^*_{br\_m}\frac{1}{h}\int_0^h \overline{P_{o\_br}(\theta_o,x,z)}dx \tag{21}$$

The average visual factor of the vegetation surface (A′ and B′ in Figure 3a) and one soil surface (C′ in Figure 3a) is defined as $\overline{P_{other}}$, and $\overline{P_{other}} = 1 - \overline{P_{open}}$. Combining Equations (18) and (21), the multiple scattering of the between-row area becomes

$$r_{br\_m} = \frac{1}{h}\int_0^h \overline{P_{o\_br}(\theta_o,x,z)}dx\left[\frac{r^*_{br\_1}k_{br}\left(1-\overline{P_{open}}\right)}{1-k_{br}\left(1-\overline{P_{open}}\right)}\right] \tag{22}$$

From Equation (22), the extinction coefficient of the between-row area is key in calculating the reflectance of the between-row area. To apply Equation (22) in the GO approach, we assumed that the extinction coefficient of the between-row area is the weight sum of the two mediums, soil and leaf. The weight is determined by the length ratio of the medium in this area, which is

$$k_{br} = \frac{2h}{A_2+2h}k + \frac{A_2}{A_2+2h}k_s = \frac{2hk+A_2k_s}{A_2+2h} \tag{23}$$

Here, k is the extinction coefficient of the canopy closure in the viewing direction (A′ and B′ in Figure 3a), and its equation is Equation (8). k_s is the extinction coefficient of the between-row soil in the viewing direction (C′ in Figure 3a). According to [30], the extinction coefficient of the between-row soil in the viewing direction is

$$k_s = 2 - \frac{4r^*_{br\_1}(\cos\theta_s + \cos\theta_o)}{p(\delta)\cos\theta_s^2\cos\theta_o^2}\left(1-\frac{b}{4}\right) \tag{24}$$

Here, $p(\delta)$ is the scattering phase function of the soil particle, which represents the second-order Legendre polynomial (an approximation of spherical function) [41], $p(\delta) = 1 + b\cos\delta + 0.5c\left(3\cos^2\delta - 1\right)$, and $\cos\delta = \cos\theta_s\cos\theta_o + \sin\theta_s\sin\theta_o\cos\varphi_{so}$. Here, b and c are the adjustment parameters for the second-order Legendre polynomial in the scattering phase function of the soil particle, which can be determined by [42].

2.2. Materials and Preparations

2.2.1. Preparations for Validation-Based Computer-Simulated Data

We used a 3D computer simulation to validate (or compare) the row models. In this study, the 3D computer simulation used an extended 3D Radiosity–Graphics Combined Model (RGM) [43,44]. The RGM model is a radiosity model based on a bilinear equation (a simple nonlinear differential equation) [2]. It uses a numerical calculation method (Gauss–Seidel) to calculate the scattering of polygons in a scene constructed by a computer graphics method with high calculation accuracy [43,44]. To calculate the sum of the reflectance of row crops (r) for the RGM model, we divided it into two steps. In the first step, we used computer graphics to establish a computer scene similar to the assumption of our row model (the turbid medium bound in the periodic box-shaped vegetation material) (abstract scenes in Figure 4). For the abstract scene, we generated four representative scenes of row crops, namely the proportion of between-row dominance (Stage_rc1), proportions of between-row and canopy closure equality (Stage_rc2), the proportion of the canopy closure dominance (Stage_rc3), and continuous crops (Stage_cc). The parameters constructed in the abstract scene are shown in Table 1. In the second step, based on the established abstract scene, we used the RGM model to calculate polygonal scattering in the abstract scene, and finally calculated the sum of the reflectance of row crops (r). Moreover, the sum of the reflectance of row crops (r) calculated by the RGM model can be used as a "true value" to validate the sum of the reflectance of row crops calculated by our model with the same parameters in Table 1. In setting the angle, the solar zenith angle was 25° and the solar azimuth angle was 130° for both models. Moreover, to keep the computer scene consistent with the periodic box-shaped plant materials assumed by our model, we used the infinite canopy of the RGM model in this study (Appendix B in [43]). This set implies that the reflectance calculated by RGM is not the reflectance of the one- or two-row cycle shown in Figure 4, but the reflectance of the scene where the row cycle is infinitely extended. The output results of reflectance are shown in Section 3.1.

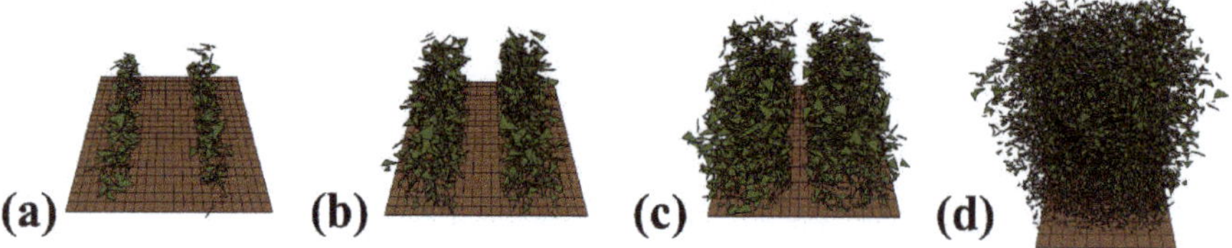

Figure 4. The abstract scenes of row crops. (**a**) The proportion of between-row dominance (Stage_rc1), (**b**) proportions of between-row and canopy closure equality (Stage_rc2), (**c**) proportion of the canopy closure dominance (Stage_rc3), and (**d**) continuous crops (Stage_cc).

Table 1. Values required when constructing abstract scenes with computer graphics.

Scenes	L (m·m^{-1})	θ_1 (°) [1]	A_1 (m)	A_2 (m)	H (m)	φ_r (°)	W_p (m) [1]
Stage_rc1	0.5	49.3	0.25	0.75	0.5	0	0.07
Stage_rc2	1.5	43.4	0.5	0.5	0.8	0	0.07
Stage_rc3	2.5	48.4	0.75	0.25	1.1	0	0.07
Stage_cc	3.5	46.7	1	0	1.4	0	0.07

[1] Here, the inclined leaf angle and azimuth leaf angle are set to a random distribution. By counting the leaf inclination angle and width of the leaves for each polygon, we obtained the average leaf inclination angle (θ_1) and the characteristic width of leaves (W_p).

2.2.2. Preparations for Validation-Based In Situ Data

The measurements include two sites in arid Northwestern China: Zhangye, Gansu Province, and Zhongwei, Ningxia Hui Autonomous Region. Field and satellite measurements were performed in Zhangye and Zhongwei, respectively.

1. Field measurement data in Zhangye

The in situ data of Zhangye came from Watershed Allied Telemetry Experiment Research (WATER) [45,46] and were measured from 20 May to 11 July 2008. A plot with an area of 180 × 180 m was selected, and the center coordinates of the plot were 38.857056°N, 100.410444°E (Figure 5). The type of crop in the plot was corn, and four quadrats were randomly selected to measure the required parameters for validation. In the measurement of optical parameters, the reflectances of the illuminated leaf (r_c), shaded leaf (r_i), illuminated soil (r_z), and shaded soil (r_g) were measured by the ASD FieldSpec Pro Spectrometer [47,48]. The sum of the reflectance of row crops (r) (blue line in Figure 9) was measured by the ASD FieldSpec Pro Spectrometer, and the distance from the sensor to the top of the canopy was about 1 m, ensuring that at least one row cycle was observed. The measured spectral curve was from 400 to 2500 nm. The measurement time was 22 May, 25 May, 1 June, 16 June, 22 June, and 1 July (Table 2) [47,48]. Since the row structure was obvious on 22 June (Table 2), we performed a multiangle observation. The sum of the reflectance of row crops (r) (blue triangle scatter and blue line in Figure in Section 3.2.2) was measured by the ASD FieldSpec Pro Spectrometer combined with the multiangle frame. The distance between the instrument and ground was 5 m, the field of view was 25°, and a complete row cycle was observed. The zenith angle was from −60° to 60° with 10° for an interval [49]. The study considered the anisotropy of reflectance. Measurements were separated by four modes in the azimuth: the principal plane (PP), orthogonal plane (OP, viewing azimuth angle perpendicular to the sun azimuth angle), along-row plane (AR, the plane along the row direction), and orthogonal row plane (OR, the plane perpendicular to the row direction). The other structure parameters used the direct measurement method [50–52]. These parameters were the leaf area index (L), average leaf inclination angle (θ_l), row width (A_1), between-row distance (A_2), canopy height (h), characteristic width of leaves (W_p), and row azimuth angle (φ_r). The measurement results are shown in Table 2. The output results of the spectral curve for the growing season are displayed in Figure 9. Corresponding output results of the distribution of reflectances in the multiangle observation (22 June, in Table 1) are displayed in Section 3.2.2.

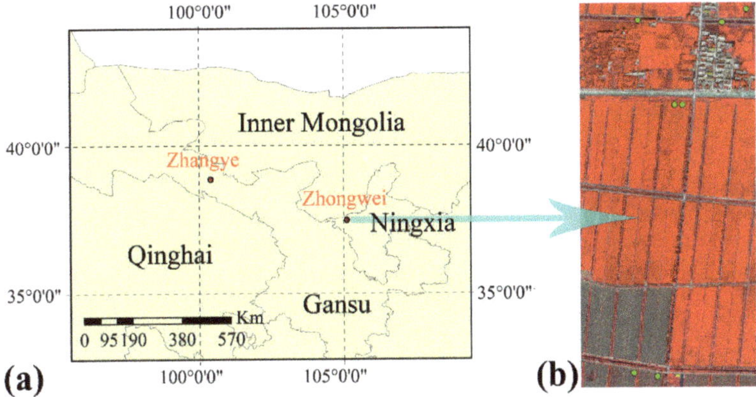

Figure 5. Geographic location of the study area (**a**) Plots in Zhangye and Zhongwe; (**b**) World-View 3 image in the Zhongwei area and its corresponding quadrats. Here, the field measurement was performed in Zhangye, hence there is no satellite image. The green points in (**b**) represent the quadrats in the field measurement with the same size as the resolution of the World-View 3 image.

2. Satellite measurement data in Zhongwei

The satellite measurement was performed on 16 July, 2016 at Zhongwei city. In the optical parameter measurements, the sum of the reflectance of row crops (r) was measured via satellite with a high spatial resolution (World-View 3 image in Figure 4b). Therefore, the measured height was about 617 km. The World-View 3 image has 8 bands from 400 to 1040 nm, and the spatial resolution is 1.24 m. To obtain the sum of the reflectance of row crops, we preprocessed the image, involving radiation calibration, geometric correction, and atmospheric correction [53]. The surface objects in the image have three types of row crops: rice, corn, and matrimony vine (strictly speaking, matrimony vine is an economic forest planted in row form). To obtain the optical parameters required for the row model, the seven typical quadrats with the same size as the resolution of the World-View 3 image (Figure 5b) were directly selected to measure the reflectance of the illuminated leaf (r_c), shaded leaf (r_i), illuminated soil (r_z), and shaded soil (r_g). For these optical parameters, we used the ASD FieldSpec Pro Spectrometer on the ground. The other structure parameters used the direct measurement method, and the measurement results are shown in Table 3. The corresponding output results simulated by our model are shown in second part of Section 3.2.1.

Table 2. Structural parameters of corn in the growing season.

Phenology	Date (2008)	L (m·m^{-1})	θ_1 (°)	A_1 (m)	A_2 (m)	H (m)	φ_r (°)	W_p (m)
Emergence	20 May	0.23	49.16	0.2	0.80	0.16	110	0.03
Emergence	25 May	0.34	48.15	0.25	0.75	0.22	110	0.04
Jointing stage	1 June	0.46	48.15	0.30	0.70	0.35	110	0.07
Jointing stage	16 June	1.76	40.83	0.5	0.5	0.87	110	0.08
Jointing stage	22 June	2.52	59.0	0. 65	0.35	0.98	110	0.13
Jointing stage	1 July	4.52	37.83	0.85	0.15	1.54	110	0.14

Table 3. Structural parameters of crops in the quadrats.

Quadrats	L (m·m^{-1})	θ_1 (°)	A_1 (m)	A_2 (m)	H (m)	φ_r (°)	W_p (m)
Corn 1	3.19	45.33	0.9	0.4	2.66	40	0.16
Corn 2	3.54	44.67	0.91	0.29	2.63	40	0.18
Corn 3	2.6	32.00	0.99	0.2	2.82	45	0.14
Rice 1	3.18	59.00	0.42	0.12	0.68	50	0.02
Rice 2	3.17	63.67	0.3	0.13	0.78	53	0.04
Matrimony vine 1	1.25	36.25	1.29	1.03	1.2	43	0.01
Matrimony vine 2	0.77	68.33	1.04	0.93	0.86	42	0.01

3. Results

3.1. Validation of Row Model Using Computer-Simulated Data

In Figure 6, the distribution of the sum of the reflectance of two models in four scenes (Stage_rc1, Stage_rc2, Stage_rc3, and Stage_cc) are shown. With an increase in LAI (leaf area index) and a change in row structure (increase in the row width and height, decrease in the between-row distance) (Table 1), the distribution of the sum of the reflectance in the red band and NIR band simulated by the row model and the distribution of the sum of the reflectance in the red band and NIR band simulated by the RGM model have high consistency in four viewing modes (PP mode, OP mode, AR mode, and OR mode). The correlation coefficients (R) are greater than 0.9281. The root mean square errors (RMSEs) are less than 0.0012 in the red band and less than 0.0095 in the NIR band (Table 4). Moreover, the differences between the reflectance simulated by the two models are less than 5%. The consistency of the red band (mean absolute deviation (MAD) = 0.25 × (1.98% + 3.08% + 4.64% + 4.03%) ≈ 3.43%) is greater than that of the near-infrared band (MAD = 0.25 × (1.87% + 2.99% + 1.68% + 1.82%) ≈ 2.08%) (Table 4). With an increase in LAI and a change in row structure, the sum of the reflectance in the red band and NIR band shows opposite trends, i.e., the sum of the reflectance in the red band is decreasing while the sum of the reflectance in the NIR band is gradually increasing. Our model is the same as the RGM model in the inverted v-shaped area where the hotspot is located (PP in Figure 6a,b). When Stage_rc1 changes to Stage_cc, the width of hotspots (slope of an inverted v-shaped area) gradually narrows.

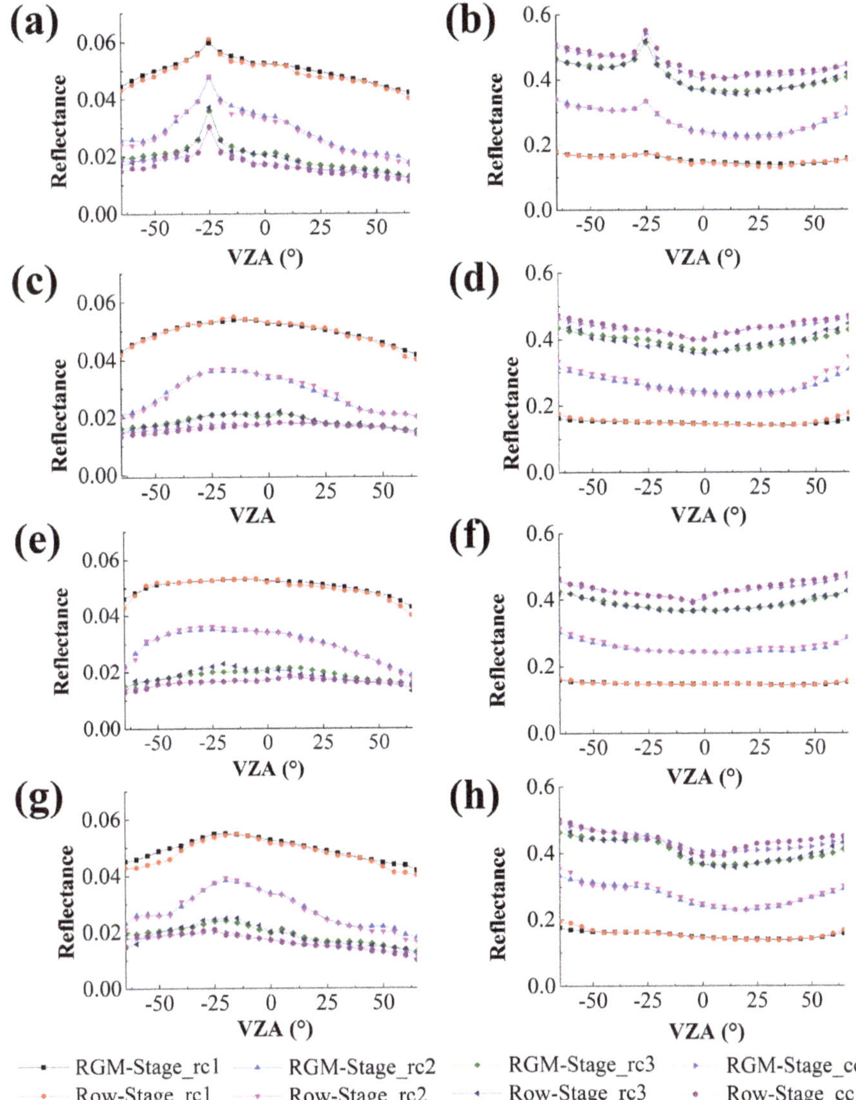

Figure 6. The distribution of the sum of the reflectance in the red band and near-infrared (NIR) band simulated by the RGM model and the row model in four viewing modes: (**a**) principal plane (PP) mode in the red band, (**b**) principal plane (PP) mode in the NIR band, (**c**) orthogonal plane (OP) mode in the red band, (**d**) orthogonal plane (OP) mode in the NIR band, (**e**) along-row plane (AR) mode in the red band, (**f**) along-row plane (AR) mode in the NIR band, (**g**) orthogonal row plane (OR) mode in the red band, and (**h**) orthogonal row plane (OR) mode in the NIR band. Here, VZA is the viewing zenith angle.

Table 4. The statistical results of the comparison of curves of reflectance simulated by the row model versus the reflectance simulated by the Radiosity–Graphics Combined Model (RGM) model.

Scenes	Statistics	R_red	R_NIR	R_NIR_1	R_NIR_m
Stage_rc1	R	0.9766	0.9282	0.8975	0.7972
	RMSE	0.0009	0.0046	0.0043	0.0029
	MAD	1.98%	1.87%	1.86%	9.82%
Stage_rc2	R	0.9902	0.9612	0.9683	0.9731
	RMSE	0.0009	0.0094	0.0056	0.0045
	MAD	3.08%	2.99%	2.58%	3.54%
Stage_rc3	R	0.9568	0.9678	0.9817	0.9754
	RMSE	0.0011	0.0087	0.0048	0.0061
	MAD	4.64%	1.68%	1.81%	1.93%
Stage_cc	R	0.9296	0.9601	0.9699	0.9850
	RMSE	0.0009	0.082	0.0056	0.0031
	MAD	4.03%	1.82%	2.64%	0.99%

Figure 7 shows the decomposition of the reflectance in the NIR band in Figure 6b,d,f,h, namely the single-scattering contribution in Figure 7a,c,e,g and multiple-scattering contribution in Figure 7b,d,f,h. In Figure 7a,c,e,g, the overall difference between the single-scattering contribution in the NIR band simulated by the row model and the single-scattering contribution in the NIR band simulated by the RGM model is small. The R-value is greater than 0.8974 and RMSEs are less than 0.0057 for single scattering in the NIR band in Table 4. Moreover, the differences between the single scattering in the NIR band simulated by the two models are less than 3% (Table 4). Figure 8 shows the dense points in the single scattering near the hotspot in PP, with our model and RGM model also having high consistency. In addition to the change in the width of the hotspot, the changing trend of the single-scattering hotspot is more obvious than the sum of the reflectance in Figure 6a,b.

In Figure 7b,d,f,h, the multiple-scattering contribution in the NIR band simulated by our model and multiple-scattering contribution in the NIR band using the RGM model have high consistency. The R-value is greater than 0.7971 and RMSEs are less than 0.0062 for multiple scattering in the NIR band in Table 4. Moreover, the differences between the multiple scattering in the NIR band simulated by the two models are less than 10% (Table 4). Comparing Figure 6b,d,f,h and Figure 7b,d,f,h, we can find that the proportion of the multiple scattering in the NIR band in the sum of the reflectance gradually changes from 20% to 50% as LAI increases and changes in row structure (from Stage_rc1 to Stage_cc).

Here, R_NIR_1 is the single scattering in the NIR band, and R_NIR_m is the multiple scattering in the NIR band. R is the correlation coefficient, RMSE is the root mean square error, and MAD is the absolute value of average difference, which represents the difference between Data A and Data B, expressed as a percentage. Its expression is $\sum_{m=1}^{m} \frac{1}{m}\left|(A-B)/B\right|$. Here, m is the number of comparison groups, A is the reflectance simulated by the row model, and B is the reflectance simulated by the RGM model.

3.2. Validation of Row Model Using In Situ Data

3.2.1. Spectral Curve

1. Validation of the sum of the reflectance during the growing season using field data

In Figure 9, the sum of the reflectance simulated by the row model and the field data have high consistency in the growing season of row crops (it can be described as an increase of LAI and change in row structure in Table 2), except for the spectrum less than 1500 nm measured on 20 May (Figure 9a). As the crop grows over time, the photosynthetically active radiation (400–700 nm) gradually decreases. This phenomenon is especially noticeable at the "red edge" (700–750 nm, the area where the vegetation reflectance from the red band to the near-infrared band increases sharply). On 25 May (Figure 9b), the "red edge" gradually formed, and it was increasingly obvious with the growth of crops (Figure 9b–f). The sum of the reflectance in the near-infrared (NIR) region (700–1100 nm) gradually increases with the growth of crops; after that, it reaches a maximum value (0.51) on 1 July (Figure 9f).

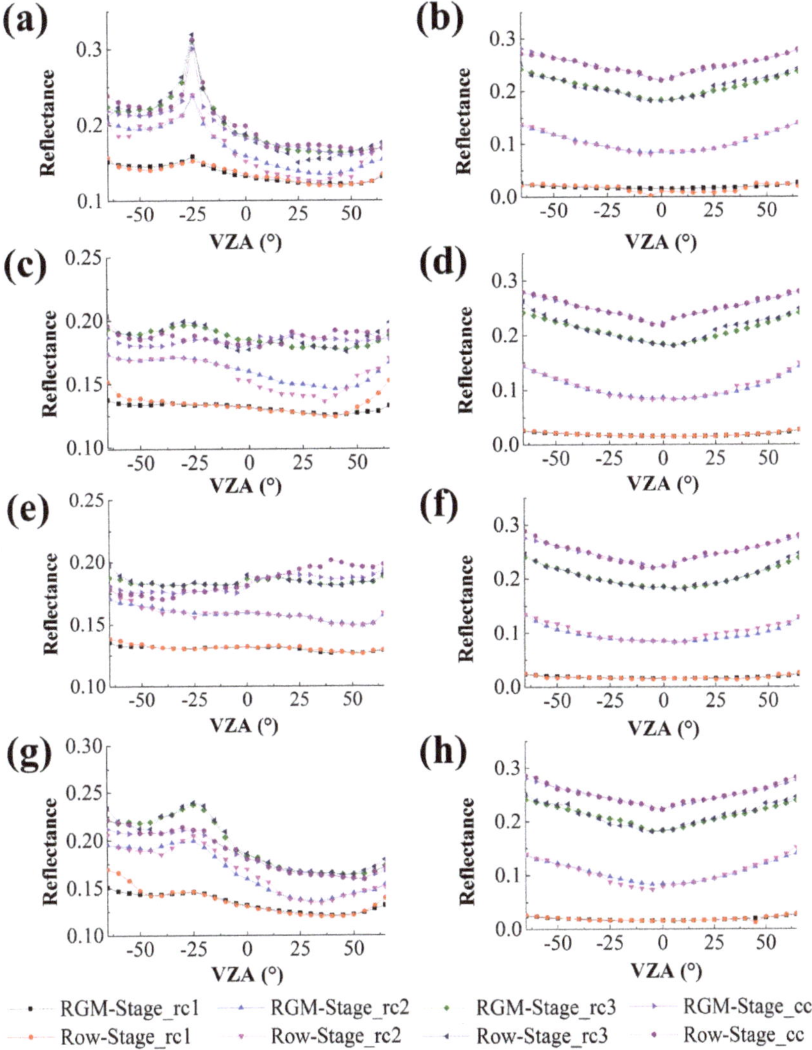

Figure 7. The distribution of reflectances in the single scattering in the NIR band and the multiple scattering in the NIR band simulated by the RGM model and the row model in four viewing modes: (**a**) principal plane (PP) mode in the single scattering in the NIR band, (**b**) principal plane (PP) mode in the multiple scattering in the NIR band, (**c**) orthogonal plane (OP) mode in the single scattering in the NIR band, (**d**) orthogonal plane (OP) mode in the multiple scattering in the NIR band, (**e**) along-row plane (AR) mode in the single scattering in the NIR band, (**f**) along-row plane (AR) mode in the multiple scattering in the NIR band, (**g**) orthogonal row plane (OR) mode in the single scattering in the NIR band, and (**h**) orthogonal row plane (OR) mode in the multiple scattering in the NIR band. Here, VZA is the viewing zenith angle.

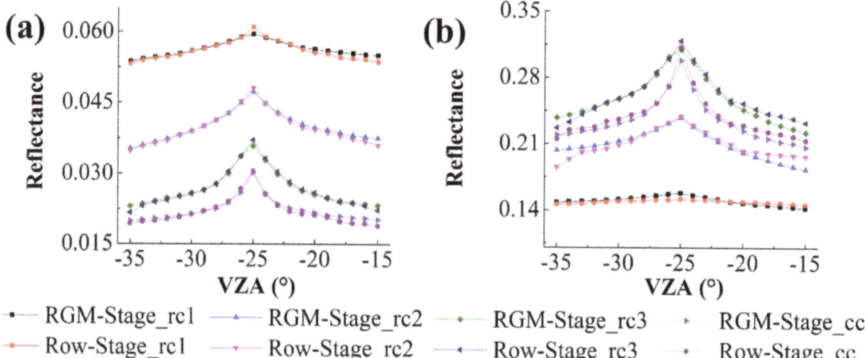

Figure 8. The distribution of reflectances for the single scattering near the hotspot (principal plane) model. (**a**) Single scattering near the hotspot in the red band, and (**b**) single scattering near the hotspot in the NIR band. Here, VZA is the viewing zenith angle.

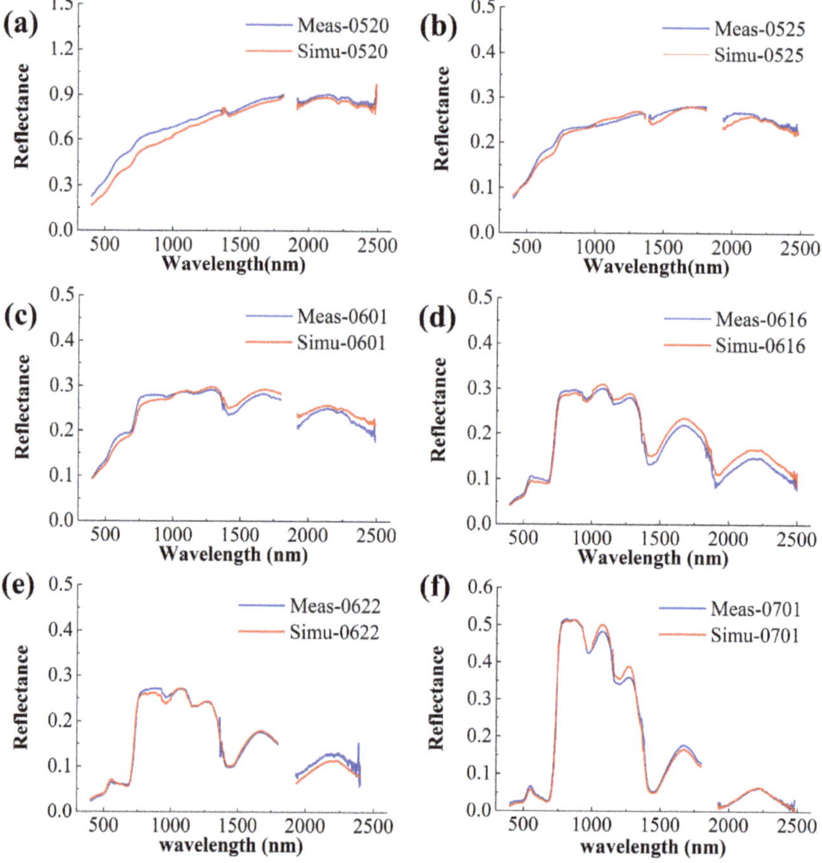

Figure 9. Comparison of the sum of the reflectance simulated by the row model and field data at the wavelength in the vertical viewing direction on (**a**) 20 May, (**b**) 25 May, (**c**) 1 June, (**d**) 16 June, (**e**) 22 June, and (**f**) 1 July. Here, the noise in the water vapor absorption was removed.

2.   Validation of the reflectance for different types of row crops using satellite data

We used remote sensing data to validate the row model. The sum of the reflectance simulated by the row model and the sum of the reflectance measured by the World-View 3 satellite have high consistency. In the simulation of each species, the accuracy of the simulation of corn and wheat is the best, while the simulations of wolfberry had a slight deviation (Figure 10).

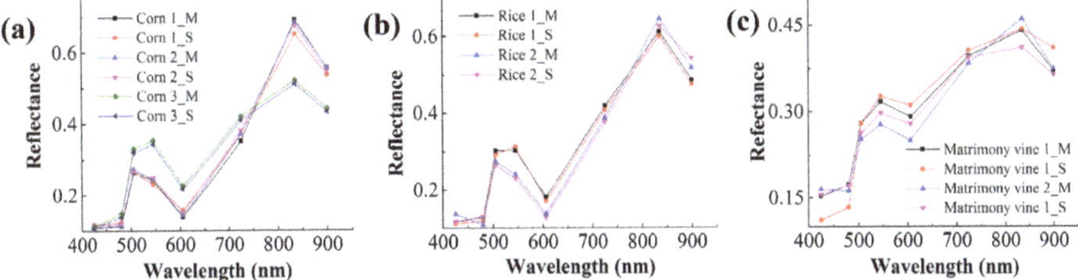

Figure 10. Comparison of the sum of the reflectance simulated by the row model and the reflectance at the top of the canopy measured by the World-View 3 satellite in the vertical viewing direction. (**a**) Corn; (**b**) rice; (**c**) matrimony vine. Here, M represents the data measured by the World-View 3 satellite and S represents the simulated data of the row model.

3.2.2. Distribution of the Sum of the Reflectance on the Multiangle Observation

In the distribution of the sum of the reflectance (Figure 11), the results simulated by the row model and the measured data have a slight systematic deviation from some angles. In the forward direction of the PP mode in the NIR band, the reflectance has the largest computational deviation between the simulation and measurement. However, from the statistical results, we can find that the R-values are greater than 0.8317 (except R = 0.5392 in the PP mode in the NIR band) and RMSE values less than 0.0403 (Figure 12). The MAD results show that the difference between the sum of the reflectance simulated by the row model and field data is less than 13%, except for the OR mode in the red band (24.11% in Figure 12). Similar to the computer simulation in Figures 6 and 7, the consistency of the red band (MAD = 0.25 × (10.71% + 12.4% + 6.29% + 24.11%) ≈ 13.38%) is greater than that in the NIR band (MAD = 0.25 × (9.65% + 5.29% + 1.61% + 3.76%) ≈ 5.08%) (Figure 12). Generally speaking, the consistency between the two is still relatively high, simulating multiangle changes in the PP mode, OP mode, AR mode, and OR mode.

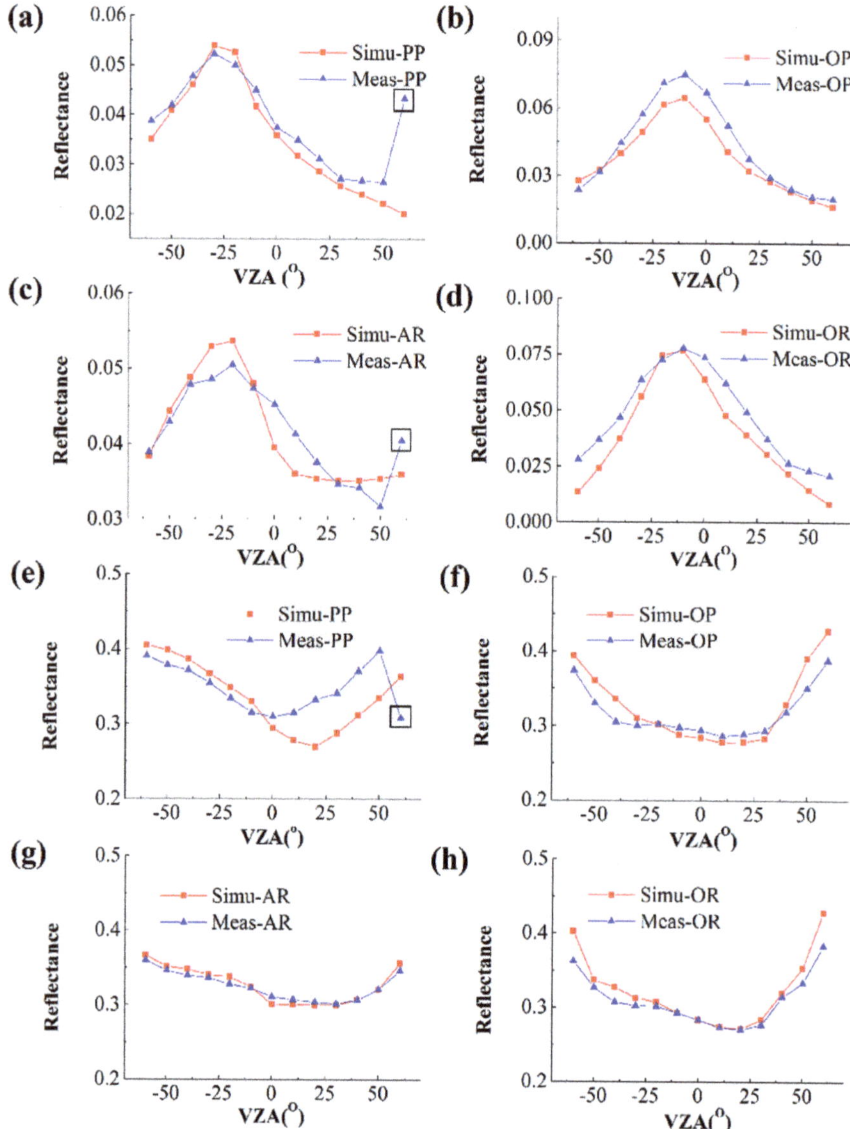

Figure 11. Comparison of the distribution of the sum of the reflectance simulated by the row model and field data in the multiangle observation for the principal plane (PP) mode (**a**,**e**), orthogonal plane (OP) mode (**b**,**f**), along-row plane (AR) mode (**c**,**g**), and orthogonal row plane (OR) mode (**d**,**h**). (**a**–**d**) Red band (670 nm); (**e**–**h**) NIR band (860 nm). VZA is the viewing zenith angle. The black box is the abnormal point in the measurement.

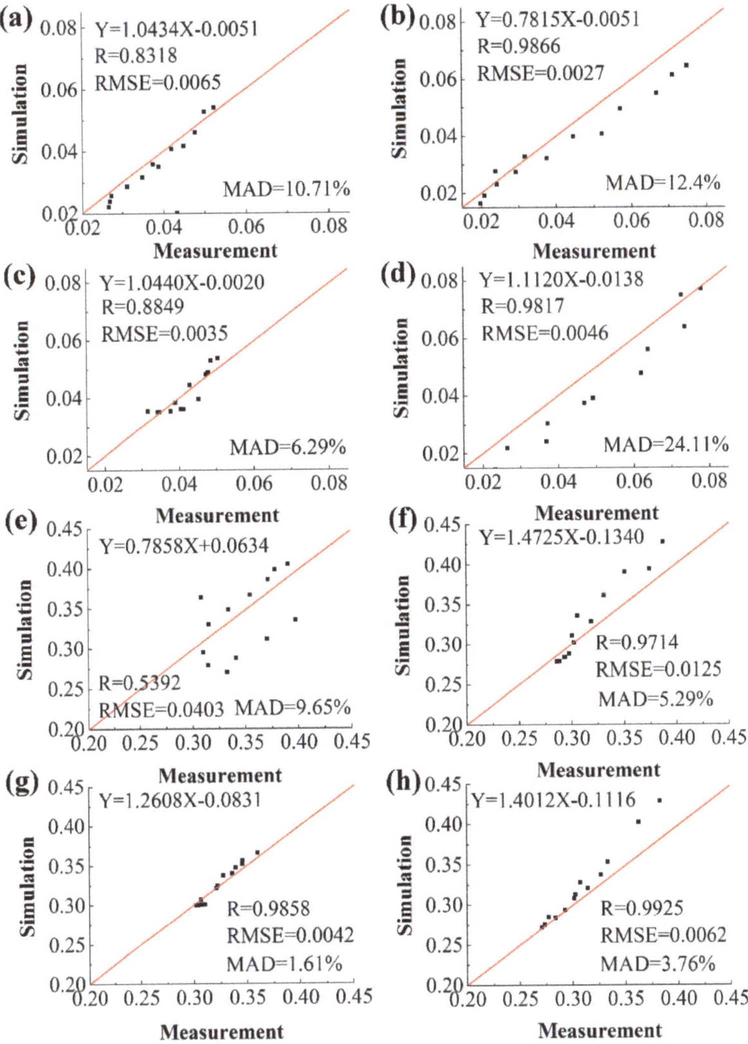

Figure 12. Statistics of the distribution of the sum of the reflectance simulated by the row model and field data in the multiangle observation for the principal plane (PP) mode (**a,e**), orthogonal plane (OP) mode (**b,f**), along-row plane (AR) mode (**c,g**) and orthogonal row plane (OR) mode (**d,h**). The red line is a 1:1 line. R is the correlation coefficient, RMSE is the root mean square error, and MAD is the mean absolute deviation, which represents the difference between Data A and Data B, expressed as a percentage. Its expression is $\sum_{m=1}^{m} \frac{1}{m}|(A-B)/B|$. Here, m is the number of comparison groups, A is the sum of the reflectance simulated by the row model, and B is field data.

4. Discussion

4.1. Multiple Scattering of Row Crops in the GO Approach

This study substantiates similar findings obtained in earlier studies by showing that it is necessary to consider the multiple-scattering contribution in the NIR band in the GO approach [13,26,27]. Moreover, we considered the multiple-scattering contribution in row crops as an extension of the previous study field (forest field) to the object of row crops. In the row modeling based on the GO approach, its initial purpose was to provide a physical basis to calculate temperature (or emissivity). However, due to the different physical properties between temperature (or emissivity) and reflectance, the previous studies rarely considered multiple scattering issues in row modeling [14,28,29]. Although subsequent studies attempted to consider multiple scattering issues in the reflectance modeling of row crops, these models usually consider a row structure to modify existing continuous crop models based on the RT approach [17,31,54,55]. Therefore, their physical basis was still the volume scattering of the radiative transfer equation, and their essence was the RT model. Different from the previous modeling, our model took the multiple-scattering equation constructed by the RT approach in GO modeling further to modify the GO approach. The multiple-scattering equation achieved a good calculation accuracy (Figure 7b,d,f,h). In the multiple-scattering contributions of row crops, the radiation energy of multiple scattering in the NIR band was very strong, accounting for a large proportion of the sum of the reflectance (comparing Figure 6b,d,f,h and Figure 7b,d,f,h). Specifically, as crops gradually grew from row crops to continuous crops, the proportion of multiple scattering in the NIR band in the sum of the reflectance gradually increased from 20% to 50% (comparing Figure 6b,d,f,h and Figure 7b,d,f,h). These results imply that the sum of the reflectance would be significantly underestimated if the multiple-scattering contribution is not considered, and the underestimation becomes more obvious as LAI gets bigger. Therefore, the multiple-scattering equation is considered in GO modeling and can be accurately calculated, which can effectively improve the accuracy of the sum of the reflectance in the NIR band. However, in Table 4, the mean absolute deviation (MAD) of R_NIR_m gradually decreased from 9.82% in the Stage_rc1 to 0.99% in the Stage_cc. This result implies that the row structure influences the calculation accuracy of multiple scattering, whereby, when the row width decreases, the between-row distance increases, and the difference in the multiple-scattering contribution between our model and the RGM model will increase. This conclusion is consistent with the simulation results in [30]. The primary reason is that our model describes Lambertian soil (soil reflectance as a single value, see Equations (13) and (22)), but the RGM model describes non-Lambertian soil (the brown squares shown in Figure 4 are approximated as an anisotropic soil, hence soil reflectance is a multivalue consisting of multiple squares). Therefore, the increase in the between-row distance will cause a difference in the multiple scattering between our model and the RGM model.

In the GO approach, we studied an adding method (Section 2.1.2) and a mathematical solution of integral radiative transfer equation (Section 2.1.3) to derive the multiple-scattering equation. This is a typical method for the multiple-scattering equation, which was established by the RT approach, to be coupled into the GO approach for a modified GO model. This method is based on the RT approach to derive the multiple-scattering equation. However, in a previous study (mainly in the heterogeneous forest), this method often ignored the geometric characteristics between individual canopies [26], which was reflected in the results as a computational deviation in the sum of the reflectance [13,27]. The between-individual canopies mentioned in [13,27] are the between-row area in row crops. Our research focused on this point in row modeling. In our model, the openness angle of the between-row area (α_1) and nonopenness angle of the between-row area (α_2) were introduced to calculate an average value of the escape probability of radiation in the between-row area ($\overline{P_{open}}$). The above treatment showed the scattering direction (or scattering angle with typical geometric characteristics) in the multiple-scattering contribution of the between-row area, and further reflected the influence of geometric characteristics (A_2, φ_r and h in Equation (20)) on the multiple-scattering equation (Equation (22)). The multiple-scattering contribution simulated by our model had higher R-values and lower RMSEs compared to the RGM model (Table 4). The modified treatments of these equations and results further illustrated that our model considered how the geometric characteristics between individual canopies can improve the calculation accuracy of the multiple-scattering contribution. We further analyzed the sum of the reflectance simulated by our model compared to those of the RGM model was of high consistency (Figure 6). The simulation results and in situ measurement of our model also have good accuracy overall (Figures 9–11). These results show that geometric characteristics considered in the multiple-scattering equation established by the RT approach are necessary for GO modeling. This consideration not only improves the accuracy of multiple scattering but also the accuracy of the sum of the reflectance.

When we analyzed the multiple-scattering equation in the previous model in the heterogeneity forest (Equation (27) in [27] and Equations (8)–(11) in [13]), we found that the single-scattering albedo of the leaf was added to derive the multiple-scattering equation. The purpose of this treatment in the previous models was to remedy the defects of the GO approach without the scattering phase function in the calculation of scattering. However, the single-scattering albedo of the leaf is difficult to determine in the GO approach without leaf transmittance [7]. Therefore, the single-scattering albedo of the leaf needs to adopt an empirical approach, which leads to the lack of a physical mechanism at the foundation of multiple scattering. In this study, we focused on the above limitation. In the multiple-scattering equation of row crops, we used a single-layer adding method (Figure 2) and mathematical solution of integral radiative transfer equation (its mathematical properties are also cumulative, Equation (22)) to avoid introducing the single-scattering albedo of the leaf to calculate multiple scattering. We tried to use gap probabilities to approximate the transmittance in canopy closure ($P_o(\theta_o,x,h)$ and $P_s(\theta_s,x,h)$ in Equation (12)) and the transmittance in the between-row area ($\overline{P_{o\_br}(\theta_o,h)}$ and $\overline{P_{open}}$ in Equation (22)). In this way, the gap probabilities become a connection that can seamlessly connect single-scattering contributions and multiple-scattering contributions together.

4.2. Hotspot of Row Crops in the Single-Scattering Contribution

The single-scattering contribution was accurately calculated, which was the basis of the multiple scattering calculations. Therefore, the issues of single scattering cannot be ignored in the calculation of the multiple-scattering contribution. Equations (12) and (22) require a single-scattering contribution as the original function to calculate the multiple-scattering contribution. In this study, we extended the modeling ideas of two-overlapping relationship (the overlapping relationship between leaves and canopy closures [21,22]) previously used for heterogeneity forest modeling to row modeling (Supplementary Materials B). By comparing single-scattering contributions near the hotspot in four canopies (the four growth stages of crops in Figure 4) simulated by our model and the RGM model, we found that the simulation results between the two models were very consistent (Figure 8), which indicates that our study achieved good accuracy. These results substantiate that our model considered the two-overlapping relationship to calculate gap probabilities (especially bidirectional gap probabilities and vegetation probabilities, Equations (B-24) and (B-32) in Supplementary Materials B-3), which effectively improved the calculation accuracy of single scattering near the hotspot in the GO approach. The mechanism for these results in Figure 8 shows that the shadow produced by the overlapping relationship between the canopy closures is gradually replaced by the shadow produced by the overlapping relationship for leaves when the row structure changes from row crops to continuous crops. In addition, the shadow area produced by the overlapping relationship between canopy closures is larger than the shadow area produced by the overlapping relationship for leaves. Therefore, the shadow near the hotspot in the crops will be reduced and single scattering near the hotspot will be enhanced. These results also imply that gap probabilities are calculated and the two-overlapping relationship is necessary.

The cause of the peak value near the hotspot involves bidirectional gap probability and bidirectional vegetation probability issues [2]. The reason for this phenomenon is that the shadow created by the overlap between the medium is not seen when the solar direction and the viewing direction are very close, and the reflectance will have a maximum value (the peak value near the hotspot) [56,57]. In continuous crops, the peak value near the hotspot is mainly influenced by LAI, leaf size, leaf orientation mode, and sun geometry [20]. However, in addition to the above factors, the peak value near the hotspot in row crops is also influenced by the row structure [31]. In our study, we found that as the row structure changes from row crops to continuous crops (from Stage_rc1 to Stage_cc), the width of the hotspot (slope of an inverted v-shaped area) gradually narrows, and the peak value becomes more obvious (Figure 8 for the single-scattering contribution and Figure 6a,b for the sum of the reflectance). Our study accurately simulated the width of the hotspot, which not only considered the two-overlapping relationship but also considered the hotspot kernel function. In order to describe the width of the hotspot, cross-correlation [19] or overlap functions [57] have been commonly used as hotspot kernel functions in the RT approach. However, our study was based on the GO approach, hence the above functions were not applicable. In our study, we used the hotspot kernel function (Equation (B-25) in Supplementary Materials B-3) proposed by Chen et al. [25] in the heterogeneity forest modeling based on the GO approach. To make it suitable for the canopy of row crops, our study analogized the internal relationship between forest parameters and row parameters to the modified C_2 in Equation (B-25), where C_2 is the control factor of the width of the hotspot. From C_2 (Equation (B-25)), it can be found that L, h, θ_s, W_p, and k (note that k is a function of A_1, A_2, h, θ_o, θ_1, and $f(\theta_1)$ in Equation (8)) were considered, which also implies that our study involved equation modeling that takes into account the main influencing factors described above, such as LAI, leaf size, leaf orientation mode, and the sun geometry, row structure. These results and equations in the modeling show that the hot kernel function is very important in simulating the peak value near the hotspot.

4.3. Analysis of the Row Model

As a GO model, our model mainly uses the geometric relationship between light and medium to calculate the sum of the reflectance. The RGM model is a computer model based on radiosity. It uses the geometric probability between leaves to calculate the sum of the reflectance. Specifically, a view factor describing the overlapping relationship between polygons (leaves) was previously introduced into the RGM model [43]. If we exclude the numerical calculation of reflectance (Gauss–Seidel algorithm [44]) in the RGM model, the geometric principles involved in our model are very similar to those of the RGM model. Therefore, based on similar physical principles, the consistency between the simulation results of our model and the RGM model is high (Figures 6–8, MAD is less than 10% in Table 4). However, our model still shows calculation deviation in the in situ validation, especially in the distribution of the sum of the reflectance on the multiangle observation (Figure 11, MAD is less than 24% in Figure 12). These results imply that the calculation deviation may come from other sources. At present, it is difficult for multiangle instruments to accurately measure directional reflectance (specifically, bidirectional reflectance distribution function (BRDF) [58]) and obtain an approximate reflectance distribution. In theory, directional reflectance (BRDF) refers to the measurement value at the same time (the same second), and there is no shadow of the instrument (note: the instrument is nontransparent) during the measurement [58], which is hard to achieve with the current instruments. In the multiangle observation, instruments take at least 10 minutes or more to complete the four modes (PP, OP, AR, and OR). Even if the measurement is performed in strict accordance with the measurement specification, changes in environmental variables over time will have a strong influence on measurement [59]. Therefore, the results in the measurements are only an approximation (abnormal points in the black box in Figure 11a,c,e). The instrument had a field of view (FOV) of 25° (common FOV of ASD spectrometer), and the observation distance was 5 m (this height is different from the current canopy reflectance model assuming that the sensor is located at infinity) in our study. Therefore, we observed the sum of the reflectance of a limited row cycle (a canopy closure plus a between-row area) in FOV (according to Table 1, we used the triangular relationship to calculate that the observed row cycle changes from 2.2 to 10.4 and from 0° to ±60° in the zenith angle), which is different from our model assumptions (periodic box-shaped plant materials). However, the validations of computer simulation have not been influenced by time in the multiangle observation. The infinite canopy is set in the RGM model in this study [43], which is the same as our model assumptions. This implies that the validation of computer simulation has more advantages than the current lack of accurate instruments. Our study reconfirmed the conclusion (or objective) from RAMI [4–6,60] and a wide range of mathematical modeling in earth sciences [59], and computer simulation is one of the most effective ways to validate the model of a nonnumerical calculation.

Since the leaves of crops are not randomly distributed in the real world, if the uncertainty in the measurement process is excluded, the most likely cause of this phenomenon is that the state of leaf aggregation on the horizontal plane has not been described, i.e., the clumping index [61]. The calculations of the clumping index and gap probabilities are inseparable. For the calculation of gap probabilities, we used a penetration function ($P = e^{-ks}$ in Supplementary Materials B-2). This function is based on the assumption that leaves are randomly distributed in the canopy. Therefore, there may be deviations between this assumption and the actual canopy, which implies that the equation describing the gap probabilities may need to be further refined considering the clumping index ($P = e^{-ks\Omega}$, where Ω is the clumping index). The clumping index is a hot topic in the field of radiation regimes and agricultural measuring equipment [62–65]. Compared with coniferous forests, the degree of leaf aggregation of crops is not very obvious. Therefore, considering the computational complexity, and whether the clumping index should be the main focus in improving the model, requires further study.

5. Conclusions

In this study, we established a row model to accurately estimate the reflectance by considering the multiple scattering framework. We validated the row model using both computer simulations and in situ measurements. The validation results show that the row model can be used to simulate crop canopy reflectance at different growth stages, with an accuracy comparable to the computer simulations. The study modified the calculation accuracy of single scattering near the hotspot by considering the calculation of the two-overlapping relationship and hot kernel function. Moreover, the row model can be successfully used to simulate the multiple-scattering contribution by considering multiple-scattering equations based on the RT approach. Therefore, this can address the underestimation problem in row crops by ignoring multiple scattering calculations in the GO approach. Our results demonstrate that the multiple-scattering contribution is a very important process that needs to be considered in row modeling based on the GO approach. This study provided a potential mechanism for remote sensing inversion based on the physical model.

Supplementary Materials: The following are available online at http://www.mdpi.com/2072-4292/12/21/3600/s1 and (https://zenodo.org/record/4171572#.X54e_1gzbIU) doi:10.5281/zenodo.4171572.

Author Contributions: Conceptualization, X.M. and Y.L.; methodology, X.M.; validation, X.M.; writing—original draft preparation, X.M.; writing—review and editing, X.M. and Y.L.; supervision, Y.L.; All authors have read and agreed to the published version of the manuscript.

Funding: This research received no external funding.

Acknowledgments: The measured data in this paper were supported by the Watershed Allied Telemetry Experiment Research (WATER). Great thanks to Guohua Huang for providing the RGM model and related computer code. Finally, great thanks to help from Tiejun Wanng.

Conflicts of Interest: The authors declare no conflict of interest.

References

1. Moran, M.S.; Inoue, Y.; Barnes, E.M. Opportunities and limitations for image-based remote sensing in precision crop management. *Remote Sens. Environ.* **1997**, *61*, 319–346. [CrossRef]
2. Liang, S.L. *Quantitative Remote Sensing of Land Surfaces*; John Wiley-Sons, Inc.: Hoboken, NJ, USA, 2004; pp. 413–415.
3. Li, X.; Strahler, A.H. Geometric-Optical Modeling of a Conifer Forest Canopy. *IEEE Trans. Geosci. Remote Sens.* **1985**, *23*, 705–721. [CrossRef]
4. Pinty, B.; Gobron, N.; Widlowski, J.L.; Gerstl, S.A.W.; Verstraete, M.M.; Antunes, M.; Bacour, C.; Gascon, F.; Gastellu, J.P.; Goel, N. Radiation transfer model intercomparison (RAMI) exercise. *J. Geophys. Res. Atmos.* **2001**, *106*, 523–538. [CrossRef]

5. Pinty, B.; Widlowski, J.L.; Taberner, M.; Gobron, N.; Verstraete, M.M.; Disney, M.; Gascon, F.; Gastellu, J.P.; Jiang, L.; Kuusk, A. Radiation Transfer Model Intercomparison (RAMI) exercise: Results from the second phase. *J. Geophys. Res. Atmos.* **2004**, *109*, 523–538. [CrossRef]

6. Widlowski, J.L.; Taberner, M.; Pinty, B.; Bruniquel-Pinel, V.; Disney, M.; Fernandes, R.; Gastellu-Etchegorry, J.P.; Gobron, N.; Kuusk, A.; Lavergne, T. The third Radiation transfer Model Intercomparison (RAMI) exercise: Documenting progress in canopy reflectance modelling. *J. Geophys. Res. Atmos.* **2007**, *112*, 139–155. [CrossRef]

7. Ross, J. *The Radiation Regime and Architecture of Plant Stands*; Springer: Berlin/Heidelberg, Germany, 1981.

8. Yang, X.; Short, T.H.; Fox, R.D.; Bauerle, W.L. Plant architectural parameters of a greenhouse cucumber row crop. *Agric. For. Meteorol.* **1990**, *51*, 93–105. [CrossRef]

9. Verbrugghe, M.; Cierniewski, J. Effects of Sun and view geometries on cotton bidirectional reflectance. Test of a geometrical model. *Remote Sens. Environ.* **1995**, *54*, 189–197. [CrossRef]

10. Annandale, J.G.; Jovanovic, N.Z.; Campbell, G.S.; Du, S.N.; Lobit, P. Two-dimensional solar radiation interception model for hedgerow fruit trees. *Agric. For. Meteorol.* **2004**, *121*, 207–225. [CrossRef]

11. Pieri, P. Modelling radiative balance in a row-crop canopy: Cross-row distribution of net radiation at the soil surface and energy available to clusters in a vineyard. *Ecol. Model.* **2010**, *221*, 802–811. [CrossRef]

12. Norman, J.M.; Welles, J.M. Radiative Transfer in an Array of Canopies. *Agron. J.* **1983**, *75*, 481–488. [CrossRef]

13. Ni, W.; Li, X.; Woodcock, C.E.; Caetano, M.R.; Strahler, A.H. An analytical hybrid GORT model for bidirectional reflectance over discontinuous plant canopies. *IEEE Trans. Geosci. Remote Sens.* **2002**, *37*, 987–999.

14. Kimes, D.S. Remote sensing of row crop structure and component temperatures using directional radiometric temperatures and inversion techniques. *Remote Sens. Environ.* **1983**, *13*, 33–55. [CrossRef]

15. Jackson, R.D.; Reginato, R.J.; Pinter, P.J.; Idso, S.B. Plant canopy information extraction from composite scene reflectance of row crops. *Appl. Opt.* **1979**, *18*, 3775–3782. [CrossRef] [PubMed]

16. Yan, B.Y.; Xu, X.R.; Fan, W.J. A unified canopy bidirectional reflectance (BRDF) model for row crops. *Sci. China Earth Sci.* **2012**, *55*, 824–836. [CrossRef]

17. Zhou, K.; Guo, Y.; Geng, Y.; Zhu, Y.; Cao, W.; Tian, Y. Development of a Novel Bidirectional Canopy Reflectance Model for Row-Planted Rice and Wheat. *Remote Sens.* **2014**, *6*, 7632–7659. [CrossRef]

18. Nilson, T. A theoretical analysis of the frequency of gaps in plant stands. *Agric. Meteorol.* **1971**, *8*, 25–38. [CrossRef]

19. Kuusk, A. The hot-spot effect of a uniform vegetative cover. *Sov. J. Remote Sens.* **1985**, *3*, 645–658.

20. Qin, W.; Goel, N.S. An evaluation of hotspot models for vegetation canopies. *Remote Sens. Rev.* **1995**, *13*, 121–159. [CrossRef]

21. Li, X.; Strahler, A.H. Modeling the gap probability of a discontinuous vegetation canopy. *IEEE Trans. Geosci. Remote Sens.* **1988**, *26*, 161–170. [CrossRef]

22. Chen, J.M.; Leblanc, S.G. A four-scale bidirectional reflectance model based on canopy architecture. *Geosci. Remote Sens. IEEE Trans.* **1997**, *35*, 1316–1337. [CrossRef]

23. Chen, J.M.; Li, X.; Nilson, T.; Strahler, A. Recent advances in geometrical optical modelling and its applications. *Urban Stud.* **2013**, *50*, 1403–1422. [CrossRef]

24. Li, X.; Strahler, A.H. Geometric-Optical Bidirectional Reflectance Modeling of a Conifer Forest Canopy. *IEEE Trans. Geosci. Remote Sens.* **1986**, *24*, 906–919. [CrossRef]

25. Chen, J.M.; Cihlar, J. A hotspot function in a simple bidirectional reflectance model for satellite applications. *J. Geophys. Res. Atmos.* **1997**, *102*, 25907–25914. [CrossRef]

26. Chen, J.M.; Leblanc, S.G. Multiple-Scattering Scheme Useful for Geometric Optical Modeling. *IEEE Trans. Geosci. Remote Sens.* **2001**, *39*, 1061–1071. [CrossRef]

27. Li, X.; Strahler, A.H.; Woodcock, C.E. A hybrid geometric optical-radiative transfer approach for modeling albedo and directional reflectance of discontinuous canopies. *IEEE Trans. Geosci. Remote Sens.* **1995**, *33*, 466–480. [CrossRef]

28. Chen, L.F.; Liu, Q.H.; Fan, W.J.; Li, X.W.; Xiao, Q.; Yan, G.J.; Tian, G.L. A bi-directional gap model for simulating the directional thermal radiance of row crops. *Sci. China Earth Sci.* **2002**, *45*, 1087–1098. [CrossRef]

29. Yan, G.J.; Jiang, L.M.; Wang, J.D.; Chen, L.F.; Li, X.W. Thermal bidirectional gap probability model for row crop canopies and validation. *Sci. China Earth Sci.* **2003**, *46*, 1241–1249. [CrossRef]

30. Ma, X.; Wang, T.; Lu, L. A Refined Four-Stream Radiative Transfer Model for Row-Planted Crops. *Remote Sens.* **2020**, *12*, 1290. [CrossRef]

31. Zhao, F.; Gu, X.; Verhoef, W.; Wang, Q.; Yu, T.; Liu, Q.; Huang, H.; Qin, W.; Chen, L.; Zhao, H. A spectral directional reflectance model of row crops. *Remote Sens. Environ.* **2010**, *114*, 265–285. [CrossRef]

32. Dorigo, W.A. Improving the Robustness of Cotton Status Characterisation by Radiative Transfer Model Inversion of Multi-Angular CHRIS/PROBA Data. *IEEE J. Sel. Top. Appl. Earth Obs. Remote Sens.* **2012**, *5*, 18–29. [CrossRef]

33. Verhoef, W. *Theory of Radiative Transfer Models Applied in Optical Remote Sensing of Vegetation Canopies*; Landbouw Universiteit Wageningen: Wageningen, The Netherlands, 1998.

34. Campbell, G.S. Derivation of an angle density function for canopies with ellipsoidal leaf angle distributions. *Agric. For. Meteorol.* **1990**, *49*, 173–176. [CrossRef]

35. Wang, W.M.; Li, Z.L.; Su, H.B. Comparison of leaf angle distribution functions: Effects on extinction coefficient and fraction of sunlit foliage. *Agric. For. Meteorol.* **2007**, *143*, 106–122. [CrossRef]

36. Liou, K.N. *An Introduction to Atmospheric Radiation*; Academic Press: London, UK, 2002; pp. 1–28.

37. Wang, Q.; Li, P. Canopy vertical heterogeneity plays a critical role in reflectance simulation. *Agric. For. Meteorol.* **2013**, *169*, 111–121. [CrossRef]

38. Verhoef, W. Earth observation modeling based on layer scattering matrices. *Remote Sens. Environ.* **1985**, *17*, 165–178. [CrossRef]

39. Lang, A.R.G.; Yueqin, X.; Norman, J.M. Crop structure and the penetration of direct sunlight. *Agric. For. Meteorol.* **1985**, *35*, 83–101. [CrossRef]

40. Antyufeev, V.S.; Marshak, A.L. Inversion of Monte Carlo model for estimating vegetation canopy parameters. *Remote Sens. Environ.* **1990**, *33*, 201–209. [CrossRef]

41. Hapke, B. Bidirectional reflectance spectroscopy 1: Theory. *J. Geophys. Res. Atmos.* **1981**, *86*, 3039–3054. [CrossRef]

42. Jacquemoud, S.; Baret, F.; Hanocq, J.F. Modeling spectral and bidirectional soil reflectance. *Remote Sens. Environ.* **1992**, *41*, 123–132. [CrossRef]

43. Goel, N.S.; Rozehnal, I.; Thompson, R.L. A computer graphics based model for scattering from objects of arbitrary shapes in the optical region. *Remote Sens. Environ.* **1991**, *36*, 73–104. [CrossRef]

44. Liu, Q.; Huang, H.; Qin, W.; Fu, K.; Li, X. An Extended 3-D Radiosity–Graphics Combined Model for Studying Thermal-Emission Directionality of Crop Canopy. *IEEE Trans. Geosci. Remote Sens.* **2007**, *45*, 2900–2918. [CrossRef]

45. Li, X.; Ma, M.G.; Wang, J.; Liu, Q.; Che, T.; Hu, Z.Y.; Xiao, Q.; Liu, Q.H.; Su, P.X.; Chu, R.Z. Simultaneous remote sensing and ground-based experiment in the Heihe River Basin: Scientific objectives and experiment design. *Adv. Earth Sci.* **2008**, *23*, 897–914.

46. Sandoval, C.; Kim, A.D. Extending generalized Kubelka-Munk to three-dimensional radiative transfer. *Appl. Opt.* **2015**, *54*, 7045–7053. [CrossRef]

47. Fan, W.; Yan, G.; Xin, X.; Tao, X.; Yan, B.; Yao, Y.; Chen, L.; Ren, H.; Wang, H.; Zhou, H.; et al. *WATER: Dataset of Spectral Reflectance Observations in the Yingke Oasis and Huazhaizi Desert Steppe Foci Experimental Areas*; National Tibetan Plateau Data Center: Beijing, China, 2013. [CrossRef]

48. Chen, L.; Yan, G.; Fan, W.; Ren, H.; Tao, X.; Zhang, W.; Wang, H.; Xin, X.; Zhang, Y. *WATER: Dataset of BRDF Observations in the Yingke Oasis and Huazhaizi Desert Steppe foci Experimental Areas*; National Tibetan Plateau Data Center: Beijing, China, 2013. [CrossRef]

49. Fan, W.; Xin, X.; Tao, X.; Liu, S.; Zhou, C.; Chen, L.; Guo, X.; Zou, J.; Tao, X. *WATER: Dataset of Ground Truth Measurement Synchronizing with PROBA CHRIS in the Yingke Oasis and Huazhaizi Desert Steppe Foci Experimental Areas on Jun 22, 2008*; National Tibetan Plateau Data Center: Beijing, China, 2014. [CrossRef]

50. Yan, G.; Zhang, W.; Wang, H.; Ren, H.; Chen, L.; Qian, Y.; Wang, J.; Wang, T. *WATER: Dataset of Vegetation Cover Fraction Observations in the Yingke Oasis, Huazhaizi Desert Steppe and Biandukou Foci Experimental Areas*; National Tibetan Plateau Data Center: Beijing, China, 2013. [CrossRef]

51. Fan, W.; Xin, X.; Yan, G.; Wang, J.; Tao, X.; Yao, Y.; Yan, B.; Shen, X.; Zhou, C.; Li, L.; et al. *WATER: Dataset of LAI Measurements in the Yingke Oasis and Huazhaizi Desert Steppe Foci Experimental Areas*; National Tibetan Plateau Data Center: Beijing, China, 2013. [CrossRef]

52. Yao, Y.J.; Fan, W.J.; Liu, Q.; Li, L.; Yao, X.; Xin, X.Z.; Liu, X.H. Improved harvesting method for corn LAI measurement in corn whole growth stages. *Trans. CSAE* **2010**, *26*, 189–194.

53. Matthew, M.W.; Adler-Golden, S.M.; Berk, A.; Richtsmeier, S.C.; Hoke, M.P. Status of Atmospheric Correction using a MODTRAN4-Based Algorithm. *Proc. Spie Int. Soc. Opt. Eng.* **2000**, *4049*, 11.

Remote Sens. **2020**, *12*, 3600

54. Goel, N.S.; Grier, T. Estimation of canopy parameters for inhomogeneous vegetation canopies from reflectance data. III. TRIM: A model for radiative transfer in heterogeneous three-dimensional canopies. *Int. J. Remote Sens.* **1986**, *25*, 255–293. [CrossRef]

55. Goel, N.S.; Grier, T. Estimation of canopy parameters of row planted vegetation canopies using reflectance data for only four view directions. *Remote Sens. Environ.* **1987**, *21*, 37–51. [CrossRef]

56. Myneni, R.B.; Kanemasu, E.T. The hot spot of vegetation canopies. *J. Quant. Spectrosc. Radiat. Transf.* **1988**, *40*, 165–168. [CrossRef]

57. Jupp, D.L.B.; Strahler, A.H. A hotspot model for leaf canopies. *Remote Sens. Environ.* **1992**, *38*, 193–210.

58. Nicodemus, F.E.; Richmond, J.C.; Hsia, J.J.; Ginsberg, I.W.; Limperis, T. *Geometrical Considerations and Nomenclature for Reflectance*; Department of Commerce, National Bureau of Standards: Washington, DC, USA, 1977.

59. Oreskes, N.; Shrader-Frechette, K.; Belitz, K. Verification, Validation, and Confirmation of Numerical Models in the Earth Sciences. *Science* **1994**, *263*, 641–646. [CrossRef]

60. Widlowski, J.-L.; Mio, C.; Disney, M.; Adams, J.; Andredakis, I.; Atzberger, C.; Brennan, J.; Busetto, L.; Chelle, M.; Ceccherini, G. The fourth phase of the radiative transfer model intercomparison (RAMI) exercise: Actual canopy scenarios and conformity testing. *Remote Sens. Environ.* **2015**, *169*, 418–437. [CrossRef]

61. Chen, J.M.; Black, T.A. Foliage area and architecture of plant canopies from sunfleck size distributions. *Agric. For. Meteorol.* **1992**, *60*, 249–266. [CrossRef]

62. Chen, J.M.; Cihlar, J. Plant canopy gap-size analysis theory for improving optical measurements of leaf-area index. *Appl. Opt.* **1995**, *34*, 6211–6222. [CrossRef]

63. Ryu, Y.; Nilson, T.; Kobayashi, H.; Sonnentag, O.; Law, B.E.; Baldocchi, D.D. On the correct estimation of effective leaf area index: Does it reveal information on clumping effects? *Agric. For. Meteorol.* **2010**, *150*, 463–472. [CrossRef]

64. Ryu, Y.; Sonnentag, O.; Nilson, T.; Vargas, R.; Kobayashi, H.; Wenk, R.; Baldocchi, D.D. How to quantify tree leaf area index in an open savanna ecosystem: A multi-instrument and multi-model approach. *Agric. For. Meteorol.* **2010**, *150*, 63–76. [CrossRef]

65. Lang, A.R.G.; Xiang, Y. Estimation of leaf area index from transmission of direct sunlight in discontinuous canopies. *Agric. For. Meteorol.* **1986**, *37*, 229–243. [CrossRef]

Publisher's Note: MDPI stays neutral with regard to jurisdictional claims in published maps and institutional affiliations.

 *remote sensing* 

Article

A Quantitative Monitoring Method for Determining Maize Lodging in Different Growth Stages

HaiXiang Guan [1], HuanJun Liu [1,2], XiangTian Meng [1], Chong Luo [2], YiLin Bao [1], YuYang Ma [1], ZiYang Yu [1] and XinLe Zhang [1,*]

[1] School of Public Administration and Law, Northeast Agricultural University, Harbin 150030, China; haixiangguan@yeah.net (H.G.); liuhuanjun@neigae.ac.cn (H.L.); mxt0123neau@yeah.net (X.M.); byl1211neau@yeah.net (Y.B.); myy18846918996@163.com (Y.M.); ziyangyu@yeah.net (Z.Y.)
[2] Northeast Institute of Geography and Agroecology, Chinese Academy of Sciences, Changchun 130012, China; luochong@iga.ac.cn
* Correspondence: xinlezhang@neau.edu.cn

Received: 18 August 2020; Accepted: 22 September 2020; Published: 25 September 2020

Abstract: Many studies have achieved efficient and accurate methods for identifying crop lodging under homogeneous field surroundings. However, under complex field conditions, such as diverse fertilization methods, different crop growth stages, and various sowing periods, the accuracy of lodging identification must be improved. Therefore, a maize plot featuring different growth stages was selected in this study to explore an applicable and accurate lodging extraction method. Based on the Akaike information criterion (AIC), we propose an effective and rapid feature screening method (AIC method) and compare its performance using indexed methods (i.e., variation coefficient and relative difference). Seven feature sets extracted from unmanned aerial vehicle (UAV) images of lodging and nonlodging maize were established using a canopy height model (CHM) and the multispectral imagery acquired from the UAV. In addition to accuracy parameters (i.e., Kappa coefficient and overall accuracy), the difference index (DI) was applied to search for the optimal window size of texture features. After screening all feature sets by applying the AIC method, binary logistic regression classification (BLRC), maximum likelihood classification (MLC), and random forest classification (RFC) were utilized to discriminate among lodging and nonlodging maize based on the selected features. The results revealed that the optimal window sizes of the gray-level cooccurrence matrix (GLCM) and the gray-level difference histogram statistical (GLDM) texture information were 17×17 and 21×21, respectively. The AIC method incorporating GLCM texture yielded satisfactory results, obtaining an average accuracy of 82.84% and an average Kappa value of 0.66 and outperforming the index screening method (59.64%, 0.19). Furthermore, the canopy structure feature (CSF) was more beneficial than other features for identifying maize lodging areas at the plot scale. Based on the AIC method, we achieved a positive maize lodging recognition result using the CSFs and BLRC. This study provides a highly robust and novel method for monitoring maize lodging in complicated plot environments.

Keywords: lodging; unmanned aerial vehicle (UAV); canopy structure feature; Akaike information criterion (AIC) method; difference index (DI); texture

1. Introduction

Maize plays an important role among the world's cereal crops; it is used in food, fodder, and bioenergy production. High and stable maize yields are crucial to global food security. Lodging is a natural plant condition that reduces the yield and quality of various crops. In terms of different displacement positions, crop lodging can be divided into stem lodging (Lodging S) and root lodging (Lodging R). Lodging S involves the bending of crop stems from their upright position, while Lodging

R refers to damage or failure to the plant's root-soil anchorage system [1]. Due to maize lodging, harvest losses could be as high as 50% [2]. Timely and exact identification of maize lodging is essential for estimating yield loss, making comprehensive production decisions and supporting insurance compensation. The traditional approaches to lodging investigation rely on manual measurements made at plots, which is time consuming, laborious, inefficient, and unsuitable for large-scale lodging surveys.

Currently, given the rapid evolution of remote sensing techniques, an increasing number of scholars have used remote sensing imagery combined with various feature-extraction tools to obtain crop lodging information. Li et al. [3] investigated the potential for maize lodging extraction and feature screening from satellite remote sensing imagery based on the difference in texture values between lodging and nonlodging areas. Based on the changes in vegetable index (VI) features under lodging and nonlodging maize, Wang et al. [2] reported that the correlation between the changes and the lodging percentage could be used as a standard to select effective features. Crop lodging is frequently accompanied by adverse weather; hence, there is no guarantee that seasonable images can be obtained successfully due to the limited capabilities of optical satellites in poor weather conditions. In contrast to optical sensors, the spaceborne synthetic aperture radar (SAR) technique is not only robust to severe weather adverse effects but also provides abundant information regarding the structure of vegetation. SAR technology is currently widely used to discriminate between lodging and nonlodging crops. By utilizing the fully polarimetric C-band radar images, Yang et al. [4] explored the difference between lodging and nonlodging wheat areas in various growth periods. Chen et al. [5] reported on polarimetric features extracted from consecutive RADARSAT-2 data to identify sugarcane lodging areas. However, due to its low spatial resolution, the SAR technique is more applicable to large and relatively homogeneous crop planting areas than to small and heterogeneous plots [6].

To precisely extract lodging areas within a small patchwork field, unmanned aerial vehicles (UAV) can acquire remote sensing images that offer both high temporal and spatial resolution and have strong operability [7]. At present, researchers usually employ UAV images and their features acquired shortly after lodging to identify crop lodging at the field scale. The identified crops are mainly in the vigorous growth period. During this period, lodging crops and nonlodging crops have similar growth conditions, but their leaf color is quite different, which can be observed in UAV images. Therefore, high-accuracy identification results of maize lodging can be achieved by applying only red-green-blue (RGB) or multispectral images.

Using UAV-collected multispectral, RGB, and thermal infrared imagery, image features such as texture features, VI, and canopy structure features can be derived from these UAV images. Based on extracted feature information, single- and multiple-class feature sets (SFS and MFS, respectively) can be constructed and used for lodging identification [8]. SFSs can be used to promptly and efficiently recognize lodging areas due to their low data dimensions and straightforward computation. RGB, multispectral imagery, and VI were introduced to extract crop lodging based on the leaf color discrepancy between lodging and nonlodging areas [9]. However, this discrepancy experiences interference from different crop varieties and fertilization treatments. To address this situation, Liu et al. [10] reported that although thermal infrared images have low spatial resolution, features extracted from them are beneficial for lodging assessment. In addition, the patterns, shapes, and sizes of vegetation in UAV imagery can be quantitatively described by analyzing texture characteristics; thus, they have been widely employed in lodging assessment research [11]. However, these features also have some defects. For instance, it is difficult to determine the optimal scale and moving direction for the texture window. More recently, the canopy structure feature (CSF) has been broadly applied to determine whether lodging has occurred and, ultimately, to estimate lodging severity, which fosters accurate monitoring of crop lodging [12]. However, these features are likely to be limited by adverse field environments and high experimental consumption; therefore, SFS often cannot fully reflect the crop lodging properties under intricate field environments, and MFSs have been used more frequently in recent years. For example, by combining the digital surface model (DSM) and texture information, crop lodging recognition can be performed with reliable accuracy [13]. Based on a feature set containing color and texture features, adding temperature information also greatly improved the recognition accuracy of indica rice lodging [10]. In addition, the probability that lodging has occurred can be predicted by using a feature set containing the canopy structure and the VI feature [14].

Nevertheless, MFSs have a data redundancy defect that causes greater consumption of computing resources and reduces the identification efficiency, necessitating optimal feature selection. The number of studies regarding screening methods for lodging identification is limited. The existing methods rely mainly on the difference evaluation index, which has less computational overhead [1,15,16]. However, these methods do not consider the interactions between features, and there is no clear standard for determining the most appropriate feature dimension. While the feature numbers can be explicitly determined by utilizing single-feature probability values based on a Bayesian network [13], this approach assumes that all features are mutually independent and that no correlation occurs between the selected features. Han et al. [14] reported that univariate and multivariate logistic regression analysis methods can be used to screen lodging features. Although this method explores the underlying relationship between the outcome and the selected factors, it is not suitable for high-dimensional feature sets because of its complex screening process.

Therefore, the main purpose of this study is to construct a simple and efficient feature screening method for lodging recognition under complicated field environments that also considers the interactions between features. We verify the feasibility and stability of the proposed method under both high- and low-dimensional circumstances. The primary objectives are (1) to analyze the changes in lodging extraction accuracy under different texture window sizes and to determine the optimal window size that results in the maximal accuracy; (2) to determine the optimal screening method by comparing the accuracy of maize lodging results based on different feature screening methods; and (3) to extract maize lodging areas by using the optimal feature screening method and various classifications.

2. Materials and Methods

2.1. Study Area

The maize field trial was conducted at Youyi Farm, Youyi County, Heilongjiang Province in China (46°44′N, 131°40′E), which covers an area of approximately 17.29 hectares. The area has a temperate continental monsoon climate with four distinct seasons, rain and heat occur during the same period, and the soil type is dark brown soil with high fertility.

The maize planted in this experimental plot was sown on 6 May 2019, and harvested on 14 September 2019. Maize lodged on 28 July 2018, and multispectral imagery (Figure 1) was acquired on 12 September 2019. In the study site, which is affected by natural factors such as high winds, difficult terrain, and soil, water, and fertilizer transportation problems, maize development is diverse and exists in both development and mature periods. In Figure 1c, the typical yellow areas show maize in the mature period, and its growth is better than that of the typical green area maize in the development period. Maize lodging can be divided into the two groups shown in Figure 2: Lodging S and Lodging R. Compared with Lodging S, Lodging R is more serious. The nonlodging maize includes three classes (Figure 2): maize with green leaves (No-Lodging G), maize with yellow leaves (No-Lodging Y), and maize with yellowish leaves (No-Lodging LY). The growth of No-Lodging LY is better than that of the other two categories. Influenced by the edge effect [17], the growth status of No-Lodging Y is the worst.

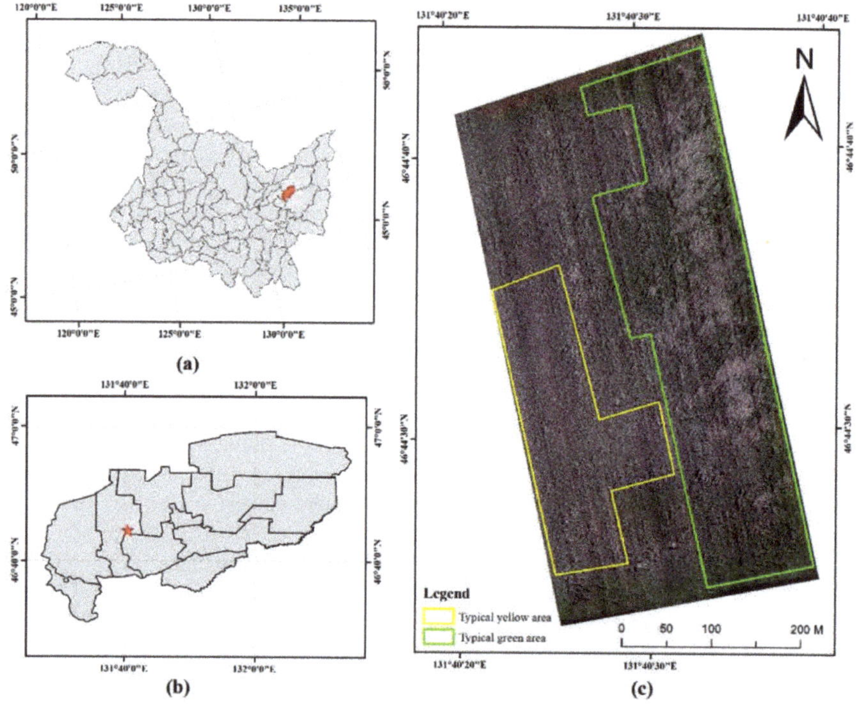

Figure 1. (**a**) The location of Youyi County in China (red area); (**b**) the location of the study site in Youyi County (red area); and (**c**) the multispectral imagery of the study site.

Figure 2. Types of lodging and nonlodging maize.

2.2. Image Collection

This study used a RedEdge 3 camera equipped with a DJI M600 Pro UAV system (SZ DJI Technology Co., Shenzhen, China) to record multispectral images with five bands (Table 1) on 12 September 2018, and 10 April 2019. The camera has a resolution of 1280 × 960 pixels and a lens focal length of 5.5 mm. By using DJI GS Pro 1.0 (version 1.0, SZ DJI Technology Co., Shenzhen, China) software, the overflight speeds were set to 7 m/s and a flight altitude 110 m above ground level. The forward and lateral overlaps were configured to 80% and 79%, respectively, which yielded UAV images with a ground sampling distance of 7.5 cm.

Table 1. The band information of RedEdge 3 camera.

Band Name	Center Wavelength (nm)	Band Width (nm)
Blue	475	20
Green	560	20
Red	668	10
Red-edge	717	10
Near-infrared	840	40

2.3. Data Processing

The agricultural multispectral processing module in Pix4Dmapper 4.4 was used to automatically generate multispectral imagery and DSM. The fundamental working flow includes (1) radiometric calibration, (2) structure-from-motion (SFM) processing, (3) spatial correction, and (4) orthoimage and DSM generation. By employing radiometric calibration images and reflectance values, radiometric calibration can be performed for every image. Using the overlapping images, SFM processing recreates their positions and orientations in a densified point cloud by using automatic aerial triangulation. Based on four ground control points (manually collected before image acquisition), the three-dimensional point cloud was corrected during the stitching process and subsequently used to construct orthoimages and DSM.

Based on the multispectral imagery, the study area was divided into 119 blocks using the created fishnet function provided by ArcGIS 10.2. Then, 89 blocks were selected as the modeling areas and 30 as the verification areas. In the modeling areas, 643 maize lodging samples and 654 nonlodging samples were collected, marked as "1" and "0," respectively. These maize samples were subjected to binary logistic regression classification (BLRC), texture information extraction and feature screening. We also selected 50 samples each of lodging and nonlodging maize from the verification areas to calculate the precision of all the lodging identification results in this study.

2.4. Feature Set Construction

Three SFSs for maize lodging recognition were extracted from multispectral imagery that included eight spectral features, 41 texture features, and nine CSFs. Moreover, we constructed all the possible feature sets of these three SFSs to build the MFSs.

2.4.1. Spectral Feature Set

In general, the spectral reflectance of vegetation changes after lodging. To take full advantage of this change, five multispectral image bands were added to the spectral feature set. The VI quantitatively reflects the growing situation of vegetation and the vegetation coverage [18,19]. In this study, three types of VI (see Table 2) were derived to expand the difference between lodging and nonlodging areas. Based on the above image features, a spectral feature set was constructed to identify maize lodging.

Table 2. The calculation formula and implication of the vegetable index (VI).

Type	Calculation Formula	Implication	References
ExG	$R = \frac{r}{r+g+b}$ $G = \frac{g}{r+g+b}$ $B = \frac{b}{r+g+b}$ $ExG = 2*G - R - B$	Reflects the chlorophyll content and growth status of vegetation	[20,21]
NDVI	$NDVI = \frac{nir-r}{nir+r}$	Describes the vegetation's cover and heath condition	[22,23]
RENDVI	$RENDVI = \frac{re-r}{re+r}$	Indicates the biomass and leaf structure of vegetation	[24,25]

Note: ExG, NDVI, and RENDVI are excess green index, normalized vegetation index, and red-edge normalized difference vegetation index, respectively. r, g, b, nir, and re represent the blue band, green band, red band, red-edge band, and near-infrared band from multispectral images, respectively.

2.4.2. Canopy Structure Feature Set

Based on the multispectral images of maize lodging and bare land collected by UAV, we constructed two DSMs of the study site. The DSM of bare land, which is not affected by maize height, can be considered as the digital elevation model (DEM) of the study area. The DSM of maize lodging was subtracted from the DEM to generate a canopy height model (CHM), which was employed to calculate nine CSFs (Table 3).

Table 3. The calculation formula and implication of canopy structure features (CSFs).

Name	Calculation Formula	Implication	References
H_{max}	$H_{max} = Max\left(h_{1,2,3\ldots n}\right)$	These characteristics can reflect the growth status of vegetation through the canopy height	[26,27]
H_{mean}	$H_{mean} = \frac{1}{n}\sum_{i=1}^{n} h_i$		
H_{cv}	$H_{cv} = \frac{H_{std}}{H_{mean}}$	Measures the vertical heterogeneity of the vegetation canopy structure	[28,29]
H_{std}	$H_{std} = \sqrt{\frac{1}{n}\sum_{i=1}^{n}(h_i - h_{mean})}$		
$H_{25}, H_{50}, H_{75}, H_{90}$	$(n-1)p\% = j+g,$ $p^{th} = h_j(g=0);$ $p^{th} = h_j * g + (1-g) * h_{j+1}$	Reflects the vertical distribution of the plant canopy height	[30]
CRR	$CRR = \frac{H_{mean}-H_{min}}{H_{max}-H_{min}}$	Describes the horizontal and vertical heterogeneity of the plant canopy structure	[31,32]

Note: $H_{1,2,3\ldots n}$ represents the canopy height of each pixel in each square meter area; n is the number of points; h_i is the height of points; p represents the percentiles of height; j and g represent the integer part and fractional part, respectively.

2.4.3. Texture Feature Set

Texture features can express the pattern, size, and shape of various objects by measuring the adjacency relationship and frequency of gray-tone changes of pixels [33]. Based on texture features, the spatial heterogeneity of remote sensing images can be expressed quantitatively [34]. The calculation approaches of texture features are usually divided into four types: structural, statistical, model-based, and transformation approaches. At present, the statistical approaches are widely used in lodging recognition research, and the high accuracy of lodging monitoring results can be derived. In addition, the statistical approaches have proved to be superior to structural and transformation approaches. Therefore, the different measurement levels in statistical approaches (the gray-level cooccurrence matrices and gray-level difference matrices) were selected to calculate the texture features of remote sensing images. The gray-level cooccurrence matrices (GLCM) and gray-level difference matrices

(GLDM) have been broadly used to quantitatively describe texture features that can be applied in image classification and parameter inversion [35].

To fully describe the details of the lodging and nonlodging maize in the image, we analyzed as many different types of texture features as possible based on the GLCMs and GLDMs. In this study, eight types of GLCM texture measures [36] (mean, variance, homogeneity, contrast, dissimilarity, entropy, second moment, and correlation) and five types of GLDM texture measures [37] (data range, mean, entropy, variance, and skewness) were extracted to construct a texture feature set to discriminate between lodging and nonlodging areas. Based on the average reflectivity of three bands (the red, green, and blue bands), red-edge, near-infrared bands and CHM, 41 texture measures were obtained. Specifically, we calculated only the mean texture features (GLCM and GLDM) of CHM to reduce data redundancy.

To speed up the processing, the texture features extracted from CHM were not used when determining the optimal texture window size. Fifteen different sizes of sliding windows were calculated based on the texture feature set without the mean texture features of CHM: $3 \times 3, 5 \times 5,$ $7 \times 7, 9 \times 9, 11 \times 11, 13 \times 13, 15 \times 15, 17 \times 17, 19 \times 19, 21 \times 21, 31 \times 31, 41 \times 41, 51 \times 51, 61 \times 61$ and 71×71. Based on maximum likelihood classification (MLC), we analyzed the variations in the Kappa coefficient and the overall accuracy with the different window sizes. The difference index (DI) value of the texture feature with the highest classification accuracy was calculated. Finally, the filter window corresponding to the minimum DI value was adopted as the optimal texture window size for maize lodging extraction. To analyze the suitability of the selected optimal texture window, the areas of single lodging and nonlodging maize in multispectral images were measured as 1.55 m^2 and 0.17 m^2, respectively. In addition, we measured the area of small-scale maize lodging as 2.21 m^2.

The formula for DI calculation is as follows:

$$DI = \frac{SD_1 + SD_2}{ABS(MN_1 - MN_2)} \tag{1}$$

where SD_1 and SD_2 are the standard deviations of the lodging and nonlodging samples, respectively, and MN_1 and MN_2 are the means of the lodging and nonlodging samples, respectively.

2.5. Feature Screening

2.5.1. Feature Screening Based on Akaike Information Criterion (AIC) Method

The Akaike information criterion (AIC), which is an information criterion based on the concept of entropy, can be used to select statistical models by evaluating the accuracy and complexity of the model [38,39]. In general, AIC must be combined with a logistic regression model to achieve feature selection. Therefore, in this study, we introduced binary logistic regression to describe the relationships between features, which can also help predict the probability of lodging.

In this study, the independent variables of the binary logistic regression model were the maize lodging recognition features, and the dependent variables were binary variables with a value of 1 or 0, representing lodging or nonlodging maize, respectively. The maximum likelihood method was used to estimate the parameters of the binary logistic regression model. Following this approach, the binary logistic regression model for predicting the occurrence probability of lodging is

$$P = \frac{e^{\beta_0 + \beta_1 z_1 + \ldots + \beta_m z_m}}{1 + e^{\beta_0 + \beta_1 z_1 + \ldots + \beta_m z_m}} \tag{2}$$

where $z_1, z_2, \ldots, z_m$ are the features to be screened, β_0 is the intercept, and $\beta_1, \beta_2 \ldots, \beta_m$ are regression coefficients.

The maize samples were divided into a training set and a testing set, and the AIC value of the model was calculated as follows:

$$AIC = -2lnL\left(\hat{\beta}_k \middle| y\right) + 2k \tag{3}$$

where *ln* stands for the natural logarithm. $L(\hat{\beta}_k|y)$ is the maximum likelihood function value of the logistic regression model, which indicates the probability that the model will result in a correct classification, and k is the number of parameters in the logistic regression model.

When calculating the AIC value of the binary logistic regression model, three-tenths of 1297 maize samples were randomly selected as the training set, and the remaining maize samples were used as the testing set. The dependent variables in the model were eliminated along the decreasing direction of the AIC value until the AIC value reached its lowest point. Finally, the parameters contained in the model with the lowest AIC values were regarded as the optimal features.

2.5.2. Feature Screening Based on Index Method

Variation coefficients and relative differences have a wide range of applications in screening and analyzing the image features of crop lodging [6,7]. The variation coefficient directly reflects the dispersion degree of lodging and nonlodging crops in the image features, while the relative difference indicates the difference between the lodging and nonlodging areas based on the image characteristics. Features suitable for lodging identification should have low variation coefficients and high relative differences.

This study adopts the variation coefficient and relative difference as two evaluation indicators for feature selection. First, the variation coefficient and relative difference of the lodging and nonlodging areas in features were calculated, and the ten features with the largest relative difference were selected as predictors for lodging recognition. Based on these predictors, the factor whose variation coefficient was less than 20.57% was selected as the optimal feature to ensure classification stability.

The formulas to calculate the variation coefficient and relative difference are as follows:

$$CV = \frac{sd}{mn} \times 100\% \tag{4}$$

$$RD = \frac{ABS(mn_1 - mn_2)}{mn_1} \times 100\% \tag{5}$$

where *CV* denotes the variation coefficient; *sd* and *mn* represent the standard deviation values of maize samples; *RD* denotes the relative difference; mn_1 and mn_2 represent the mean values of maize samples; and *ABS* is the absolute value algorithm.

2.6. Lodging Maize Identification

BLRC, MLC, and random forest classification (RFC) are specialized for maize lodging identification using image features. The samples used as inputs to the MLC and RFC were randomly collected by using the "region of interest" (ROI) tools provided by ENVI version 5.1. The distribution of the ROI regions was uniform, which made the samples highly representative of the lodging and nonlodging maize in the entire study area. Then, 27 and 28 ROI regions were selected for lodging and nonlodging maize samples, respectively. The total numbers of sample pixels representing lodging and nonlodging maize were 58,819 and 61,747 pixels, respectively. In terms of shadow, according to the predecessor's processing method [9], shadows with areas of more than 5.41 m^2 were placed into the lodging class to reduce the influence of shadows on the classification process.

2.6.1. Binary Logistic Regression Classification (BLRC)

Logistic regression is a generalized linear regression model that can explain the relationships between variables. This model has been extensively applied to classification and regression problems. Because lodging recognition is a binary classification problem in this study, we adopted a binary logistic regression model as a classifier [40].

In this binary logistic regression model, the maize sample values extracted from different image features form the independent variables. The predicted dependent variable is a function of the probability that maize lodging will occur. Moreover, there is a basic assumption about the dependent variable: the probability of the dependent variable takes the value of 1 (positive response). According to the logistic curve, the value of the dependent variable can be calculated using the following formula [41]:

$$P(y = 1|X) = \frac{\exp(\sum CX)}{1 + \exp(\sum CX)} \tag{6}$$

where P represents the probability of maize lodging (the dependent variable); X indicates the independent variables; and C represents the estimated parameters.

The logit transformation was performed to linearize the above model and eliminate the 0/1 boundaries for the original dependent variable, which is the probability. The new dependent variable is boundless and continuous within the range from 0 to 1. After logit transformation, the above model is expressed by the following equation:

$$\ln\left(\frac{P}{(1-P)}\right) = C_0 + C_1 X_1 + C_2 X_2 + C_3 X_3 + \cdots + C_k X_k \tag{7}$$

where P represents the probability of maize lodging (the dependent variable), $X_1, X_2, X_3, \ldots, X_k$ are the independent variables, and $C_0, C_1, C_2, \ldots, C_k$ represent the estimated parameters.

The binary logistic regression model for predicting maize lodging probability encoded by R programming language was trained using R Studio software. The encoded program automatically determined the coefficients of all independent variables according to the relationship between independent and dependent variables. The fitted model was inputted into the "Band Math" module of ENVI software to obtain the lodging probability map of the study area. Finally, a uniform and optimal probability threshold was determined for all pixels to judge whether they represented lodging.

2.6.2. Maximum Likelihood Classification (MLC)

MLC is a nonlinear discriminant function based on the Bayesian rule that is suitable for processing low-dimensional data [42]. MLC assumes that a hyper ellipsoid decision volume can be applied to approximate the shapes of the data clusters. The mean vector and covariance matrix of specified unknown pixels in each class is calculated. The probability of each pixel's class is then analyzed, and the pixel is predicted to belong to the class with the maximum probability [43]. In this study, the MLC module from the ENVI software was used to identify maize lodging. The data scale factor, which is a major parameter in the classifier, was set to 1.0 in each classification. This is mainly because the UAV images and their characteristics involved in this study are floating-point data. The UAV image features and ROI regions were provided as the input data for the MLC to generate the maize lodging recognition results.

2.6.3. Random Forest Classification (RFC)

The random forest algorithm is an ensemble learning method based on the decision tree method that has been applied to both object classification and regression analysis of high-dimensional data [44]. In the random forest algorithm, all the decision trees are trained on a bootstrapped sample of the original training data. To completely divide the variable space, the split node in each decision tree is randomly selected from the input samples. Only two-thirds of the input samples are involved in the building of each decision tree. The remaining one-third of the samples, which are called out-of-bag (OOB) samples, are used to validate the accuracy of the prediction. Finally, the prediction result is obtained through a majority voting strategy of the individual decision trees [45–47]. In this study, the RFC program [48] embedded in ENVI software was employed to identify maize lodging, and two parameters were tuned to generate the optimal RFC model: the number of split nodes and the number of decision trees. The optimal parameters were determined according to the accuracy of the lodging classification results. In each classification, both were reset to adapt to different feature sets.

2.7. Accuracy Assessment of Lodging Detection Results

The Kappa coefficient is an index used to measure the spatial consistency of classification results; it explicitly reveals the spatial change in the classification results [49]. The overall accuracy (OA) reflects the quality of the classification result in terms of quantity [50]. In this study, the above indicators were chosen to evaluate the correctness of maize lodging detection results.

Some scholars believe that the Kappa coefficient is discontinuous and marginal in the evaluation of remote sensing image classification results [51]. However, these views are held only at the theoretical level and lack sufficient practical proof, especially in the study of binary classification based on remote sensing images. In general, remote sensing images and features are not considered to be independent, which is the main reason for the failure of confusion matrix and kappa coefficient [52]. The nonoverlapping ROI or sample points of lodging and nonlodging maize needed by model training were selected before image classification. Consequently, in each image feature, the values of lodging and nonlodging maize were independent and had no influence on each other. Supported by a large number of remote sensing application practices [53,54], this study uses the Kappa coefficient as the evaluation index for lodging recognition results.

2.8. Traditional Crop Lodging Identification Method

The main process of traditional crop lodging identification is as follows. First, different types of image features—color, spectral reflectance, vegetation index, GLCM texture, and canopy structure—are extracted from RGB and multispectral images. Certain image features are selected, and the optimal features are obtained by using the index method or prior knowledge. Finally, the optimal features and manually selected ROIs are input into the supervised classification model to obtain the recognition results of crop lodging in the plot.

There are three main differences between our new proposed method and the traditional method: (1) more image features (compared to less than 20 features in previous studies) are extracted to determine whether they are suitable for maize lodging identification under complicated field environments. (2) Both GLCM texture and GLDM texture are used for maize lodging recognition. In addition, the optimal GLCM and GLDM texture window sizes are determined. (3) The AIC method is used to screen the maize lodging identification features.

3. Results and Analysis

3.1. Optimal Window for Texture Features

Based on the identification results obtained by using GLCM and GLDM textures with different window scales, the variations in the Kappa coefficient and OA are shown in Figure 3, where the overall accuracy and Kappa coefficient clearly have the same trend using both GLCM and GLDM textures. The Kappa coefficient and OA of the GLCM texture are higher than those of the GLDM texture at window scales from 3 to 51, but the Kappa coefficient and OA of the GLCM texture become lower than those of GLDM at window scales from 51 to 71. The highest accuracies of the GLCM texture window sizes are 15, 17, 19, and 21 (the OA is 88% and the Kappa coefficient is 0.76), and their DI values are 14.68, 12.69, 14.11, and 13.76, respectively. When using the GLDM texture, the most accurate window sizes are 7, 21, and 31 (the OA is 85% and the Kappa coefficient is 0.70), and their DI values are 56.39, 11.18, and 12.64, respectively. Therefore, GLDM texture with a window size of 21 and GLCM texture with a window size of 17 are adopted for subsequent maize lodging recognition.

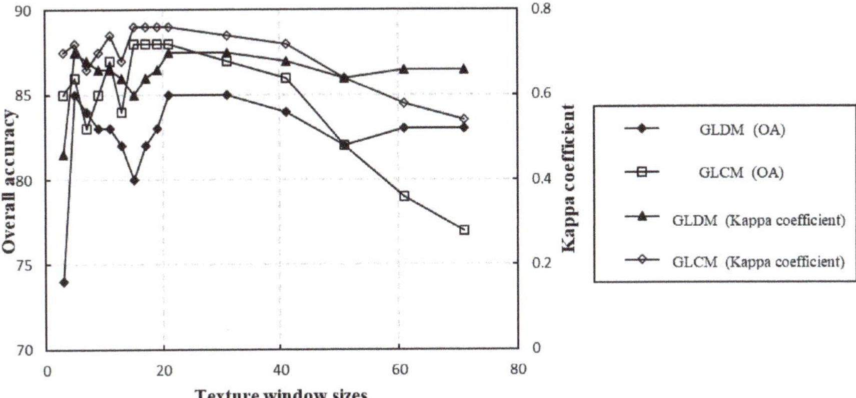

Figure 3. The changes in maize lodging recognition accuracy with gray-level cooccurrence matrix (GLCM) and gray-level difference matrix (GLDM) textures under different window sizes.

3.2. Performances of Different Feature Screening Methods

In this paper, two feature sets are established to compare the performance of the AIC method and index method: the GLCM texture features extracted from the blue, green, and red bands (RGB_GLCM) and the GLCM texture features extracted from blue, green, red, red-edge, and near-infrared bands (All_band_GLCM). When utilizing the AIC method to screen RGB_GLCM and All_band_GLCM, we obtained fourteen and sixteen satisfactory features, respectively. The optimal texture characteristics obtained by employing the index method are shown in Table 4.

Table 4. The selected features using index method-based GLCM texture.

Feature Set	Selected Feature	Relative Difference	Variation Coefficient of Lodging Maize	Variation Coefficient of Nonlodging Maize
	GLCM_B_Correlation	9.45%	8.20%	10.15%
RGB_GLCM	GLCM_G_mean	16.92%	14.28%	13.97%
	GLCM_B_mean	23.38%	17.17%	23.95%
	GLCM_G_mean	16.92%	14.28%	13.97%
All_band_GLCM	GLCM_B_mean	23.38%	17.17%	23.95%
	GLCM_Nir_Dissimilarity	20.22%	18.84%	20.25%

Note: GLCM_G_mean and GLCM_B_mean are GLCM mean textures derived from the green and blue bands, respectively. GLCM_B_Correlation and GLCM_Nir_Dissimilarity are the GLCM correlation texture of the blue band and GLCM correlation texture of the near-infrared band, respectively.

Tables 5 and 6 display the recognition accuracy of AIC and the index method under different classification algorithms. The Kappa coefficient and OA of the AIC method are generally higher than those of the index method. The AIC method combined with RGB_GLCM enabled estimation of maize lodging with higher accuracy than that of the index method (average Kappa coefficient is 0.71, average OA is 85.67%). In contrast, when the All_band_GLCM is used, the average Kappa coefficient and average OA of the index method fall to 0.15 and 57.67%, respectively, considerably lower than those of the AIC method. Furthermore, comparing the AIC and index methods clarified that the index method seriously overestimated the lodging area (Figure 4).

Table 5. The Kappa coefficient of the Akaike information criterion (AIC) and index methods.

Feature Set	Screening Method	Classification Algorithm			Mean
		MLC	RFC	BLRC	
RGB_GLCM	AIC	0.78	0.54	0.82	0.71
	Index	0.52	0.06	0.12	0.23
All_band_GLCM	AIC	0.70	0.40	0.70	0.60
	Index	0.30	0.06	0.10	0.15

Table 6. The overall accuracy (OA) of AIC and index methods.

Feature Set	Screening Method	Classification Algorithm			Mean
		MLC	RFC	BLRC	
RGB_GLCM	AIC	89.00%	77.00%	91.00%	85.67%
	Index	76.00%	53.00%	56.00%	61.67%
All_band_GLCM	AIC	85.00%	70.00%	85.00%	80.00%
	Index	65.00%	53.00%	55.00%	57.67%

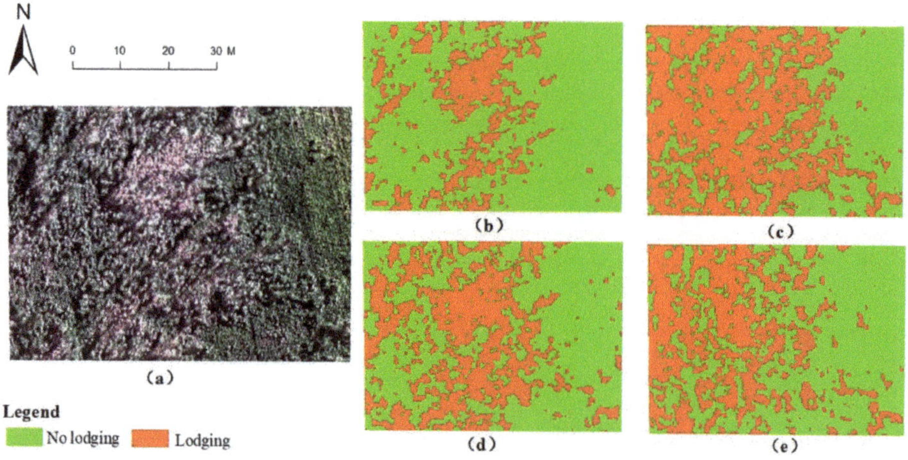

Figure 4. Maps of the distribution of lodging and nonlodging maize based on the AIC and index methods. (**a**) Part of the multispectral image of the study area; (**b**) Maize lodging area using AIC method and binary logistic regression classification (BLRC) under RGB_GLCM; (**c**) Maize lodging area using index method and maximum likelihood classification (MLC) under RGB_GLCM; (**d**) Maize lodging area using AIC method and MLC under All_band_GLCM; (**e**) Maize lodging area using index method and MLC under All_band_GLCM.

The analyzed data are displayed in Table 5, and Table 6 clearly shows that RGB_GLCM achieves higher classification accuracy than All_band_GLCM for each screening method. During feature selection, we found that although RGB_GLCM is included in All_band_GLCM, the features derived from one screening method are not the same. Consequently, under the condition of using the feature selection method, it is necessary to construct as many feature sets as possible to obtain the optimal lodging recognition features.

Table 7 illustrates the performance of the AIC and index method for discriminating lodging and nonlodging areas in different regions within the study location under the maximum likelihood classification method. The accuracy of the two methods differs only slightly in the green region. On the one hand, using RGB_GLCM, the Kappa coefficient and OA of the index method were 0.034 and 1.00% higher than those of the AIC method, respectively. The same precision can be obtained from these methods when employing RGB_GLCM. On the other hand, the AIC method resulted in a smaller error than the index method in the yellow area. In particular, when using the All_band_GLCM, the Kappa coefficient and OA of the AIC method are 0.53 and 16.00% higher than those of the index method, respectively.

Table 7. The accuracy assessment of AIC and index methods within different regions.

Feature Set	Screening Method	Yellow Area		Green Area	
		Kappa Coefficient	OA	Kappa Coefficient	OA
RGB_GLCM	AIC	0.53	86.00%	0.73	92.00%
	Index	0.50	85.00%	0.76	93.00%
All_band_GLCM	AIC	0.73	92.00%	0.73	92.00%
	Index	0.20	76.00%	0.73	92.00%

3.3. Identification Accuracy of Different Screening Methods

The number of features, AIC value, and Kappa coefficient of MLC before and after screening are presented in Table 8. The dimensions and AIC values of all feature sets are reduced by varying degrees after applying the AIC method. When the AIC method was used to screen the high-dimensional feature sets (number of features > 40), the feature dimensions and AIC values decreased significantly: the feature dimension decreased by 90.25% on average, and the AIC value decreased by 88.44% on average. However, for the low-dimensional feature sets, the number of features and the AIC values decreased only slightly after employing the AIC method: the feature dimension decreased by only 31.21% on average, and the AIC value decreased by only 5.83% on average. According to the reference data in Table 8, the average Kappa coefficient of each feature set before screening was 0.03 higher than that after screening. Moreover, the set with the highest Kappa coefficient (0.92) was Texture + canopy structure (CS) + Spectral.

Table 8. Feature number, AIC value, and Kappa coefficient before and after using AIC method.

Feature Set	Feature Number		AIC Value		Kappa Coefficient	
	Before Screening	After Screening	Before Screening	After Screening	Before Screening	After Screening
Texture	41	5	84.00	12.00	0.68	0.72
CS	9	6	117.22	113.11	0.76	0.72
Spectral	8	6	1038.10	1034.60	0.04	0.02
Texture + CS	50	5	102.00	12.00	0.86	0.86
Texture + Spectral	49	4	100.00	10.00	0.68	0.74
Spectral + CS	17	11	92.35	82.47	0.66	0.74
Texture + CS + Spectral	58	5	118.00	12.00	0.84	0.92

Note: Texture, CS, and Spectral represent the texture feature set, canopy structure feature set, and spectral feature set, respectively.

Here, we show the selected maize lodging features from the Texture + CS + Spectral set for further discussion, which includes the red band, the GLCM mean texture, and dissimilarity texture of the average reflectivity of the red, green, and blue bands; the GLCM mean texture of CHM; and the GLDM mean texture of the average reflectivity of the red, green, and blue bands.

3.4. Maize Lodging Identification Result

Based on the features after AIC method screening, the verification results under different classification methods are given in Table 9. The Texture + CS + Spectral set showed the best performance compared with the other feature sets in the MLC and RFC algorithms. However, MLC provided higher classification accuracy (Kappa coefficient = 0.92, OA = 96%) than RFC. As shown in Figure 5a, RFC overlooks more maize lodging pixels in the UAV image. A favorable lodging identification result

(Kappa coefficient = 0.86, OA = 93.00%), which is shown in Figure 5c, can be obtained by utilizing BLRC combined with the CS feature set. Overall, the optimal lodging detection result (Kappa coefficient = 0.92, OA = 96.00%) was generated when using MLC and Texture + CS + Spectral (Table 9). In comparison to other feature sets, the Texture + CS + Spectral had the highest average Kappa coefficient and average OA; thus, it is more suitable for accurately extracting the lodging area.

Table 9. Lodging identification accuracy under different classification methods and various feature sets.

Feature Set	MLC		RFC		BLRC		Mean	
	Kappa Coefficient	OA	Kappa Coefficient	OA	Kappa Coefficient	OA	Kappa Coefficient	OA
Texture	0.72	86.00%	0.64	82.00%	0.62	81.00%	0.66	83.00%
CS	0.72	86.00%	0.64	82.00%	0.86	93.00%	0.74	87.00%
Spectral	0.02	51.00%	0.02	51.00%	0.10	55.00%	0.05	52.33%
Texture + CS	0.86	93.00%	0.76	88.00%	0.84	92.00%	0.82	91.00%
Texture + Spectral	0.74	87.00%	0.62	81.00%	0.56	78.00%	0.64	82.00%
Spectral + CS	0.74	87.00%	0.66	83.00%	0.68	84.00%	0.69	84.67%
Texture + CS + Spectral	0.92	96.00%	0.78	89.00%	0.78	89.00%	0.83	91.33%

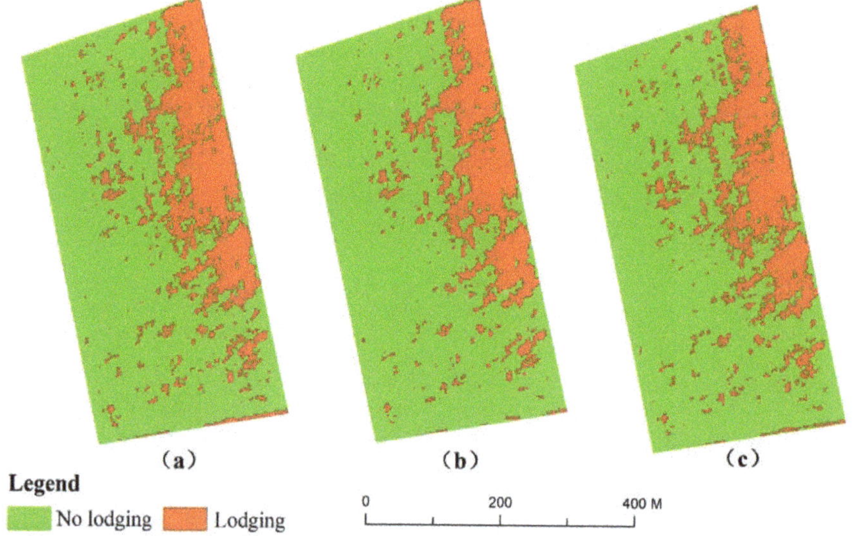

Figure 5. The lodging identification results with the highest accuracies under MLC, BLRC, and random forest classification (RFC) (**a**) MLC combined with Texture + CS + Spectral; (**b**) RFC combined with Texture + CS + Spectral; (**c**) BLRC combined with CS.

In terms of the SFS in Table 9, CS had the best verification results (average Kappa coefficient = 0.74, average OA = 87.00%). Moreover, the Kappa coefficient and OA of texture + spectra are clearly the lowest among MFS (Table 9). The results of this study show that the CS still provides powerful support for accurate extraction of maize lodging, whether in SFS or MFS.

4. Discussion

In this study, nine CSFs, 41 texture features, and eight spectral features of maize in the study area were extracted based on UAV multispectral images and CHM. The different types of image characteristics can not only describe the discrepancies between lodging and nonlodging areas from various aspects but also establish information complementarity during classification. Therefore, we constructed all possible feature sets (SFS and MFS) of the above features as potential factors for the identification of lodging maize. After screening the above feature sets by the AIC method, MLC, BLRC, and RFC were implemented to identify maize lodging. Finally, we obtained the optimal maize lodging recognition result by analyzing the verification results. The present study offers a novel method for monitoring maize lodging. The results revealed that the AIC method is effective and helpful for discriminating among maize lodging and nonlodging areas.

4.1. Optimal Texture Window Size

The precision of image classification based on texture measures relies primarily on the size of the texture window and the size of the target object [34]. At present, accuracy evaluation indexes such as the Kappa coefficient and OA are usually used to determine the optimal texture window size for target recognition in images [55,56]. However, in this study, we found several texture window sizes with the highest accuracy when we used these indexes (Figure 3). Generally, higher separability between the image features of two objects indicates a greater likelihood that they will be correctly classified. Consequently, based on the Kappa coefficient and OA, we utilized DI to analyze the optimal window sizes for the GLCM and GLDM textures (which were 21 and 17, respectively) to ensure the suitability of the window size. Because the spatial resolution of the UAV image is 7.5 cm, the actual areas of the texture features at the 17 and 21 window scales are 1.59 m^2 and 2.40 m^2, respectively. The former area (1.59 m^2) is close to the measured area (1.55 m^2) of single lodging maize in the orthographic multispectral image, while the latter (2.40 m^2) is quite different from the measured area (0.17 m^2) of single nonlodging maize, but it is close to the area of small-scale maize lodging. Therefore, the texture measures of a single lodging plant and a small-scale lodging area are well described under the optimal window size in this study, which avoids missing maize lodging areas.

As shown in Figure 3, the accuracy of the GLDM texture measures increases sharply at window scales from 3 to 5 and decreases slowly at window scales from 5 to 7. This result occurs principally because the actual area of the 5 window size in the UAV image (0.14 m^2) is the closest to the measured area of single nonlodging maize compared to the areas at window sizes of 3 and 7. In this way, the GLDM texture with a window size of 5 maintains good separability between lodging and nonlodging maize. In addition, the randomness of sample selection may cause large fluctuations in the accuracy curve.

4.2. Generalizability of the AIC Method in Selecting Lodging Features

As the fundamental method of identifying the feature selection of lodging crops, the index method [7,15,16] is unable to directly determine the optimal number of features. Conversely, the AIC method not only can quickly determine the optimal number of features but also has higher recognition accuracy (Tables 5–7). The AIC method maintains high recognition accuracy (Table 7) even in two maize regions in different growth stages; therefore, it also achieves good generalizability. Due to the complexity of the plot environment in the study area, we used multiple features to identify maize lodging. Nevertheless, the screening criterion used in the index method—which determines whether a single feature is suitable for lodging identification—does not consider the interaction between multiple features. In contrast, the AIC method establishes the relationships between multiple features using a logistic regression model, thereby enhancing the complementarity of selected features. Therefore, the accuracy of the AIC method is generally higher than that of the index method.

At present, scholars use UAV images to identify lodging crops mainly during the vigorous growth period of crops [12,57]; thus, the method for selecting lodging recognition features is greatly affected by other crop growth periods and field environments. In contrast, the study area in this study includes both lodging and nonlodging maize in different growth stages. Therefore, the AIC method can be extensively applied to various situations (different varieties, growth periods, leaf colors, etc.) In summary, the AIC screening method is more effective for identifying lodging maize in complex field environments or in plots with different fertilization rates, varieties, maturity dates, and sowing periods.

4.3. Suitability of Canopy Structure Feature for Extraction of Maize Lodging in Complex Environments

Texture, spectral, and color features extracted from remote sensing images are frequently used as effective factors to estimate crop lodging [3,11,13]. In addition to these characteristics, CSFs were calculated according to CHM. Among the SFS, the CS had the highest accuracy; its average Kappa coefficient and overall accuracy were 0.74 and 87.00%, respectively (Table 9). Lodging maize in the study area is dominated by Lodging R. The differences in CSFs between Lodging R and No-Lodging LR are significant, but their leaf color and spectral reflectance are similar. Therefore, compared with the texture and spectral features, the CSFs make it easier to identify lodging maize. Although both spectral and texture features are effective for assessing crop lodging [1], different crop types and growing periods still affect the recognition accuracy and stability of the features.

The MFS had the largest difference in the average Kappa coefficient and OA between Texture + Spectral and Texture + CS + Spectral: 0.19 and 9.33% (Table 9), respectively. These results further indicated that CSF is the primary factor that can improve the accuracy of maize lodging detection. Although CSF was able to directly reflect the height difference between lodging and nonlodging maize, their heights were sometimes similar due to the spatial heterogeneity of soil and terrain. Consequently, to compensate for this defect of CSF and accurately distinguish lodging and nonlodging areas, CSF must be combined with other image features such as texture and multispectral reflectance.

4.4. Comparison of Optimal Classification Results under Different Classification Algorithms

Among MLC, RFC, and BLRC, the most accurate respective feature sets were Texture + CS + Spectral, Texture + CS + Spectral, and CS (Table 9). Based on the same feature set, the accuracy of MLC was higher than that of RFC because MLC is more suitable for classifying low-dimensional data than RFC. The BLRC algorithm is very suitable for addressing the issue of binary classification in imagery used for object detection. However, both the lodging and nonlodging maize involved in this study can have multiple forms (Figure 2). Therefore, although high-precision lodging recognition results can be obtained by utilizing BLRC combined with CSF, the results were still lower than those of MLC.

4.5. Comparison of Traditional and New Methods for Crop Lodging Identification

Compared with the traditional crop lodging recognition methods, the new method proposed in this study describes the difference between lodging and nonlodging crops from more aspects—because it extracts more image features. In addition, the optimal and unique texture window size that fully expresses the morphological characteristics of crop lodging in the image was determined, thus improving the probability of successful crop lodging identification. However, extracting a large number of image features requires more computation and consumes more time. In this paper, the AIC method, which can obtain the optimal features more quickly and directly, was applied to feature selection while maintaining a strong relationship between features. Therefore, the AIC method is more suitable for identifying crop lodging using multiple image features. In contrast, the traditional crop lodging identification methods are based on homogeneous field environments in which the crops are all in the same growth period, and their growth conditions are similar. Nevertheless, different maize growth periods and growth statuses existed in the study area. Therefore, compared with traditional methods, the new crop lodging identification method is better able to identify crop lodging in a complex field environment, and it has stronger universality.

4.6. Analysis of Optimal Feature Set under Different Data Availability

In practice, the recognition of lodging maize must consider both data availability and the accuracy of the results. Therefore, based on the reference results in Table 9, we analyzed the optimal feature set of maize lodging extraction features using the AIC method under different data conditions. Compared with multispectral images, the CHM acquisition requires more manpower, material resources, and time. Therefore, we were able to employ only the texture and spectral features derived from the multispectral imagery to save data collection costs. Based on this approach, the optimal feature set and classification algorithm were Texture + Spectral and MLC, respectively. The Texture + Spectral feature set is suitable for situations in which the color discrepancy between leaves and stems is significant. In the absence of multispectral images, the optimal feature set and classification method were CS and BLRC, respectively. The CS feature set is suitable when there are large differences in lodging and nonlodging maize. When both CHM and multispectral imagery can be obtained, the most suitable feature set and classification method were Texture + CS + Spectral and MLC for lodging detection. In a complex field environment, the Texture + CS + Spectral feature set can accurately discriminate between lodging and nonlodging areas, but the time and economic costs of this approach are high; therefore, it is not practical for rapid monitoring of crop lodging.

4.7. Shortcoming and Prospects of Research

Although the new crop lodging identification method proposed in this paper has good stability and high accuracy in complicated field conditions, its universality at the regional scale requires further exploration and verification. When applied to satellite images, the accuracy of the new method may decrease due to the following factors: (1) the spatial resolution and image quality of satellite images are lower than those of UAV images; (2) satellite images include more non-vegetation objects, including houses, other lodgings, and lakes; and (3) detailed CSF cannot be obtained from satellite images.

Compared with the multispectral camera, the three-dimensional cloud points obtained by LiDAR have higher density and accuracy, which would improve the accuracy of CSF and lodging identification. In the future, we could use UAVs equipped with hyperspectral sensors to obtain hyperspectral images with abundant spectral information, allowing more vegetation index features to be constructed. Moreover, the relationship between background factors and lodging crops should be fully analyzed to determine the main factors that cause lodging.

5. Summary and Conclusions

In this study, the texture measures, spectral features, and CSFs of maize in the study area were extracted from UAV-acquired multispectral images, and the optimal texture window size was determined for lodging identification. Based on the above image features, the performance of the two screening methods was analyzed under the MLC, RFC, and BLRC algorithms. The main outcomes were as follows. (1) The optimal texture window sizes of GLCM and GLDM texture were 17 and 21, respectively, as determined by the Kappa coefficient, OA and DI. The area of the former size (1.59 m^2) was close to the measured area (1.55 m^2) of single lodging maize in the orthographic multispectral image, while the area of the latter (2.40 m^2) was quite different from the measured area (0.17 m^2) of single nonlodging maize but quite close to the maize lodging area with a small size. (2) Compared with the index method, the AIC method had a higher Kappa coefficient and OA; thus, this method is more suitable for screening lodging recognition features. In a complex field environment, the AIC method has strong generalizability when using MFS. (3) The accuracy of the detection result based on the Texture + CS + Spectral and AIC methods was the highest compared with the other feature sets, and CSF played an important role in lodging recognition. The Texture + CS + Spectral features, screened by the AIC method and classified by the MLC, achieved the highest lodging recognition accuracy: the Kappa coefficient and OA of this combination were 0.92% and 96.00%, respectively. (4) The new crop lodging identification method proposed in this paper can adapt to complex field environments, especially in crop fields with different growth periods. (5) In practice, researchers can obtain the CSF of maize first and then combine it with the BLRC algorithm to achieve fast extraction of crop lodging areas.

Author Contributions: Conceptualization, H.G. and H.L.; methodology, H.G.; software, X.M.; validation, Y.M. and Z.Y.; formal analysis, C.L.; investigation, H.G., Y.B., and Z.Y.; resources, X.Z.; data curation, Y.B.; writing—original draft preparation, H.G.; writing—review and editing, H.G.; visualization, H.G.; supervision, H.L.; project administration, H.L.; funding acquisition, H.L. and X.Z. All authors have read and agreed to the published version of the manuscript.

Remote Sens. **2020**, *12*, 3149

Funding: This study was supported by the National Natural Science Foundation of China (41671438) and the "Academic Backbone" Project of Northeast Agricultural University (54935112).

Acknowledgments: We thank Baiwen Jiang for helping in selection of survey fields and Qiang Ye for helping collect the field data. We are also grateful to the anonymous reviewers for their valuable comments and recommendations.

Conflicts of Interest: The authors declare no conflict of interest.

References

1. Chauhan, S.; Darvishzadeh, R.; Boschetti, M.; Pepe, M.; Nelson, A. Remote sensing-based crop lodging assessment: Current status and perspectives. *ISPRS J. Photogramm. Remote Sens.* **2019**, *151*, 124–140. [CrossRef]
2. Wang, L.Z.; Gu, X.H.; Hu, S.W.; Yang, G.J.; Wang, L.; Fan, Y.B.; Wang, Y.J. Remote sensing monitoring of maize lodging disaster with multi-temporal HJ-1B CCD image. *Sci. Agric. Sin.* **2016**, *49*, 4120–4129. [CrossRef]
3. Li, Z.N.; Chen, Z.X.; Ren, G.Y.; Li, Z.C.; Wang, X. Estimation of maize lodging are based on Worldview-2 image. *Trans. CSAE* **2016**, *32*, 1–5. [CrossRef]
4. Yang, H.; Chen, E.X.; Li, Z.Y.; Zhao, C.J.; Yang, G.J.; Pignatti, S.; Casa, R.; Zhao, L. Wheat lodging monitoring using polarimetric index from RADARSAT-2 data. *Int. J. Appl. Earth Obs. Geoinf.* **2015**, *34*, 157–166. [CrossRef]
5. Chen, J.; Li, H.; Han, Y. Potential of RADARSAT-2 data on identifying sugarcane lodging caused by typhoon. In Proceedings of the Fifth International Conference on Agro-geoinformatics (Agro-Geoinformatics 2016), Tianjin, China, 18–20 July 2016; pp. 36–42. [CrossRef]
6. Somers, B.; Asner, G.P.; Tits, L.; Coppin, P. Endmember variability in spectral mixture analysis: A review. *Remote Sens. Environ.* **2011**, *115*, 1603–1616. [CrossRef]
7. Li, Z.N.; Chen, Z.X.; Wang, L.M.; Liu, J.; Zhou, Q.B. Area extraction of maize lodging based on remote sensing by small unmanned aerial vehicle. *Trans. CSAE* **2014**, *30*, 207–213. [CrossRef]
8. Yang, M.D.; Tseng, H.H.; Hsu, Y.C.; Tsai, H.P. Semantic segmentation using deep learning with vegetation indices for rice lodging identification in multi-date UAV visible images. *Remote Sens.* **2020**, *12*, 633. [CrossRef]
9. Zhao, X.; Yuan, Y.T.; Song, M.D.; Ding, Y.; Lin, F.F.; Liang, D.; Zhang, D.Y. Use of unmanned aerial vehicle imagery and deep learning Unet to extract rice lodging. *Sensors* **2019**, *19*, 3859. [CrossRef]
10. Liu, T.; Li, R.; Zhong, X.; Jiang, M.; Jin, X.; Zhou, P.; Liu, S.; Sun, C.; Guo, W. Estimates of rice lodging using indices derived from UAV visible and thermal infrared images. *Agric. For. Meteorol.* **2018**, *252*, 144–154. [CrossRef]
11. Zhang, X.L.; Guan, H.X.; Liu, H.J.; Meng, X.T.; Yang, H.X.; Ye, Q.; Yu, W.; Zhang, H.S. Extraction of maize lodging area in mature period based on UAV multispectral image. *Trans. CSAE* **2019**, *35*, 98–106. [CrossRef]
12. Chu, T.; Starek, M.; Brewer, M.; Murray, S.; Pruter, L. Assessing lodging severity over an experimental maize (Zea mays L.) field using UAS images. *Remote Sens.* **2017**, *9*, 923. [CrossRef]
13. Yang, M.-D.; Huang, K.-S.; Kuo, Y.-H.; Tsai, H.; Lin, L.-M. Spatial and spectral hybrid image classification for rice lodging assessment through UAV imagery. *Remote Sens.* **2017**, *9*, 583. [CrossRef]
14. Han, L.; Yang, G.; Feng, H.; Zhou, C.; Yang, H.; Xu, B.; Li, Z.; Yang, X. Quantitative identification of maize lodging-causing feature factors using unmanned aerial vehicle images and a nomogram computation. *Remote Sens.* **2018**, *10*, 1528. [CrossRef]
15. Wang, J.-J.; Ge, H.; Dai, Q.; Ahmad, I.; Dai, Q.; Zhou, G.; Qin, M.; Gu, C. Unsupervised discrimination between lodged and non-lodged winter wheat: A case study using a low-cost unmanned aerial vehicle. *Int. J. Remote Sens.* **2018**, *39*, 2079–2088. [CrossRef]
16. Dai, J.G.; Zhang, G.S.; Guo, P.; Zeng, T.J.; Cui, M.; Xue, J.L. Information extraction of cotton lodging based on multi-spectral image from UAV remote sensing. *Trans. CSAE* **2019**, *35*, 63–70. [CrossRef]
17. Orlowski, G.; Karg, J.; Kaminski, P.; Baszynski, J.; Szady-Grad, M.; Ziomek, K.; Klawe, J.J. Edge effect imprint on elemental traits of plant-invertebrate food web components of oilseed rape fields. *Sci. Total Environ.* **2019**, *687*, 1285–1294. [CrossRef]
18. Meyer, G.E.; Neto, J.C. Verification of color vegetation indices for automated crop imaging applications. *Comput. Electron. Agric.* **2008**, *63*, 282–293. [CrossRef]
19. Patel, N.K.; Patnaik, C.; Dutta, S.; Shekh, A.M.; Dave, A.J. Study of crop growth parameters using Airborne Imaging Spectrometer data. *Int. J. Remote Sens.* **2010**, *22*, 2401–2411. [CrossRef]

20. Hufkens, K.; Friedl, M.; Sonnentag, O.; Braswell, B.H.; Milliman, T.; Richardson, A.D. Linking near-surface and satellite remote sensing measurements of deciduous broadleaf forest phenology. *Remote Sens. Environ.* **2012**, *117*, 307–321. [CrossRef]

21. Sonnentag, O.; Hufkens, K.; Teshera-Sterne, C.; Young, A.M.; Friedl, M.; Braswell, B.H.; Milliman, T.; O'Keefe, J.; Richardson, A.D. Digital repeat photography for phenological research in forest ecosystems. *Agric. For. Meteorol.* **2012**, *152*, 159–177. [CrossRef]

22. Guan, S.; Fukami, K.; Matsunaka, H.; Okami, M.; Tanaka, R.; Nakano, H.; Sakai, T.; Nakano, K.; Ohdan, H.; Takahashi, K. Assessing correlation of high-resolution NDVI with fertilizer application level and yield of rice and wheat crops using small UAVs. *Remote Sens.* **2019**, *11*, 112. [CrossRef]

23. Tucker, C.J. Red and photographic infrared linear combinations for monitoring vegetation. *Remote Sens. Environ.* **1979**, *8*, 127–150. [CrossRef]

24. Kanke, Y.; Tubaña, B.; Dalen, M.; Harrell, D. Evaluation of red and red-edge reflectance-based vegetation indices for rice biomass and grain yield prediction models in paddy fields. *Precision Agric.* **2016**, *17*, 507–530. [CrossRef]

25. Bai, G.; Ge, Y.; Hussain, W.; Baenziger, P.S.; Graef, G. A multi-sensor system for high throughput field phenotyping in soybean and wheat breeding. *Comput. Electron. Agric.* **2016**, *128*, 181–192. [CrossRef]

26. Næsset, E.; Bjerknes, K.-O. Estimating tree heights and number of stems in young forest stands using airborne laser scanner data. *Remote Sens. Environ.* **2001**, *78*, 328–340. [CrossRef]

27. Li, W.; Niu, Z.; Chen, H.; Li, D.; Wu, M.; Zhao, W. Remote estimation of canopy height and aboveground biomass of maize using high-resolution stereo images from a low-cost unmanned aerial vehicle system. *Ecol. Indic.* **2016**, *67*, 637–648. [CrossRef]

28. Ciceu, A.; Garcia-Duro, J.; Seceleanu, I.; Badea, O. A generalized nonlinear mixed-effects height-diameter model for Norway spruce in mixed-uneven aged stands. *For. Ecol. Manag.* **2020**, *477*, 118507. [CrossRef]

29. Wang, L.; Zheng, N.; Gao, S.; Huang, N.; Chen, H. Correlating the horizontal and vertical distribution of LiDAR point clouds with components of biomass in a Picea crassifolia forest. *Forests* **2014**, *5*, 1910–1930. [CrossRef]

30. Kane, V.; Bakker, J.; McGaughey, R.; Lutz, J.; Gersonde, R.; Franklin, J. Examining conifer canopy structural complexity across forest ages and elevations with LiDAR data. *Can. J. For. Res.* **2010**, *40*, 774–787. [CrossRef]

31. Li, W.; Niu, Z.; Chen, H.; Li, D. Characterizing canopy structural complexity for the estimation of maize LAI based on ALS data and UAV stereo images. *Int. J. Remote Sens.* **2016**, *38*, 2106–2116. [CrossRef]

32. Parker, G.; Harmon, M.; Lefsky, M.; Chen, J.; Van Pelt, R.; Weiss, S.; Thomas, S.; Winner, W.; Shaw, D.; Franklin, J. Three-dimensional structure of an old-growth Pseudotsuga-Tsuga canopy and its implications for radiation balance, microclimate, and gas exchange. *Ecosystems* **2004**, *7*, 440–453. [CrossRef]

33. Coburn, C.; Roberts, A.C. A multiscale texture analysis procedure for improved forest stand classification. *Int. J. Remote Sens.* **2004**, *25*, 4287–4308. [CrossRef]

34. Feng, Q.; Liu, J.; Gong, J. UAV remote sensing for urban vegetation mapping using random forest and texture analysis. *Remote Sens.* **2015**, *7*, 1074–1094. [CrossRef]

35. Backes, A.R.; Casanova, D.; Bruno, O.M. Texture analysis and classification: A complex network-based approach. *Inf. Sci.* **2013**, *219*, 168–180. [CrossRef]

36. Weszka, J.; Dyer, C.; Rosenfeld, A. A comparative study of texture measures for terrain classification. *IEEE Trans. Syst. Man Cybern.* **1976**, *6*, 269–285. [CrossRef]

37. Conners, R.W.; Harlow, C.A. A Theoretical comparison of texture algorithms. *IEEE Trans. Pattern Anal.* **1980**, *PAMI-2*, 204–222. [CrossRef]

38. Nishimura, K.; Matsuura, S.; Suzuki, H. Multivariate EWMA control chart based on a variable selection using AIC for multivariate statistical process monitoring. *Stat. Probab. Lett.* **2015**, *104*, 7–13. [CrossRef]

39. Hughes, A.; King, M. Model selection using AIC in the presence of one-sided information. *J. Stat. Plan Inference* **2003**, *115*, 397–411. [CrossRef]

40. Ye, H.; Huang, W.; Huang, S.; Cui, B.; Dong, Y.; Guo, A.; Ren, Y.; Jin, Y. Recognition of banana fusarium wilt based on UAV remote sensing. *Remote Sens.* **2020**, *12*, 938. [CrossRef]

41. Jokar Arsanjani, J.; Helbich, M.; Kainz, W.; Darvishi Booloorani, A. Integration of logistic regression, Markov chain and cellular automata models to simulate urban expansion. *Int. J. Appl. Earth Obs. Geoinf.* **2013**, *21*, 265–275. [CrossRef]

42. Sun, J.; Yang, J.; Zhang, C.; Yun, W.; Qu, J. Automatic remotely sensed image classification in a grid environment based on the maximum likelihood method. *Math. Comput. Model.* **2013**, *58*, 573–581. [CrossRef]
43. Jia, K.; Liang, S.; Zhang, L.; Wei, X.; Yao, Y.; Xie, X. Forest cover classification using Landsat ETM+ data and time series MODIS NDVI data. *Int. J. Appl. Earth Obs. Geoinf.* **2014**, *33*, 32–38. [CrossRef]
44. Li, J.; Zhong, P.-A.; Yang, M.; Zhu, F.; Chen, J.; Liu, W.; Xu, S. Intelligent identification of effective reservoirs based on the random forest classification model. *J. Hydrol.* **2020**, *591*, 125324. [CrossRef]
45. Breiman, L. Random forest. *Mach. Learn.* **2001**, *45*, 5–32. [CrossRef]
46. Bao, Y.; Meng, X.; Ustin, S.; Wang, X.; Zhang, X.; Liu, H.; Tang, H. Vis-SWIR spectral prediction model for soil organic matter with different grouping strategies. *Catena* **2020**, *195*, 104703. [CrossRef]
47. Meng, X.; Bao, Y.; Liu, J.; Liu, H.; Zhang, X.; Zhang, Y.; Wang, P.; Tang, H.; Kong, F. Regional soil organic carbon prediction model based on a discrete wavelet analysis of hyperspectral satellite data. *Int. J. Appl. Earth Obs. Geoinf.* **2020**, *89*, 102111. [CrossRef]
48. Van der Linden, S.; Rabe, A.; Held, M.; Jakimow, B.; Leitão, P.; Okujeni, A.; Schwieder, M.; Suess, S.; Hostert, P. The enmap-box—A toolbox and application programming interface for enmap data processing. *Remote Sens.* **2015**, *7*, 11249. [CrossRef]
49. Congalton, R.G. A review of assessing the accuracy of classifications of remotely sensed data. *Remote Sens. Environ.* **1991**, *37*, 35–46. [CrossRef]
50. Foody, G.M. Classification accuracy comparison: Hypothesis tests and the use of confidence intervals in evaluations of difference, equivalence and non-inferiority. *Remote Sens. Environ.* **2009**, *113*, 1658–1663. [CrossRef]
51. Pontius, R.; Millones, M. Death to kappa: Birth of quantity disagreement and allocation disagreement for accuracy assessment. *Int. J. Remote Sens.* **2011**, *32*, 4407–4429. [CrossRef]
52. Foody, G.M. Explaining the unsuitability of the kappa coefficient in the assessment and comparison of the accuracy of thematic maps obtained by image classification. *Remote Sens. Environ.* **2020**, *239*, 111630. [CrossRef]
53. Potnis, A.; Panchasara, C.; Shetty, A.; Nayak, U.K. Assessing the accuracy of remotely sensed data: Principles and practices. *Photogram Rec.* **2010**, *25*, 204–205. [CrossRef]
54. Warrens, M.J. Category kappas for agreement between fuzzy classifications. *Neurocomputing* **2016**, *194*, 385–388. [CrossRef]
55. Chen, D.; Stow, D.A.; Gong, P. Examining the effect of spatial resolution and texture window size on classification accuracy: An urban environment case. *Int. J. Remote Sens.* **2010**, *25*, 2177–2192. [CrossRef]
56. Puissant, A.; Hirsch, J.; Weber, C. The utility of texture analysis to improve per-pixel classification for high to very high spatial resolution imagery. *Int. J. Remote Sens.* **2005**, *26*, 733–745. [CrossRef]
57. Sun, Q.; Sun, L.; Shu, M.; Gu, X.; Yang, G.; Zhou, L. Monitoring maize lodging grades via unmanned aerial vehicle multispectral image. *Plant Phenomics* **2019**, *2019*, 1–16. [CrossRef]

 *remote sensing*

Article

Prediction of the Kiwifruit Decline Syndrome in Diseased Orchards by Remote Sensing

Francesco Savian [1,2,*], Marta Martini [1], Paolo Ermacora [1], Stefan Paulus [3] and Anne-Katrin Mahlein [3]

1   Department of Agricultural, Food, Environmental and Animal Sciences (DI4A), University of Udine, Via delle Scienze 206, 33100 Udine, Italy; marta.martini@uniud.it (M.M.); paolo.ermacora@uniud.it (P.E.)
2   CREA - Council for Agricultural Research and Economics, Research Centre for Agriculture and Environment, I-40128 Bologna, Italy
3   Institute of Sugar Beet Research (IfZ), Holtenser Landstraße 77, 37079 Göttingen, Germany; paulus@ifz-goettingen.de (S.P.); mahlein@ifz-goettingen.de (A.-K.M.)
*   Correspondence: savian.francesco@spes.uniud.it

Received: 15 May 2020; Accepted: 7 July 2020; Published: 9 July 2020

Abstract: Eight years after the first record in Italy, Kiwifruit Decline (KD), a destructive disease causing root rot, has already affected more than 25% of the area under kiwifruit cultivation in Italy. Diseased plants are characterised by severe decay of the fine roots and sudden wilting of the canopy, which is only visible after the season's first period of heat (July–August). The swiftness of symptom appearance prevents correct timing and positioning for sampling of the disease, and is therefore a barrier to aetiological studies. The aim of this study is to test the feasibility of thermal and multispectral imaging for the detection of KD using an unsupervised classifier. Thus, RGB, multispectral and thermal data from a kiwifruit orchard, with healthy and diseased plants, were acquired simultaneously during two consecutive growing seasons (2017–2018) using an Unmanned Aerial Vehicle (UAV) platform. Data reduction was applied to the clipped areas of the multispectral and thermal data from the 2017 survey. Reduced data were then classified with two unsupervised algorithms, a K-means and a hierarchical method. The plant vigour (canopy size and presence/absence of wilted leaves) and the health shifts exhibited by asymptomatic plants between 2017 and 2018 were evaluated from RGB data via expert assessment and used as the ground truth for cluster interpretation. Multispectral data showed a high correlation with plant vigour, while temperature data demonstrated a good potential use in predicting health shifts, especially in highly vigorous plants that were asymptomatic in 2017 and became symptomatic in 2018. The accuracy of plant vigour assessment was above 73% when using multispectral data, while clustering of the temperature data allowed the prediction of disease outbreak one year in advance, with an accuracy of 71%. Based on our results, the unsupervised clustering of remote sensing data could be a reliable tool for the identification of sampling areas, and can greatly improve aetiological studies of this new disease in kiwifruit.

Keywords: disease detection; outbreak prediction; sensor fusion; unsupervised clustering; multispectral imaging; thermal imaging; unmanned aerial vehicle; UAV

1. Introduction

Italy is, beside New Zealand, one of the two major kiwifruit producers and exporters worldwide, accounting for 15% of the total world demand on its own [1]. Since 2012, a major new disease known as Kiwifruit Decline (KD) has been threatening the future of kiwifruit production in Italy [2]. Up to now, KD has already destroyed more than 25% (~6600 ha) of the Italian kiwifruit growing area, and it has become a serious threat to the future of the crop in this country (Tacconi, personal communication updating [3]). To our knowledge, KD outbreaks have only been reported in Italy, but similar diseases have also been described in other countries, including New Zealand [4], Japan [5], and more recently in Turkey [6–8].

The main damage caused by KD is the almost complete degradation of the white feeding roots. Coarse roots usually show red discolouration under the cortex, and the external cylinder easily detaches from the centre, creating the so-called "rat tail" symptom [2]. Symptoms on the root might not immediately affect the canopy, which may remain completely asymptomatic for an unknown amount of time, even if the root system is considerably deteriorated [3]. Usually, diseased plants go through an irreversible and fast wilting process, which is suddenly visible late in the vegetative season (July, August, northern hemisphere) when heat waves occur [9]. Once the wilting appears, plants usually lose all their leaves and die within a few weeks. In the following years, plants that manage to survive have considerably reduced canopy vigour, may show different degrees of nutrition deficiencies, and never recover from their weakened condition [3]. KD spreads following an oil-spot pattern, with an astonishing speed that can compromise the whole orchard in only one or two growing seasons [2]. Currently, studies on KD are mostly focused on understanding the aetiology of the disease, which has not yet been fully clarified, but it seems that an interaction between waterlogging and several soil-borne pathogens is necessary to induce the disease [2,10–13].

The timing and positioning of sampling are key factors for all aetiological studies. The lack of knowledge about the dynamics of root degradation severely compromises the quality of sampling activities, which cannot be properly scheduled. In particular, the core samples (feeding roots) that are collected for laboratory analysis of KD are usually irreversibly deteriorated, and almost useless for aetiological investigations once canopy symptoms appear [10]. Currently, there are no efficient methods to survey orchards. Field surveying presents several limitations, mostly associated with the obstruction of inter-row spaces by shoots that reduce the field of view (One to three metres only, depending on plant vigour) and can increase the time required even just for observation of the canopies. In addition, there are no methods to predict disease outbreaks, because wilting appears abruptly, even within healthy plants that are apparently devoid of any warning signs [2,3,10]. The only way to confirm the presence of the disease before plant dieback is to observe the symptoms in the roots. This strategy, though, is highly unfeasible; firstly, because there are no reliable methods to select which plant must be uprooted, and secondly because it is time-consuming and requires hard work.

Remote sensing can be a real aid in monitoring activities, especially for field surveys. The high impact of the disease on plant vigour, if approached appropriately, can be exploited non-invasively to assess the spread of KD. Indeed, multispectral imagery has been widely exploited for the evaluation of canopy vigour indices [14], drought stress indices [15], plant pathogen detection [16] and, on kiwifruit, to predict the dry matter content of fruits from satellite data [17].

RGB cameras have also been largely adopted to estimate geometrical parameters of field crops [18–21], but their performance on plant vigour estimation is generally secondary to multispectral sensors, due to the higher correlation of near-infrared (NIR) wavelengths with plant biophysical and biochemical traits (e.g., chlorophyll content) [22,23]. However, for KD studies, RGB cameras have the advantage of representing the scene exactly as observed by the human eye, allowing experts to recognise the wilting symptoms, evaluate the disease spread and develop reference data for machine learning algorithms.

Finally, since KD causes fine root decay, the water absorption capabilities of kiwi vines are highly compromised [10]. Consequently, it is highly probable that the physiological response of the plant is similar to that induced by a drought stress: closure of the stomata and rising of the leaf temperature. It has been reported that kiwifruit vines are sensible to water-related stresses, and quickly close their stomata in response to drought conditions [24], severe root pruning [25] and even waterlogging [26,27]. The increase in leaf temperature can be significant in drought-stressed kiwi vines, and reported to be between +1 °C and +6 °C, depending on the exposure of the leaf to the sun, the sampling time and the intensity of the stress [28,29]. Leaf temperature might be the key to detect a compromised plant, which would be otherwise undetectable by the naked eye. Since leaves emit radiation according to their temperature in the thermal infrared band (TIR; 3 to 15 μm) [23,30], this wavelength range has been widely used to estimate the water status of several plants [31–33], and to detect a number of plant diseases that influence leaf stomatal conductance and transpiration rates directly or indirectly [34–36]. Regarding kiwifruit, thermal imaging has been previously adopted for the early detection of another vascular disease caused by the bacterium *Pseudomonas syringae* pv. *actinidiae* [37].

Among the remote sensing technologies, drone-based surveys provide several advantages in KD studies: (i) they allow us to generate high resolution images that can be registered to image the whole field and quickly evaluate the health status of the plants; (ii) they can be easily equipped with different sensors, collecting data in non-visible wavelengths such as the near infrared region (NIR) and thermal infrared; (iii) they can fly at lower altitudes, reducing the influence of the atmosphere; (iv) they are cost efficient; and (v) they have a high scheduling flexibility, which is crucial for surveying plant diseases that are strictly related to climatic condition [38,39]. Considering KD, the latter aspect is probably the most important, since the survey must be performed not only with favourable weather conditions (radiation, clouds, wind), but also with optimal soil water content, since both waterlogging and drought stress can affect the transpiration rates [26,27,29].

Images acquired with these systems can be classified using either supervised or unsupervised machine learning algorithms. However, the application of supervised algorithms is limited, since it is difficult to produce sufficient high-quality labelled data to create a robust supervised classifier that is able to account for all the variability associated with KD and kiwifruit orchards. The best method for labelling the data is the observation of the disease spread between two consecutive growing seasons, since protocols to rapidly and objectively detect diseased plant are still missing [2]. However, the swiftness of the disease spread make the use of supervised algorithms impractical. Indeed, by the time the classifier has been trained, the field could have already been irreversibly compromised. Unsupervised learning may prove to be an effective alternative, since it can reveal the underlying patterns and structure of the data without the need for labelling [40].

Thus, this study aims to understand the feasibility and reliability of thermal and multispectral imaging in predicting and assessing disease outbreaks, via an unsupervised classification approach. To the best of our knowledge, the present study is the first one that introduces remote sensing as a reliable tool for addressing aetiological and epidemiological studies of Kiwifruit Decline.

2. Materials and Methods

2.1. Experimental Site

UAV-based surveys were performed over one hectare of *Actinidia deliciosa* (A. Chev.) CF Liang et AR Ferguson cultivar 'Hayward' in San Giorgio della Richinvelda (Friuli Venezia Giulia, NE Italy, 46.039°N 12.864°E). The orchard was 20 years old, grown in a 5 × 5-m T-bar plantation without anti-hail netting, with a micro-irrigation system (micro-sprinkler, 20 mm/hour) and with the following soil properties: 8% gravel, 36% sand, 59% silt, 5% clay and 1.04% organic matter (w/w). The first symptoms of KD were recorded in 2015 after an irrigation pipe broke in the central rows of the orchard. Later in 2016, the disease rapidly spread towards the border of the field, compromising approximately 30% of the field. From 2017 the field was regularly inspected to check the pattern of disease spread. In that year, the disease affected almost the entire orchard, compromising approximately 70% of the field, and by 2018 only a few plant rows on the eastern boundary were still apparently healthy. At the end of 2018, the field was definitively explanted. From the first appearance of the disease, the orchard was irrigated by following the usual agronomic practices and avoiding both drought stress and excessive irrigation.

2.2. Image Acquisition

The UAV surveys were carried out on 8 August 2017 and 18 July 2018, respectively at 11.00 a.m. and 10.00 a.m. on hot sunny days. The timing of the flight was based on a previous report suggesting this time frame as one that would maximise the differences between well-fed and drought-stressed plants [28]. Weather condition were favourable for the acquisition: no clouds, wind speed <2 km/h, relative air humidity ~44%, radiation ~2900 MJ/m^2 and temperature ~29 °C. The acquisitions were performed with a single flight lasting less than 20 min, including take-off and landing. The flights were undertaken, respectively, two days after an irrigation event in 2017 and three days after a heavy rainfall (20 mm) in 2018, to exclude water deficit from the variables influencing the canopy temperature and to make sure that the soil water content was not a limiting factor. The gravimetric soil moisture was measured the day before the flight at nine locations in the surveyed area. Three sampling areas were selected inside an apparently healthy area, three in a heavily compromised area and three in a transition area between these two. The surveys were performed if the soil humidity was between 70% and 80% of the field capacity, representing an optimal condition for kiwifruit growth [41]. The field capacity value was derived from the soil texture using soil-plant-atmosphere-water field and pond hydrology (SPAW) models [42]. SPAW models are pedo-transfer functions, which use soil texture (percentage of sand, silt and cay) and percentages of soil organic matter to estimate the soil hydrodynamics properties relevant to agronomical practices, such as volumetric field capacity and wilting point [43].

The UAV flights were performed with a hexa-copter using Real Time Kinematic–Global Navigation Satellite System (RTK-GNSS) and equipped with a gimbal system on two axes (Adorn-technologies, Italy). Thermal, Multispectral and RGB images were acquired simultaneously at a flying height of 35 m, at a flying speed of 5 m/s. Frame rate was adjusted for each sensor in order to achieve a forward image overlap of 80%, while sidelap overlap was 80%, 88% and 90% between thermal, RGB and multispectral images, respectively. In 2017, the acquisition combined the three sensors: GoPro Hero 4 (GoPro, San Mateo, CA, USA) as the RGB camera, Tetracam ADC snap (Tetracam, Chatsworth, LA, USA) as multispectral sensor and FLIR TAU2 640 for thermal imaging (FLIR System, Wilsonville, OR, USA). In 2018 the RGB and multispectral sensors were exchanged for Sequoia+ sensors (Parrot, Paris, France). For detailed information about the sensors used see Table 1. Tetracam ADC snap is a low-cost multispectral sensor, modified from an RGB sensor equipped with a blue absorbing glass filter to eliminate the blue sensitivity, and to use the blue pixels in the sensor to measure NIR bands. FLIR TAU2 640 is a thermal camera capable of detecting temperature differences of ±0.05 °C with a sensibility of 2 °C. It has the advantage of performing automatic flat-field corrections (FFC) while recording, to account for microbolometer variation and remove artefacts from the 2D images. For this study, the FFC was performed every 100 frames. FLIR TAU2 640 also measures the internal temperature of the camera to frequently update the non-uniformity coefficients used to convert raw data into radiometric values [44]. Air temperature and relative humidity were measured before the flight and inserted in the camera settings, together with the plant emissivity value, which was set to 0.98 as suggested in Maes et al. (2014) [37]. Sequoia+ is a camera equipped with a sunshine sensor for image calibration, an RGB camera and four single-band cameras measuring reflectance in the green, red, red-edge and near-infrared (NIR) bands.

2.3. Pre-Processing: Creation of Orthomosaics and Alignment Checking

An overall visualisation of the workflow is presented in Figure 1. The analysis was performed using all data (RGB, multispectral and thermal) acquired in 2017, while in 2018, only RGB data were used to evaluate the disease spread, as detailed below. After the acquisition, raw RGB and multispectral data were converted to reflectance and thus normalised from 0 to 1 using Pix4D software (Pix4D., Lausanne, Switzerland). In 2017, calibration was performed using a white Teflon plate included in the Tetracam ADC snap packaging as white reference, and as dark reference a picture acquired with the shutter closed. White and dark references were acquired at the beginning of the flight. Raw measurements were then calibrated as reported in Elmasry et al. (2012) [45]. In 2018, RGB images were instead normalised following the manufacture-recommended methods for the Sequoia+ sensor, using the one-point calibration plus sunshine sensor method as reported in Poncet et al. (2019) [46]. Thermal data was not calibrated; however, rather than absolute leaf temperature, it was of higher interest in the relative difference occurring between plants with different health statuses. Temperature values (°C) were derived from the TIR wavelengths using FLIR-studio (FLIR System Inc., USA). Stitching, georeferencing and orthorectification were performed independently for each set of data (RGB, multispectral and thermal data from 2017 and RGB from 2018) with Pix4D. Image processing is based on structure from motion (SfM) algorithms to reconstruct the three-dimensional scene based on shared features detected across the images. The four orthomosaics (multispectral and thermal from 2017; RGB from 2017 and 2018) were combined in Qgis [47], and planting lines and control points were drawn separately over each orthomosaic to check alignment of the datasets. A total of 11 control points were selected as follows: four points in both the northern and southern part of the field, and three in the orchard. Errors between planting line slope were minimal (on average 0.01°), while the mean error between control points was 4.97 cm horizontally and 4.51 cm vertically, suggesting that no major deformation occurred during the stitching. Alignment was deemed good enough for the analysis purposes, therefore co-registration of the orthomosaics was not performed. Errors linked to temperature drifts were checked at two pairs of points: two points diagonally collected on a grass-free compacted portion of the earth roads delimiting the field on the northeast and southwest sides; and two points were collected on the water contained in an irrigation canal which flanks the north side of the field. Temperature values for each point were extracted from the thermal orthomosaic and respectively compared to check for temperature drifts, which were 0.2 °C on the bare soil and 0.04 °C on the water.

Table 1. Manufacturer technical specifications of the sensors used in the study.

	Year of the Survey			
	2017			**2018**
Sensor Type	RGB	Multispectral	Thermal	RGB
Sensor Model	Hero 4	ADC snap	Thermal Capture TAU 2.0 640	Sequoia$^+$
Company	GoPro	Tetracam	FLIR Systems	Parrot
Country	USA	USA	USA	France
Channel	RED, GREEN, BLUE	NIR, RED, GREEN	TIR	RED, GREEN, BLUE
Central Band (μm)	0.66, 0.55, 0.47	0.79, 0.66, 0.55	10.5	0.66, 0.55, 0.47
Wavelength Range (μm)	NA	0.52–0.90	7.5–13.5	NA
Spatial Res. [1] (cm/pixel)	1	4	7	1
Image Resolution [2] (px)	1920 × 1080	1280 × 1024	640 × 512	4608 × 3456
Focal Length (mm)	3	8.43	13	4.88
Focal-Ratio (-)	f/2.8	f/3.2	f/1.25	f/2.3
Pixel Size (μm)	1.55	5	17	1.34
Sensor Dimension (mm)	4.19 × 2.36	6.59 × 4.9	10.88 × 8.70	6.10 × 4.58
FOV2 (°)	120(H) × 90(V)	37.67(H) × 28.75(V)	45(H) × 37(V)	64(H) × 50(V)

2.4. Post-Processing: Clustering and Reference Data

Single-band orthomosaics were extracted from the multispectral orthomosaic and tested for band correlation. Green and red bands were de-correlated from the NIR band using the formula $R_{adj} = R_{ini} - R_{NIR} * 0.8$, where R_{adj} is the corrected reflectance value, R_{ini} is the initial value of the red or green bands and R_{NIR} is the reflectance value in the NIR spectrum. This correction reduced the R^2 correlation from 0.89 to 0.64 between the red and NIR band, and from 0.92 to 0.81 between the green and NIR band. Temperature and RGB values were not modified after the stitching.

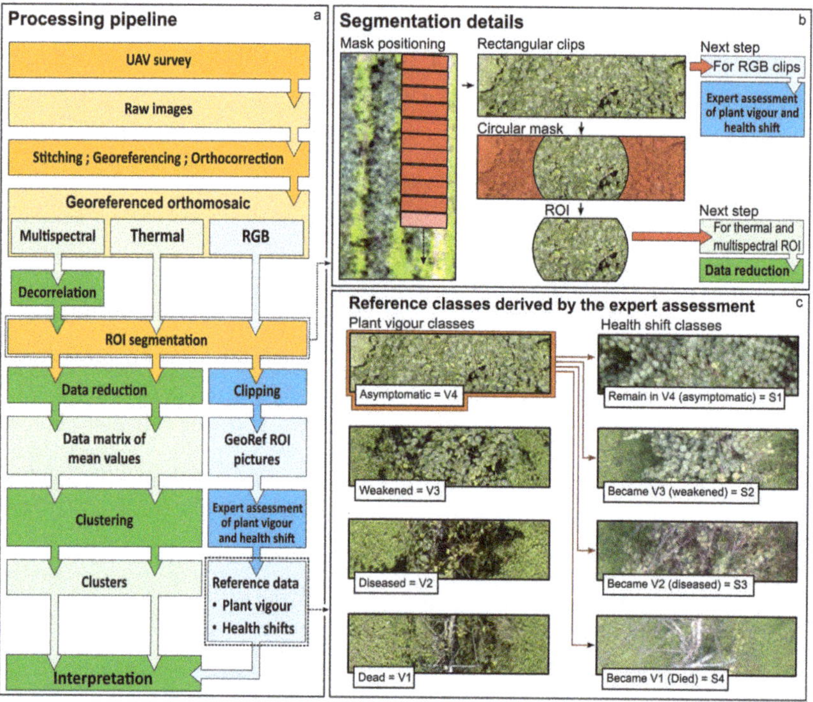

Figure 1. Overall processing pipeline. In (**a**) processes are represented with a darker background, while data have a light background. The green steps represent data and processes, used for the unsupervised classification of multispectral and thermal data; the blue steps represent factors used for reference data estimation based on RGB clips; the yellow steps are common between the two processing pipelines. In (**b**) details of the segmentation process are depicted: starting with mask positioning along planting lines, proceeding with rectangular shape clipping and concluding with circular mask extraction. In (**c**) the classes derived by the expert assessment are shown for each vigour and plant health shift class. On the left, the four vigour classes are represented from highly vigorous and asymptomatic (V4) to dead (V1). On the right, the classification of the health shifts that occurred between 2017 and 2018 in plants that were initially highly vigorous and asymptomatic (V4) is reported: from those that remain highly vigorous and asymptomatic in the second year (S1), to the ones that in 2018 were weakened (S2), diseased (S3) or dead (S4). Abbreviations: ROI, region of interest; GeoRef, georeferenced.

Identification of the stump positions and determination of the canopy boundaries is very difficult to achieve when performing in-field UAV surveys over kiwi vines because, firstly, T-bar or pergola training systems create a flat and dense canopy preventing stump identification, and secondly, the canopies of two neighbouring plants usually overlap due to the disordered growth of shoots. For these reasons, a rough segmentation algorithm was applied to the RGB, thermal, and each single-band multispectral orthomosaic. Firstly, the orthomosaics were clipped along the planting lines with rectangular masks. These digital masks were 5 m wide and 1.5 m long and were placed with the longer side perpendicular to the direction of planting (Figure 1b). The rectangles were placed 1.5 m apart from each other to avoid overlapping. Secondly, a circular shape with a radius of 1 m was applied to the centre of each rectangle clip, retaining only the area within the circle. The shape resulting from the rectangle's and circle's intersection was selected to reduce the border effect, while the mask dimensions and the distances between the mask centres were set to minimise the interference from background pixels of the inter-row grass, and to reduce the canopy overlap within the same clip. Segmentation of RGB data was stopped at the rectangular shape clipping, while multispectral and thermal data were subjected to the complete segmentation workflow. The RGB clips were used to estimate reference values for plant vigour and plant health shifts via expert assessment, while multispectral and thermal data were used for unsupervised clustering. Maintaining the entire rectangular clips for RGB data improved expert awareness of the canopy dimensions, and consequently the precision of plant vigour classification. On the other hand, multispectral and thermal data was reduced to the assumed canopy extent to reduce background bias, since the central portion of the canopy is usually flat and orthogonal to the acquisition of the images (Figure 1b).

Two types of reference data were evaluated from the rectangular RGB clips via expert assessment: plant vigour and plant health shifts (Figure 1c). Plant vigour is a good indicator to distinguish asymptomatic and symptomatic plants, but it is not directly associated with the status of the root system [2]. Therefore, the a posteriori evaluations of vigour shifts and the death rates between years were the only available means to perform realistic evaluations of plant health status in 2017 without uprooting plants.

Reference data for canopy vigour was estimated based on RGB clips derived from the segmentation process of the 2017 orthomosaics. Each clip was evaluated individually and independently by two experts and ranked in four classes based on canopy vigour, considering: (i) the presence/absence of symptoms on the leaves (yellowing, scorching or wilting), (ii) disease severity (percentage of canopy with wilting symptoms), and (iii) fractional vegetation coverage (Table 2 and Figure 1c).

Table 2. Criteria used by experts to evaluate plant vigour and health shift classes. Plant vigour was estimated from RGB clips for 2017 and 2018 surveys. Plant health shifts between the two years were then derived only for plants that were highly vigorous (V4) in 2017.

Plant Vigour Assessment			
Vigour Class	**V4**	**V3**	**V2**	**V1**
Symptoms	None	None	Wilting; yellowing; scorching	No leaf; wilting; yellowing; scorching
Disease Severity	0	0	25–50%	75–100%
Fractional Vegetation Cover	80–100%	60–80%	20–50%	0–20%
Description	High vigour, asymptomatic	Medium vigour, asymptomatic	Low vigour, symptomatic	Dead or nearly dead
Plant Health Shift Assessment			
Health Shift Class	**S1**	**S2**	**S3**	**S4**
Plant Health Shifts (2017 -> 2018)	V4 -> V4	V4 -> V3	V4 -> V2	V4 -> V1
Shift Descriptions	No vigour loss and still asymptomatic	Moderate vigour loss but still asymptomatic	Severe vigour loss and wilting appearance	Died

Reference data for plant health shifts were derived by comparing plant vigour classes in 2017 with those observed in 2018. Vigour classes for 2018 were evaluated via the same procedure described above. Shifts that occurred in highly vigorous plants from 2017 (V4) were the only ones considered due to their relevance for aetiological studies (Table 2 and Figure 1c). Indeed, plants with canopies resembling the diseased classes (V2 and V1) usually possessed a heavily deteriorated root system that may have already been colonised by secondary invaders, while the vigour reduction of plants in class V3 could have been caused by several aspects unrelated to the disease being studied.

Data reduction was applied to multispectral and thermal data by averaging the pixel values within each masked area and storing the results in a matrix associated with the coordinates of the masking centres. K-means [48] and Ward's hierarchical [49] clustering were then performed separately for multispectral (NIR and red bands) and thermal data, using two, three and four clusters. The highest number of clusters was selected based on the capability of the experts to correctly classify the reference data, while lower numbers were tested to better understand the clustering prediction capability (see below for details).

2.5. Cluster Interpretation

Finally, clustering results were compared with the reference classes of plant vigour in 2017, and the plant health shifts between 2017 and 2018, via a two-step analysis: firstly, mosaic plots were used to visualise and explore possible correlations between clusters and reference data; secondly, predictive statistics were evaluated based on confusion matrices considering the result of the correlation analysis.

Mosaic plots can display the relationship between categorical variables using rectangles whose areas represent the proportion of cases for any given combination of the multivariate categorical data [50]. Paired with residual analysis, such as Pearson X^2, the significance of such a correlation can be estimated [51]. A cluster can be considered representative of a reference class if it is positively correlated with only one class and negatively correlated with all the others.

Predictive statistics (sensitivity, accuracy, precision and F1 score) were then evaluated for combinations (reference x cluster) whose correlation was significant ($p > 0.05$) and meaningful (Figure 4). Therefore, vigour classes were compared to the clusters obtained from multispectral data, while thermal data were used for disease outbreak prediction. If the number of clusters was inferior to the number of reference classes, the latter was reduced to match the former, aggregating classes of the reference data (Figure 4). Reclassification was performed by merging the neighbour classes reported in Table 2 only if they fit the following criteria: (i) they were correlated (or at least highly represented) with the same cluster, (ii) the overall accuracy of the confusion matrix increased after class aggregation, and (iii) the new aggregated classes preserved an internal logic. Therefore, the four reference classes for plant vigour (Table 2 and Figure 4) were reduced to three and two by respectively merging the V3 and V2 classes (weakened plants) and V4, V3 and V2 classes (plants with leaves). The new aggregated reference data for vigour were then compared via multispectral clustering using three or two clusters respectively (Figure 4).

The same criteria adopted for the aggregation of vigour classes were applied also for the plant health shifts, which were compared with two clusters derived from temperature data. The plants that remained asymptomatic without changing their vigour in 2018 (S1) were merged with those displaying slightly reduced vigour without showing symptoms on the leaves (S2), creating a new class characterised by plants that remained unharmed, whereas the plants that died in 2018 (S4) or were heavily compromised (S3) were grouped in a new class composed of plants that most likely were already diseased in 2018 (Figure 4).

All the analyses regarding decorrelation, segmentation, clustering and statistics of the clusters were performed with RStudio [52].

3. Results

3.1. Multispectral and Thermal Data as an Indicator of Plant Vigour and Disease Outbreaks

For both methods tested, multispectral data were mostly correlated with plant vigour (Figure 2a–f), but it was found to not be a good predictor for disease outbreaks (Figure 2g–l). Indeed, the spatial distribution of multispectral clusters resembles that of the reference data for vigour classes (Figures 3c–h and 3a, respectively). With four clusters, it was possible to estimate all the vigour classes needed for the evaluation of the disease. With three, the weakened plants (V3 and V2) were separated from the asymptomatic and the dead ones, while with two clusters, plants with leaves were distinguished from those already dead. Overall, K-means performed better than the hierarchical method, since the differences between the middle vigour classes V3 (weakened) and V2 (diseased) could be better identified. However, hierarchical clustering showed higher correlations towards the highly vigorous plants (V4).

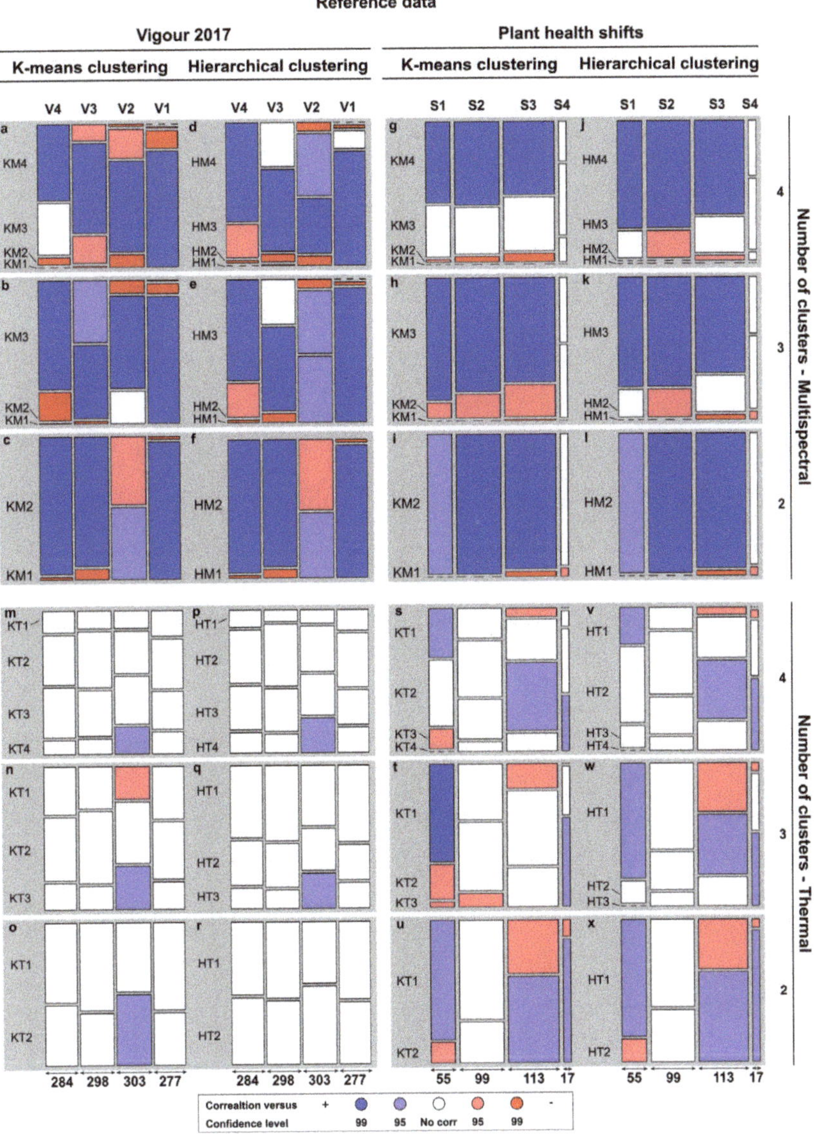

Figure 2. Correlation between clusters and reference data. (**a–f**) and (**g–l**) show the correlations of multispectral clusters with plant vigour and health shift classes, respectively. Results obtained from the K-means method are reported in (**a–c**) and (**g–i**), while those from hierarchical clustering are in (**d–f**) and (**j–l**). (**m–r**) and (**s–x**) picture the correlation of thermal clusters with plant vigour and health shift classes, respectively. (**m–o**) and (**s–u**) report the results from K-mean clustering, while (**p–r**) and (**v–x**) report those from the hierarchical method. White represents non-significant correlation. Positive and negative correlations are highlighted in blue and red, respectively, and shaded in relation to their confidence level (Pearson test): 95% confidence in brighter colours, 99% in darker tones. The bases of the rectangles are proportional to the total number of images within each reference class (numbers at the bottom of the graph), whereas heights are the proportion of reference data within the specific cluster.

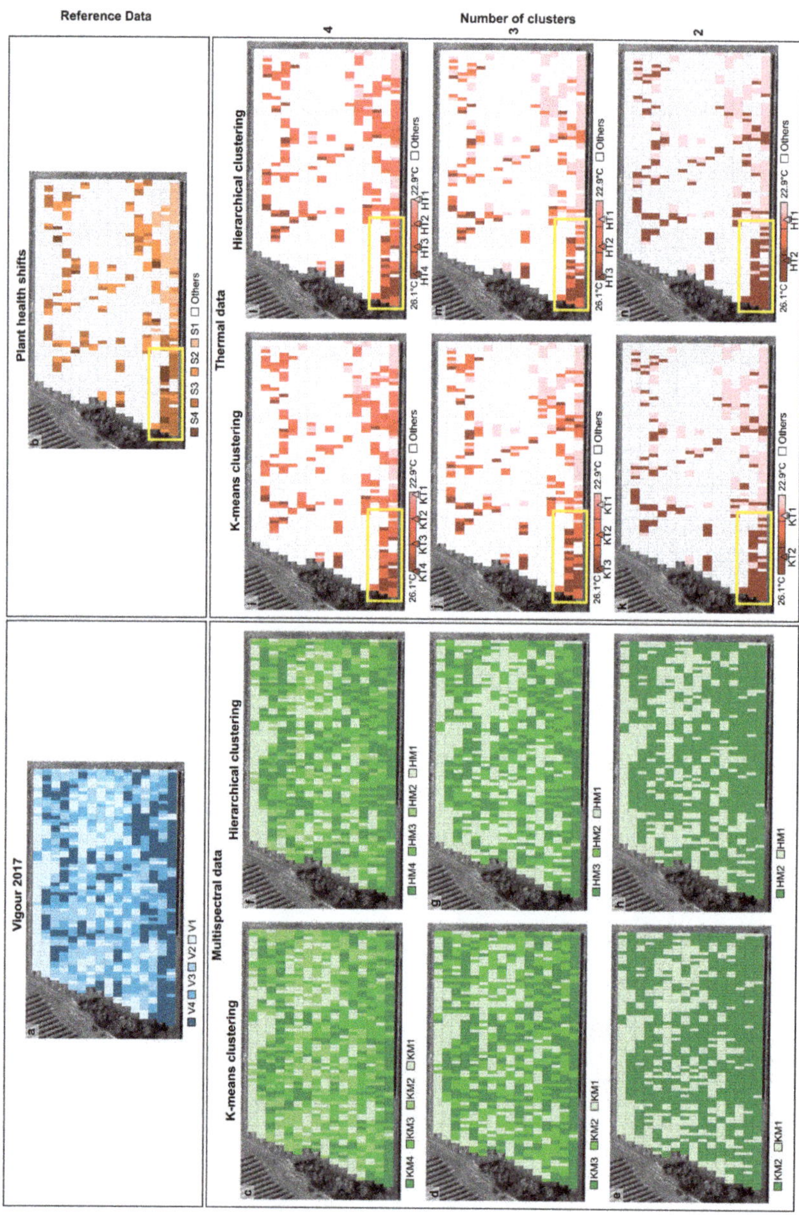

Figure 3. Spatial distribution of the results obtained after clustering and the references used for the interpretation. (**a**) In the blue scale, expert assessment ratings for the vigour classes (V4 high, V3 medium, V2 diseased, V1 dead). (**b**) In the brown scale, plant health shifts between 2017 and 2018 (S1 asymptomatic plants that preserved high vigour, S2 asymptomatic but weakened plants, S3 diseased plants but still alive, S4 plants that died in 2018). In the green scale, K-means (**c**–**e**) and hierarchical (**f**–**g**) clustering derived from multispectral bands. In the red scale, K-means (**i**–**k**) and hierarchical (**l**–**n**) clusters resulting from the analysis of thermal images: triangles indicate the temperature of the clusters'

centres. In (**b**) and (**i–n**) white represents plants that were not highly vigorous in 2017 and were therefore removed from the plant health shift analysis. Yellow rectangles in (**b**) and (**f-h**) identify the hottest area within the orchard that happened to have the highest disease incidence in 2018.

When K-means was applied to four clusters to classify multispectral orthomosaics (Figure 2a), each cluster was positively correlated with only one vigour class, and negatively correlated with all the others. Almost all the correlations were significant or highly significant, suggesting that cluster KM4 can be used for the detection of apparently asymptomatic plants with high vigour (V4), cluster KM3 for asymptomatic plants with low vigour (V3), cluster KM2 for plants that are heavily compromised but still alive (V2), and KM1 for the dead plants (V1). Misclassification errors occurred mostly between the narrower cluster classes KM3 and KM4. Hierarchical clustering showed a higher correlation between the V4 class and the HM4 cluster, and performed similarly regarding the HM1–V1 and the HM2–V2 associations (Figure 2d). However, the cluster HM3 was not clearly correlated with the V3 class, since it also contained some diseased plants belonging to the V2 class.

Using K-means with three clusters on the multispectral data (Figure 2b), the extreme vigour classes V4 and V1 were still well discriminated by KM3 and KM1, respectively, while cluster KM2 was mostly correlated with classes V2 and V3, corresponding to plants with a weakened status. Considering this association, errors can occur when assuming cluster KM3 to be solely associated with the highly vigorous plants in V4. By using hierarchical clustering, the association HM3–V4 and HM2 (V2 and V3) was maintained (Figure 2e), but most of the misclassifications occurred within the cluster HM1, which was not highly correlated solely with dead plants (V1) and was also correlated with the V2 class.

Finally, with two clusters only, both methods performed equally (Figure 2c,f). The plants with an asymptomatic canopy (V4 and V3) were clearly separated from the dead plants (V1), and highly correlated with clusters KM2 or HM2, and KM1 or HM1, respectively. Major errors occurred in the classification of plants that were diseased but still alive (V3), which seemed to be better correlated with clusters KM1 and HM1.

For both methods tested, thermal data were mostly correlated with health shifts (Figure 2s–v), but could not be associated with any vigour classes (Figure 2m–r). The hottest area in 2017 within the orchard happened to be the one with the highest disease incidence in 2018 (Figure 3b,j–n). Canopy temperature ranged between 22 °C and 30 °C, with the clusters averaging between 23 °C and 26 °C, depending on the number of clusters used. K-means centres were on average slightly higher (+0.3 °C) than the means evaluated for each hierarchical cluster. The mosaic plots suggested that the temperature data were able to predict disease outbreaks by clustering the plants into two groups: one where disease symptoms appeared (S3 and S4) and another where the plants remained asymptomatic (S1 and S2). Indeed, among the tested combinations, the two-cluster model showed the best discrimination between the plant health shift classes. Both algorithms performed similarly, with K-means manifesting slightly better correlations between the plants that remained highly vigorous in 2018 (S1).

Nevertheless, the plants that died in 2018 (S4) were always positively associated with the hottest clusters, regardless of the number of clusters used (Figure 2s–v). Furthermore, plants that were heavily compromised, but still alive, in 2018 (S3) were usually abundant in the hottest cluster, although not always with a significant correlation for all numbers of clusters tested (Figure 2s–v). Lastly, the coldest clusters, KT1 and HT1, were consistently positively associated with plants that also maintained high vigour in 2018 (S1), and negatively correlated with plants that became diseased (S3 and S4) (Figure 2s–v).

3.2. Clustering Interpretation

Based on the mosaic plot results, the accuracy in estimating the vigour classes or plant health shifts was tested via a confusion matrix (Figure 4). The clustering of multispectral data performed similarly for both methods, with better statistics for K-means compared to hierarchical clustering. Dead plants (V1) were the class with the highest predictive statistics for all the numbers of clusters tested in both methods, obtaining the best score set when four clusters were used. By reducing the number of clusters, the precision dropped to a minimum of 60%, but the sensitivity increased greatly (98%). The same behaviour was observed for plants with high vigour (V4) that were correctly classified into clusters KM4 or HM4 and KM3 or HM3, when the data were grouped into four or three clusters, respectively.

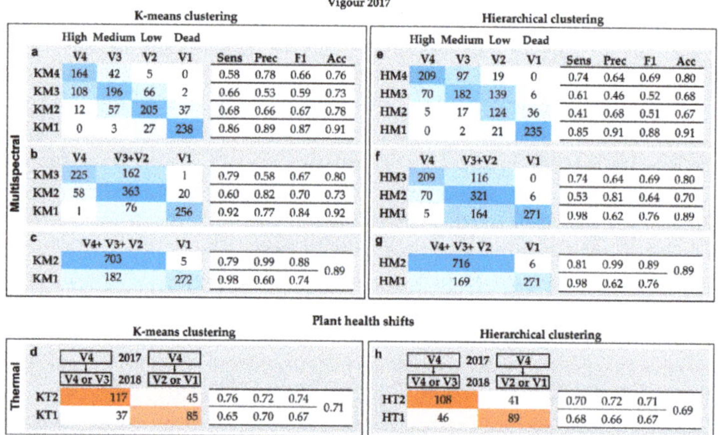

Figure 4. Confusion matrices of unsupervised clustered data for the assessment of disease spreading (plant vigour) and the prediction of future outbreak (plant health shifts). K-means (**a–c**) and hierarchical clustering (**e–g**) performance in assessing disease spreading using 4, 3 or 2 clusters, and using multispectral data. (**d**) and (**e**), performance of thermal clusters in predicting future outbreaks using 2 clusters, for K-means and hierarchical clustering, respectively. In (**b–d**) and (**f–h**) reference data were merged if the number of clusters was inferior to the number of reference classes. In (**b**) and (**f**), a class was created aggregating weakened and diseased plants (V3 and V2). In (**c**) and (**g**), the three classes with leaves (V4, V3 and V2) were grouped together. In (**d**) and (**h**), plants that remained asymptomatic until 2018 (left) were split apart from those that showed wilting symptoms in 2018 (right). Abbreviation: Sesn, Sensitivity; Prec, Precision; F1, F1-score; Acc, accuracy; KM1-KM4 and KT1, KT2 multispectral and thermal clusters.

For the K-means method, using four clusters was the approach that showed the best association with plant vigour. Each cluster was associated with a single class with an accuracy above 73% (Figure 4a). In particular, the extreme classes (V1 and V4) were precisely identified (precision above 78%), with a low number of false positives. Misclassification errors mostly occurred between neighbour classes, mostly due to false positives that occurred in the association of cluster KM3 with the vigour class V3. Reducing the number of clusters to three, the central cluster KM2 was associated with weakened plants (V3 and V2) with an accuracy of 73% and a precision above 82% (Figure 4b). Finally, using only two clusters, it was possible to distinguish between dead plants with no canopy and plants with leaves (Figure 4c).

Hierarchical clustering was more sensible than K-means clustering for highly vigorous plants (V4), especially when four clusters were used (Figure 4e). The classification of the middle classes V3 and V2 was the major reason for errors. Indeed, with four clusters, this caused a reduction of the accuracy by as much as 0.68% and 0.67%, when V3 and V2 classes were respectively associated with HM3 and HM2 (Figure 4e). By reducing the number of clusters to three (Figure 4f), the association between weakened plants (V3+V2) and the middle cluster HM2 still existed (precision 81%), but the accuracy was reduced by the association of some plants of class V3 (diseased) with the HM1 cluster that was highly correlated with dead plants as well (V1) (Figures 2e and 4f). Finally, the predictive statistics obtained with two clusters were almost equal to those obtained by the K-means method (Figure 4g).

The predictive capability of thermal data was tested for health shifts, which occurred in plants that were highly vigorous in 2017 (Figure 4d,h). The best predictive capability for disease spread was observed when clustering was performed with only two clusters, which allowed discrimination between plants that showed wilting and those that did not. The performances of the two methods were similar, with K-means being slightly better than hierarchical clustering (Figure 4d,h).

4. Discussion

Our results suggest that UAVs combined with multispectral and thermal imaging can be useful tools for scouting activities at a field scale in kiwifruit orchards, and can greatly improve monitoring activities. For instance, the sole observation of an RGB map is highly valuable, because it provides experts with an overview of the entire field, improving their awareness of the disease spread (Figure A1 in Appendix A).

Multispectral data can be used to speed up the assessment of the symptomatic plants due to their strong correlation with plant vigour. By identifying dead or highly compromised plants, it was certainly possible to precisely delimit and monitor the spread of the disease, while the detection of very vigorous plants made it possible to identify homogeneous areas, where other analyses should be focused for the prediction of disease outbreaks. Our findings are in line with those of other studies that used NIR (850–1700 nm), green (495–570 nm) and red (620–750 nm) wavelengths to develop vegetation indices for the estimation of plant biophysical traits [14,53–55]. Indeed, healthy plants are usually characterised by a higher reflectance in the NIR bands [56]. Studies of multispectral data regarding soil-borne diseases identified NIR bands as an important factor for the detection of plant vigour and the estimation of disease severity indices [57–61]. However, as observed in our study, multispectral data seem to have a poor predictive capability for this kind of diseases, and they are usually useful once symptoms are visible [34,36,57–61].

The lack of predictive capability within the multispectral data can be offset by thermal data. The canopy temperature was found to be a reliable predictor of plant health status, being able to indicate the spread of the disease one year before the actual outbreak. It is interesting to note that the best prediction metrics were obtained precisely for plants that were completely asymptomatic in 2017, far exceeding the predictive capability of any expert. Indeed, these plants were still apparently healthy in both October 2017 and in the first half of the vegetative season in 2018. Until June 2018, they produced a wide canopy, but suddenly after the first heat waves of July, they showed scorching, leaf drop and, in the worst cases, complete defoliation. Results obtained from the clustering of thermal data confirmed that plant responses to KD might be similar to those induced by drought stress, as also observed for other soil-borne diseases [23,34,36]. Indeed, just like abiotic factors, pathogens may affect the stomatal response by influencing the temperature gradient between the plant tissue and the air [62]. Soil-borne diseases cause the reduction of water absorption translocation and transpiration functions as a net result of their infection [63]. These alterations induce a closure of the stomata, and consequently the increase of leaf temperature as a consequence of the reduced evaporative cooling effect [24,63,64]. Indeed, for soil-borne diseases, canopy temperature has been shown to be particularly useful in detecting compromised plants, even during the early stages of infection when visual symptoms were invisible [23,34,36,63,64]. It can be speculated that the multispectral data probably failed to predict disease outbreaks because KD does not cause internal structure modifications until a critical point is reached, at which time the root system can no longer support the plant's transpiration rate [65]. It seems that the kiwifruit has a very high root/canopy ratio [25,66], and thus can cope with a substantial (80%) loss of the root system before the growth of shoots and leaves is affected [25]. Conversely, a justification for the predictive capability of canopy temperature data may reside in the physiological response of kiwi vines to drought stress or root loss, which induce a rapid reduction in gas exchange fluxes and consequently an increase in leaf temperature [24,25,67]. Finally, it should also be noted that even flooding conditions can induce a reduction in the transpiration rates of kiwi vines [27], and thus the water content of the soil cannot be overlooked during temperature data acquisition for KD detection.

Based on our results, the use of unsupervised clustering can be a reliable method for quickly exploring and identifying sampling areas that are suitable for aetiological studies. By using multispectral data, symptomatic plants can be easily identified, simplifying the disease assessment process. The clustering of thermal data is even more useful, since the early detection of diseased plants allows us not only to identify the best areas where samples for laboratory analysis should be collected, but also to observe the evolution of symptoms over time, and improve the understanding of root degradation dynamics. However, we believe that the unsupervised clustering of remote sensing data might also be implemented for other diseases. To the best of our knowledge, unsupervised clustering has not been used previously for forecasting purposes, although it has been used for the assessment of cotton root rot [68], for structuring the aggressiveness of fungal infections on peas [69], and for the segmentation of plant backgrounds for weed detection [70]. With our experimental set up, K-means gave better results than hierarchical clustering in both assessing and predicting the disease spread. However, it cannot be ignored that under other conditions the performances of these two methods might differ. The biggest drawback of the unsupervised classification methods is the need for an expert to assign meaning to each cluster. Nevertheless, for KD, cluster interpretation can be easily achieved by overlapping cluster results with an RGB orthomosaic. The use of supervised machine learning algorithms might be a solution to the problem of directly classifying plant health status, but when the aetiological background is not clear or the labelling is impractical (such as in the KD syndrome), the development of a reliable training dataset is difficult because rapid tests to discriminate between diseased and healthy plants are currently unavailable. Therefore, unsupervised clustering is an appropriate method of seeing or finding groups within unknown data when labels are not available, especially when we do not know what kind of sensor data is needed to find patterns and groups [71].

The study was limited by low cost technologies, and it was performed by only analysing shifts that occurred between two seasons, so several future improvements can be envisaged with this approach. A better segmentation algorithm, based on sensor fusion processes and improved geometric accuracy of the orthomosaics, is needed to remove background noise and to improve the quality of the analysis. In this regard, a similar approach was successfully adopted to detect olive plants infected by *Verticillium* wilt, using airborne data captured simultaneously with thermal, fluorescence and hyperspectral detectors [34]. Moreover, the advantages derived from sensor fusion are already evident in fruit safety and quality control studies [72–74]. Segmentation could also be improved by using automatic tree segmentation with object-based image analysis [75]. In our study, the low-cost multispectral sensors available precluded the possibility of correctly segmenting the inter-row grass from the kiwifruit canopy. Nevertheless, testing flexible and low-cost methods is important to ensure the real application of remote sensing technologies to the field scale. A better understanding of the relationship between temperature and plant health status is also needed. Models that are able to predict leaf temperature using data from weather stations (e.g., air temperature, air relative humidity, radiation, soil water content) should be developed to set more objective thresholds for discriminating healthy from diseased plants, and thus use temperature as a more reliable predictor. Temporal analysis of simple vegetation indices might also improve our understanding of the disease dynamics and increase the detection accuracy of KD. Studies on a wider range of wavelengths are also needed to identify the best bands for disease detection and prediction. In particular, it would be interesting to study the possible correlations between KD appearance and short-wave infrared wavelengths (1100 to 2500 nm), as these bands are influenced by leaf chemical composition and water content [76,77].

5. Conclusions

This study is the very first to demonstrate the prediction of the spread of Kiwifruit Decline using remote sensing technologies. Our results indicate that unsupervised clustering can be a reliable algorithm for characterising the disease in its early stages, in order to identify homogenous areas out in the field. Multispectral data can be used to discriminate symptomatic from asymptomatic plants, allowing a quick estimation of disease spread. On the other hand, thermal data have been shown to be effective in predicting future outbreaks of the disease, by providing an informative tool for directing the sampling activities for aetiological and epidemiological studies.

Author Contributions: Conceptualisation, F.S., M.M., P.E., S.P. and A.-K.M.; Data curation, F.S. and S.P.; Formal analysis, F.S. and S.P.; Funding acquisition, M.M., P.E. and A.-K.M.; Methodology, F.S., M.M., P.E., S.P. and A.-K.M.; Supervision, A.-K.M.; Writing: original draft, F.S.; Writing: review and editing, M.M., P.E., S.P. and A.-K.M. All authors have read and agreed to the published version of the manuscript.

Funding: SP and AKM were partially funded by the Deutsche Forschungsgemeinschaft (DFG, German Research Foundation) under Germany's Excellence Strategy, EXC 2070 – 390732324. FS, PE, MM were partially funded by the Friuli Venezia Giulia region (Italy) CUP: G24I19000810002.

Acknowledgments: Simone Saro, from ERSA—Phytosanitary Service of the Friuli Venezia Giulia region (Italy)—for his collaboration and support in field-scouting activities. Luca Zuliani, from Adron Technology srl, for his help in the pre-processing of images and generation of the orthomosaics. Gianni Tacconi from CREA—the Council for Agricultural Research and Agricultural Economics Analysis—for his insights regarding the spread of Kiwifruit Decline in Italy.

Conflicts of Interest: The authors declare no conflict of interest.

Appendix A

Figure A1. Low resolution RGB orthomosaics in 2017 (**a**) and 2018 (**b**).

References

1. FAOSTAT Food and Agriculture Organization of the United Nations, Rome, Italy. Available online: http://www.fao.org/faostat/en/#home (accessed on 4 September 2019).
2. Tacconi, G.; Paltrinieri, S.; Mejia, J.F.; Fuentealba, S.P.; Bertaccini, A.; Tosi, L.; Giacopini, A.; Mazzucchi, U.; Favaron, F.; Sella, L.; et al. Vine decline in kiwifruit: Climate change and effect on waterlogging and *Phytophthora* in north Italy. *Acta Hortic.* **2015**, 93–97. [CrossRef]
3. Sorrenti, G.; Tacconi, G.; Tosi, L.; Vittone, G.; Nari, L.; Savian, F.; Saro, S.; Ermacora, P.; Graziani, S.; Toselli, M. Avanza la "moria del kiwi": Evoluzione e primi riscontri della ricerca. *Frutticoltura* **2019**, *2*, 34–42.
4. Reid, J.B.; Tate, K.G.; Brown, N.S. Effects of flooding and alluvium deposition on kiwifruit (Actinidia deliciosa). *N. Z. J. Crop Hortic. Sci.* **1992**, *20*, 283–288. [CrossRef]
5. Huang, Y.; Qi, P. Studies on the cause of root rot of kiwifruit in Guangdong Province. *J. South China Agric. Univ.* **1998**, *19*, 19–22.
6. Akilli, S.; Serçe, Ç.U.; Zekaİ Katircioğlu, Y.; Karakaya, A.; Maden, S. Involvement of *Phytophthora citrophthora* in kiwifruit decline in Turkey. *J. Phytopathol.* **2011**, *159*, 579–581. [CrossRef]
7. Kurbetli, İ.; Ozan, S. Occurrence of *Phytophthora* root and stem rot of kiwifruit in Turkey. *J. Phytopathol.* **2013**, *161*, 887–889. [CrossRef]
8. Polat, Z.; Awan, Q.N.; Hussain, M.; Akgül, D.S. First report of *Phytopythium vexans* causing root and collar rot of kiwifruit in turkey. *Plant Dis.* **2017**, *101*, 1058. [CrossRef]
9. Savian, F.; Martini, M.; Borselli, S.; Saro, S.; Musetti, R.; Loi, N.; Firrao, G. Studies on Kiwifruit Decline, an emerging issue even for Friuli Venezia Giulia (Eastern Italy). *J. Plant Pathol.* **2017**, *99*, 18.
10. Savian, F.; Musetti, R.; Sandrin, N.; Ermacora, P.; Martini, M. Etiological studies over Kiwifruit Decline reveal the involvement of both flooding and biotic factors. In Proceedings of the Abstracts of Presentations at the XXV Congress of the Italian Phytopathological Society (SIPaV). *J. Plant Pathol.* **2019**, *101*, 811–848.
11. Prencipe, S.; Savian, F.; Nari, L.; Ermacora, P.; Spadaro, D.; Martini, M. First Report of *Phytopythium vexans* Causing Decline Syndrome of *Actinidia deliciosa* 'Hayward' in Italy. *Plant Dis.* **2020**, *104*, 2032. [CrossRef]
12. Spigaglia, P.; Barbanti, F.; Marocchi, F.; Mastroleo, M.; Baretta, M.; Ferrante, P.; Caboni, E.; Lucioli, S.; Scortichini, M. *Clostridium bifermentans* and *C. subterminale* are associated with kiwifruit vine decline, known as *moria*, in Italy. *Plant Pathol.* **2020**, *69*, 765–774. [CrossRef]
13. Donati, I.; Cellini, A.; Sangiorgio, D.; Caldera, E.; Sorrenti, G.; Spinelli, F. Pathogens Associated to Kiwifruit Vine Decline in Italy. *Agriculture* **2020**, *10*, 119. [CrossRef]
14. Xue, J.; Su, B. Significant remote sensing vegetation indices: A review of developments and applications. *J. Sens.* **2017**, *2017*, 1–17. [CrossRef]
15. Li, L.; Zhang, Q.; Huang, D. A review of imaging techniques for plant phenotyping. *Sensors* **2014**, *14*, 20078–20111. [CrossRef] [PubMed]
16. Sankaran, S.; Mishra, A.; Ehsani, R.; Davis, C. A review of advanced techniques for detecting plant diseases. *Comput. Electron. Agric.* **2010**, *72*, 1–13. [CrossRef]
17. Mills, L.; Flemmer, R.; Flemmer, C.; Bakker, H. Prediction of kiwifruit orchard characteristics from satellite images. *Precis. Agric.* **2019**, *20*, 911–925. [CrossRef]
18. Walter, J.; Edwards, J.; Cai, J.; McDonald, G.; Miklavcic, S.J.; Kuchel, H. High-throughput field imaging and basic image analysis in a wheat breeding programme. *Front. Plant Sci.* **2019**, *10*, 449. [CrossRef] [PubMed]
19. Mullan, D.J.; Reynolds, M.P. Quantifying genetic effects of ground cover on soil water evaporation using digital imaging. *Funct. Plant Biol.* **2010**, *37*, 703. [CrossRef]
20. Liu, J.; Pattey, E. Retrieval of leaf area index from top-of-canopy digital photography over agricultural crops. *Agric. For. Meteorol.* **2010**, *150*, 1485–1490. [CrossRef]
21. Lukina, E.V.; Stone, M.L.; Raun, W.R. Estimating vegetation coverage in wheat using digital images. *J. Plant Nutr.* **1999**, *22*, 341–350. [CrossRef]
22. Carter, G.A.; Knapp, A.K. Leaf optical properties in higher plants: Linking spectral characteristics to stress and chlorophyll concentration. *Am. J. Bot.* **2001**, *88*, 677–684. [CrossRef] [PubMed]
23. West, J.S.; Bravo, C.; Oberti, R.; Lemaire, D.; Moshou, D.; McCartney, H.A. The potential of optical canopy measurement for targeted control of field crop diseases. *Annu. Rev. Phytopathol.* **2003**, *41*, 593–614. [CrossRef] [PubMed]

24. Gucci, R.; Massai, R.; Piccotino, D.; Xiloyannis, C. Gas exchange characteristics and water relations of kiwifruit vines during drought cycles. *Acta Hortic.* **1993**, 213–218. [CrossRef]

25. Black, M.Z. The Kiwifruit (Actinidia sp.) Vine Root System: Responses to Vine Manipulations. Ph.D. Thesis, University of Waikato, Hamilton, New Zealand, 2012.

26. Savé, R.; Serrano, L. Some physiological and growth responses of kiwi fruit (*Actinidia chinensis*) to flooding. *Physiol. Plant.* **1986**, *66*, 75–78. [CrossRef]

27. Smith, G.S.; Judd, M.J.; Miller, S.A.; Buwalda, J.G. Recovery of kiwifruit vines from transient waterlogging of the root system. *New Phytol.* **1990**, *115*, 325–333. [CrossRef]

28. Nuzzo, V.; Dichio, B.; Montanaro, G.; Xiloyannis, C. Risposta di piante di actinidia in piena produzione alle limitate disponibilità idriche del suolo. Atti del Convegno Nazionale. In Proceedings of the Atti del Convegno Nazionale "La coltura dell'actinidia", Faenza, Italy, 10–12 October 1996.

29. Mills, T.M.; Li, J.; Behboudian, M.H. Physiological responses of gold kiwifruit (*Actinidia chinensis*) to reduced irrigation. *J. Am. Soc. Hortic. Sci.* **2009**, *134*, 677–683. [CrossRef]

30. Mahlein, A.-K. Plant disease detection by imaging sensors—parallels and specific demands for precision agriculture and plant phenotyping. *Plant Dis.* **2015**, *100*, 241–251. [CrossRef]

31. Jones, H.; Schofield, P. Thermal and other remote sensing of plant stress. *Gen. Appl. Plant Physiol.* **2008**, *34*, 19–32.

32. Maes, W.H.; Steppe, K. Estimating evapotranspiration and drought stress with ground-based thermal remote sensing in agriculture: A review. *J. Exp. Bot.* **2012**, *63*, 4671–4712. [CrossRef]

33. Mangus, D.L.; Sharda, A.; Zhang, N. Development and evaluation of thermal infrared imaging system for high spatial and temporal resolution crop water stress monitoring of corn within a greenhouse. *Comput. Electron. Agric.* **2016**, *121*, 149–159. [CrossRef]

34. Calderón, R.; Navas-Cortés, J.A.; Lucena, C.; Zarco-Tejada, P.J. High-resolution airborne hyperspectral and thermal imagery for early detection of Verticillium wilt of olive using fluorescence, temperature and narrow-band spectral indices. *Remote Sens. Environ.* **2013**, *139*, 231–245. [CrossRef]

35. Ortiz-Bustos, C.M.; Pérez-Bueno, M.L.; Barón, M.; Molinero-Ruiz, L. Use of Blue-Green Fluorescence and Thermal Imaging in the Early Detection of Sunflower Infection by the Root Parasitic Weed Orobanche cumana Wallr. *Front. Plant Sci.* **2017**, *8*, 833. [CrossRef] [PubMed]

36. Zarco-Tejada, P.J.; Camino, C.; Beck, P.S.A.; Calderon, R.; Hornero, A.; Hernández-Clemente, R.; Kattenborn, T.; Montes-Borrego, M.; Susca, L.; Morelli, M. Previsual symptoms of *Xylella fastidiosa* infection revealed in spectral plant-trait alterations. *Nat. Plants* **2018**, *4*, 432. [CrossRef] [PubMed]

37. Maes, W.H.; Minchin, P.E.H.; Snelgar, W.P.; Steppe, K. Early detection of Psa infection in kiwifruit by means of infrared thermography at leaf and orchard scale. *Funct. Plant Biol.* **2014**, *41*, 1207. [CrossRef] [PubMed]

38. Sankaran, S.; Khot, L.R.; Espinoza, C.Z.; Jarolmasjed, S.; Sathuvalli, V.R.; Vandemark, G.J.; Miklas, P.N.; Carter, A.H.; Pumphrey, M.O.; Knowles, N.R.; et al. Low-altitude, high-resolution aerial imaging systems for row and field crop phenotyping: A review. *Eur. J. Agron.* **2015**, *70*, 112–123. [CrossRef]

39. Manfreda, S.; McCabe, M.; Miller, P.; Lucas, R.; Pajuelo Madrigal, V.; Mallinis, G.; Ben Dor, E.; Helman, D.; Estes, L.; Ciraolo, G.; et al. On the Use of Unmanned Aerial Systems for Environmental Monitoring. *Remote Sens.* **2018**, *10*, 641. [CrossRef]

40. Kanmani, A.P.; Obringer, R.; Rachunok, B.; Nateghi, R. Assessing Global Environmental Sustainability Via an Unsupervised Clustering Framework. *Sustainability* **2020**, *12*, 563. [CrossRef]

41. Huang, H. *Kiwifruit: The Genus Actinidia*, 1st ed.; Academic Press: Cambridge, MA, USA, 2016; ISBN 978-0-12-803066-0.

42. Saxton, K.E.; Willey, P.H. The SPAW model for agricultural field and pond hydrologic simulation. In *Watershed Models*; Frevert, D., Singh, V., Eds.; CRC Press: Boca Raton, FL, USA, 2005; pp. 400–435, ISBN 978-1-4200-3743-2.

43. Baghdadi, N.; Zribi, M. Characterization of Soil Surface Properties Using Radar Remote Sensing. In *Land Surface Remote Sensing in Continental Hydrology*; Elsevier: Amsterdam, The Netherlands, 2016; pp. 1–39, ISBN 978-1-78548-104-8.

44. Malbéteau, Y.; Parkes, S.; Aragon, B.; Rosas, J.; McCabe, M. Capturing the Diurnal Cycle of Land Surface Temperature Using an Unmanned Aerial Vehicle. *Remote Sens.* **2018**, *10*, 1407. [CrossRef]

45. Elmasry, G.; Kamruzzaman, M.; Sun, D.-W.; Allen, P. Principles and Applications of Hyperspectral Imaging in Quality Evaluation of Agro-Food Products: A Review. *Crit. Rev. Food Sci. Nutr.* **2012**, *52*, 999–1023. [CrossRef]
46. Poncet, A.M.; Knappenberger, T.; Brodbeck, C.; Fogle, M.; Shaw, J.N.; Ortiz, B.V. Multispectral UAS Data Accuracy for Different Radiometric Calibration Methods. *Remote Sens.* **2019**, *11*, 1917. [CrossRef]
47. QGIS development Team. QGIS Geographic Information System. Open Source Geospatial Found. Proj. 2016. Available online: http://qgis.osgeo.org (accessed on 1 July 2020).
48. Hartigan, J.A.; Wong, M.A. Algorithm AS 136: A K-Means Clustering Algorithm. *Appl. Stat.* **1979**, *28*, 100. [CrossRef]
49. Murtagh, F.; Legendre, P. Ward's Hierarchical Agglomerative Clustering Method: Which Algorithms Implement Ward's Criterion? *J. Classif.* **2014**, *31*, 274–295. [CrossRef]
50. Friendly, M. Mosaic displays for multi-way contingency tables. *J. Am. Stat. Assoc.* **1994**, *89*, 190–200. [CrossRef]
51. Mayer, D.; Zeileis, A.; Hornik, K. The strucplot framework: Visualizing multi-way contingency tables with vcd. *J. Stat. Softw.* **2006**, *17*, 1–48. [CrossRef]
52. Core Team R: A language and environment for statistical computing. R Foundation for Statistical Computing, Vienna, Austria. 2013. Available online: http://www.R-project.org/ (accessed on 1 July 2020).
53. Thenkabail, P.S.; Smith, R.B.; De Pauw, E. Hyperspectral Vegetation Indices and Their Relationships with Agricultural Crop Characteristics. *Remote Sens. Environ.* **2000**, *71*, 158–182. [CrossRef]
54. Boelman, N.T.; Stieglitz, M.; Rueth, H.M.; Sommerkorn, M.; Griffin, K.L.; Shaver, G.R.; Gamon, J.A. Response of NDVI, biomass, and ecosystem gas exchange to long-term warming and fertilization in wet sedge tundra. *Oecologia* **2003**, *135*, 414–421. [CrossRef]
55. Quirós Vargas, J.J.; Zhang, C.; Smitchger, J.A.; McGee, R.J.; Sankaran, S. Phenotyping of plant biomass and performance traits using remote sensing techniques in pea (*Pisum sativum*, l.). *Sensors* **2019**, *19*, 2031. [CrossRef]
56. Woolley, J.T. Reflectance and transmittance of light by leaves. *Plant Physiol.* **1971**, *47*, 656–662. [CrossRef]
57. Wang, D.; Kurle, J.E.; Estevez de Jensen, C.; Percich, J.A. Radiometric assessment of tillage and seed treatment effect on soybean root rot caused by *Fusarium* spp. in central Minnesota. *Plant Soil* **2004**, *258*, 319–331. [CrossRef]
58. Raikes, C.; Burpee, L.L. Use of multispectral radiometry for assessment of rhizoctonia blight in creeping bentgrass. *Phytopathology* **1998**, *88*, 446–449. [CrossRef]
59. Duarte-Carvajalino, J.M.; Alzate, D.F.; Ramirez, A.A.; Santa-Sepulveda, J.D.; Fajardo-Rojas, A.E.; Soto-Suárez, M. Evaluating Late Blight Severity in Potato Crops Using Unmanned Aerial Vehicles and Machine Learning Algorithms. *Remote Sens.* **2018**, *10*, 1513. [CrossRef]
60. Hill, R.J.; Wilson, B.A.; Rookes, J.E.; Cahill, D.M. Use of high resolution digital multi-spectral imagery to assess the distribution of disease caused by *Phytophthora cinnamomi* on heathland at Anglesea, Victoria. *Australas. Plant Pathol.* **2009**, *38*, 110–119. [CrossRef]
61. Salgadoe, A.S.A.; Robson, A.J.; Lamb, D.W.; Dann, E.K.; Searle, C. Quantifying the Severity of Phytophthora Root Rot Disease in Avocado Trees Using Image Analysis. *Remote Sens.* **2018**, *10*, 226. [CrossRef]
62. Prashar, A.; Jones, H. Infra-red thermography as a high-throughput tool for field phenotyping. *Agronomy* **2014**, *4*, 397–417. [CrossRef]
63. Hernández-Clemente, R.; Hornero, A.; Mottus, M.; Penuelas, J.; González-Dugo, V.; Jiménez, J.C.; Suárez, L.; Alonso, L.; Zarco-Tejada, P.J. Early diagnosis of vegetation health from high-resolution hyperspectral and thermal imagery: Lessons learned from empirical relationships and radiative transfer modelling. *Curr. For. Rep.* **2019**, *5*, 169–183. [CrossRef]
64. Pinter, P.J.; Stanghellini, M.E.; Reginato, R.J.; Idso, S.B.; Jenkins, A.D.; Jackson, R.D. Remote Detection of Biological Stresses in Plants with Infrared Thermometry. *Science* **1979**, *205*, 585–587. [CrossRef]
65. Sorrenti, G.; Toselli, M.; Reggidori, G.; Spinelli, F.; Tosi, L.; Giacopini, A.; Tacconi, G. Implicazioni della gestione idrica nella "Moria del kiwi" del veronese. *Frutticoltura* **2016**, *3*, 1–7.
66. Reid, J.B.; Petrie, R.A. Effects of soil aeration on root demography in kiwifruit. *N. Z. J. Crop Hortic. Sci.* **1991**, *19*, 423–432. [CrossRef]
67. Xiloyannis, C.; Massai, R.; Piccotino, D.; Baroni, G.; Bovo, M. Method and technique of irrigation in relation to root system characteristics in fruit growing. *Acta Hortic.* **1993**, 505–510. [CrossRef]

68. Yang, C.; Odvody, G.N.; Fernandez, C.J.; Landivar, J.A.; Minzenmayer, R.R.; Nichols, R.L. Evaluating unsupervised and supervised image classification methods for mapping cotton root rot. *Precis. Agric.* **2015**, *16*, 201–215. [CrossRef]

69. Setti, B.; Bencheikh, M.; Henni, J.; Neema, C. Comparative aggressiveness of *Mycosphaerella pinodes*on peas from different regions in western algeria. *Phytopathol. Mediterr.* **2009**, *48*, 195–204.

70. Tang, L.; Tian, L.F.; Steward, B.L. Color image segmentation with genetic algorithm for in-field weed sensing. *Trans. ASAE* **2000**, *43*, 1019. [CrossRef]

71. Behmann, J.; Mahlein, A.-K.; Rumpf, T.; Römer, C.; Plümer, L. A review of advanced machine learning methods for the detection of biotic stress in precision crop protection. *Precis. Agric.* **2015**, *16*, 239–260. [CrossRef]

72. Blasco, J.; Aleixos, N.; Gómez, J.; Moltó, E. *Citrus* sorting by identification of the most common defects using multispectral computer vision. *J. Food Eng.* **2007**, *83*, 384–393. [CrossRef]

73. Gowen, A.; Odonnell, C.; Cullen, P.; Downey, G.; Frias, J. Hyperspectral imaging – an emerging process analytical tool for food quality and safety control. *Trends Food Sci. Technol.* **2007**, *18*, 590–598. [CrossRef]

74. Sighicelli, M.; Colao, F.; Lai, A.; Patsaeva, S. Monitoring post-harvest orange fruit disease by fluorescence and reflectance hyperspectral imaging. *Acta Hortic.* **2009**, 277–284. [CrossRef]

75. Zhen, Z.; Quackenbush, L.; Zhang, L. Trends in Automatic Individual Tree Crown Detection and Delineation—Evolution of LiDAR Data. *Remote Sens.* **2016**, *8*, 333. [CrossRef]

76. Jacquemoud, S.; Ustin, S.L. Leaf optical properties: A state of the art. In Proceedings of the 8th International Symposium of Physical Measurements & Signatures in Remote Sensing, Aussois, France, 8–12 January 2001; pp. 223–332.

77. Carter, G.A. Ratios of leaf reflectances in narrow wavebands as indicators of plant stress. *Int. J. Remote Sens.* **1994**, *15*, 697–703. [CrossRef]

 *remote sensing*

Article

Practical Recommendations for Hyperspectral and Thermal Proximal Disease Sensing in Potato and Leek Fields

Simon Appeltans [1], Angela Guerrero [1], Said Nawar [1], Jan Pieters [2] and Abdul M. Mouazen [1],*

[1]   Department of Environment, Faculty of Bioscience Engineering, Ghent University, 9000 Ghent, Belgium; Simon.Appeltans@UGent.be (S.A.); Angela.Guerrero@UGent.be (A.G.); said.nawar@ugent.be (S.N.)

[2]   Department of Plants and Crops, Faculty of Bioscience Engineering, Ghent University, 9000 Ghent, Belgium; Jan.Pieters@UGent.be

*   Correspondence: Abdul.Mouazen@UGent.be; Tel.: +32-9-264-6037

Received: 26 May 2020; Accepted: 12 June 2020; Published: 15 June 2020

Abstract: Thermal and hyperspectral proximal disease sensing are valuable tools towards increasing pesticide use efficiency. However, some practical aspects of the implementation of these sensors remain poorly understood. We studied an optimal measurement setup combining both sensors for disease detection in leek and potato. This was achieved by optimising the signal-to-noise ratio (SNR) based on the height of measurement above the crop canopy, off-zenith camera angle and exposure time (ET) of the sensor. Our results indicated a clear increase in SNR with increasing ET for potato. Taking into account practical constraints, the suggested setup for a hyperspectral sensor in our experiment involves (for both leek and potato) an off-zenith angle of 17°, height of 30 cm above crop canopy and ET of 1 ms, which differs from the optimal setup of the same sensor for wheat. Artificial light proved important to counteract the effect of cloud cover on hyperspectral measurements. The interference of these lamps with thermal measurements was minimal for a young leek crop but increased in older leek and after long exposure. These results indicate the importance of optimising the setup before measurements, for each type of crop.

Keywords: hyperspectral; thermal; proximal sensing; disease detection; signal-to-noise ratio

1. Introduction

The agricultural sector is under constant pressure to produce more efficiently and sustainably [1]. One of the most important factors for sustainable food production has always been disease management [2]. Now, globalisation has facilitated the spread of plant pathogens [3–5]. This, combined with changing climate conditions, poses great challenges for modern crop protection [6–9]. Disease typically appears in patches, after which it starts spreading to the rest of the crop [10]. The ability to detect these infections and manage them site-specifically, i.e., in a precision agriculture approach, has the potential to significantly increase pesticide use efficiency and thereby reduce economic and environmental costs, compared to current full-field 'homogeneous' applications [8]. Thermal and hyperspectral sensors have been proposed as useful tools for site-specific crop management, in both aerial and ground-based measurements [11–16]. However, many practical aspects of using these sensors in proximal or remote disease detection in field conditions are not fully understood [17–19]. Researchers have compared the efficacy of disease detection for these sensors in general [11] or specifically for one disease [20]. It was found that the ability to measure dozens up to hundreds of wavebands in the visible and the near-infrared region of the spectrum is an important advantage of hyperspectral sensors. This makes it possible to analyse subtle changes in the spectrum related to for example leaf structure, cell components and photosynthetic capacity, making them a very versatile disease detection tool [21]. Thermal cameras are more specifically aimed at measuring one parameter, for example temperature (through emission of radiation). They have been mainly used to detect abiotic stresses related to irrigation scheduling, although their use for biotic stress detection has also been shown, even post-harvest [11,22]. The combination of these two sensors in a data fusion approach could further increase disease detection capability [23]. For pharmaceutical applications, some research has been done to investigate practical difficulties of the use of hyperspectral imagery and practical solutions have been proposed [24]. Such a practical guideline does not yet exist—to the best of our knowledge—for disease detection, particularly for the fusion of thermal and hyperspectral sensors. Researchers have instead focused on determining the optimal setup for a single sensor for in-field measurement conditions. Franceschini et al. (2017) for example compared a handheld multispectral sensor to an airborne hyperspectral sensor to determine which of these setups performed best for measuring a series of vegetation indices [25]. Garzonio et al. (2017) and Vargas et al. (2020) further discuss the setup of a hyperspectral sensor for drone-based measurements [17,26]. Another recent example is the work of Thompson and Puntel (2020), which discusses the development of a practical decision support system based on drone-based multispectral measurements [27].

From literature, we see that several factors affect the quality of hyperspectral reflectance measurements. Factors related to incident solar radiation include the position of the sun, related to the crop and the viewing angle of the sensor, and cloud cover [28–30]. It is advised by some authors to work under cloudy conditions if possible, to counteract solar position variations [31], but this is not practical in most climates around the world. Plant-related properties affecting reflectance measurements include plant species, biotic and abiotic stresses, including drought, nutrients shortage and disease presence, within-crop shading and plant growth stage [19,28,32–35]. Finally, measurement height, exposure time (ET) of the sensor and angle and distance of artificial lights also affect reflectance measurements. Based on these interfering factors, Whetton et al. (2017) established a method to optimize the setup of a hyperspectral pushbroom camera for measuring a wheat canopy in field conditions based on maximising the signal-to-noise ratio (SNR), which is defined as the ratio of the mean to the standard deviation of the measurement data [19]. It was found for a wheat canopy that the most important parameters affecting SNR (that can be set at the start of the experiment) are the height of scanning, the off-zenith angle of the sensor and the ET [19]. The best SNR in their experiment was found for a height of 30 cm, a camera angle of 10° and an ET of 50 ms. The question remains whether the setup found for hyperspectral measurements of a wheat canopy can be extrapolated to other crops, e.g., potato and leek. In the interest of data fusion, it is also necessary to determine how to merge hyperspectral and thermal sensors into a single setup and subsequently analyse the data. Some research has been conducted towards combining thermal and hyperspectral sensors for nitrogen and irrigation management in a wheat canopy [23]. In this work, the sensors were placed side-by-side at a height of 2.5 m at a fixed angle, without artificial light. It is unclear what the effect would be if, similar to Fitzgerald et al. (2006), a thermal camera is added to the setup of Whetton et al. (2017) where artificial lights flank the camera [19,23].

This paper aims at providing practical recommendations for the use of hyperspectral and thermal proximal sensing side-by-side for canopy measurement in potato and leek. We first conducted a market study to find the optimal combination of a hyperspectral and thermal sensor for measuring crop diseases. Then, we applied the hyperspectral setup optimisation methodology [19] to potato and leek crops, taking the first step towards identifying similarities/differences between the optimal setup of three completely different crop types (broad leaves, narrow leaves and cereals). Finally, we studied the effect of artificial light on hyperspectral and thermal measurements under both sunny and cloudy conditions.

2. Materials and Methods

2.1. Materials

Based on the setup used by Fitzgerald et al. (2006) and Whetton et al. (2017), we designed a new setup for disease detection in leek and potato that combines a thermal (Flir Systems, USA) and a hyperspectral sensor with a spectral range of 400–1000 nm (Specim, Finland) into one portable, easy to use sensor box that can be placed on a variety of platforms (e.g., tractors, rovers, spray boom and fork lift) (Figure 1) [19,23].

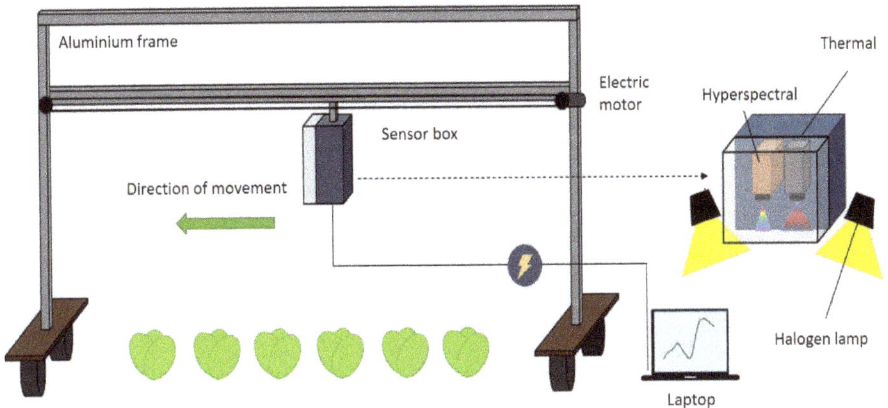

Figure 1. Aluminium frame with free-moving wheels designed to move a sensor box over leek and potato rows in field conditions. The sensor box contains a pushbroom hyperspectral camera and a thermal camera flanked by two 500 W halogen lights.

Both sensors were placed inside a waterproof box and connected to a laptop (Panasonic Belgium, Asse). The sensors were placed side by side, in the plane perpendicular to the direction of the crop row. To ensure both sensors were capable of scanning the same point directly beneath the sensor box, an 80° wide-angle lens was selected for the thermal camera so it could be placed next to the hyperspectral camera, whose narrower field of view required it to be above the crop row. The sensor box was installed in one of two ways: on an aluminium frame (Figure 1) or on a mini shovel (MultiOne, The Netherlands) (Figure 2).

The aluminium frame formed an upturned U-shaped structure, which rested on four wheels. Between the supports of the frame, an aluminium bar with a length of 3 m was positioned at adjustable height. The sensor box was attached to this bar by a wheels-and-rail system. The aluminium frame pieces and the aluminium bar could be dismounted and broken up into several smaller pieces for transport. The bar contained an electric motor that was powered by a 12 V battery. This motor moved the sensor box along the aluminium bar (in the direction of the crop row) at a pre-set constant speed by means of a rubber driver belt, which was attached to the sensor box. Two 500 W halogen lamps (Powerplus, The Netherlands) were attached on both sides of the sensor box to provide additional illumination for the hyperspectral camera. It was essential that these lamps were halogen, to ensure they emitted radiation with a similar spectral profile to that of the sun [19]. A generator was used to power the sensors and artificial lamps. The mini shovel was fitted with a 3 m aluminium beam that supported the sensor box. The laptop and power generator were attached to the mini shovel so that one person could operate both the platform and the sensor. A battery-powered pyranometer (Skye Instruments, UK) was used to measure incoming solar radiation for both the mini shovel and frame setups.

Figure 2. Close-up of mini shovel with aluminium beam supporting the sensor box. The box contained the hyperspectral and thermal sensor, flanked by two 500 W halogen lights and was moved over the row of potato test plots. A laptop was mounted on the side to allow control of the sensors. A generator mounted on the back of the mini shovel provided power for the lights, laptop and sensors.

2.2. Methods

2.2.1. Test Field

Field experiments were conducted at the Bottelare experimental farm (Merelbeke, Belgium) of Universiteit Gent and Hogeschool Gent, between December 2018 and September 2019. The geographical coordinates are 50°57′45.2″N, 3°45′36.3″E. Leek plants of cultivar Pluston were pregerminated and grown in pots, after which they were transplanted to the field. Potato plants of cultivar Agria were also pregerminated but planted directly into the field. Both potato and leek plants were planted in ridges with a width of 75 cm and a height of 30 cm. Leek plants were planted at a within-row distance of 12 cm, while potatoes were spaced at 34 cm in the row. Leek plots were designed to be 3 by 3 m, whereas potatoes were planted in 3 by 5 m plots. All data storage was done on external solid-state drives, which are more resistant to vibrations and have a higher writing speed (necessary for the storage of large amounts of hyperspectral and thermal data) compared to older storage drive models.

2.2.2. Availability of Optical Sensors in the Market

A market study of commercially available sensors was carried out to guide in the selection of the best combination of hyperspectral and thermal sensors. Although it was not needed for the set-up optimisation, we included fluorescence sensors in this market evaluation as a comparison, since it has also been proposed as a promising sensor for proximal disease sensing [11,36]. We focused only on portable sensors that can be used in field conditions and are not clip-on, so they can work from a proximal sensing perspective.

2.2.3. SNR Based Setup Optimisation

To determine the optimal setup, four ETs (1 ms, 10 ms, 30 ms and 50 ms), three heights (30 cm, 70 cm and 110 cm) and three camera off-zenith angles (0°, 8° and 17°) were tested for scanning in conditions of 300–400 W/m^2 for the leek canopy, which are fully sunny conditions during winter, and 800–900 W/m^2 for the potato canopy, which are fully sunny conditions during summer. For each of the 36 configurations, around 1000 linescans were taken with the hyperspectral camera at a framerate of 60 Hz. These scans were then stored in one data cube per setup through Lumo software (Specim, Finland). These cubes were subsequently corrected using Matlab software (The Mathworks, Inc., USA) using a white and dark reference value (Equation (1)). The pyranometer was placed in a fixed position before the measurements began and indicated whether a new white reference sample needed to be taken. The white reference target (SphereOptics, Germany, Alucore reflectance target, 500×500 mm, 95% reflectance, calibrated) was measured with the hyperspectral camera to obtain a reference value at the start of each measurement and during measurements if the solar radiation varied more than 75 W/m^2. This measurement was done in exactly the same conditions as those of the crop. The dark reference value was measured at the start of the experiment by closing the hyperspectral camera shutter. These reference values were later used to correct the data using following equation:

$$R\_cor = (R\_raw - R\_dr)/(R\_wr - R\_dr) \tag{1}$$

where R_cor is the corrected reflectance value of the measured sample, R_wr is the white reference value, or 'maximum' reflectance value, R_raw is the raw reflectance of the sample measured and R_dr is the dark, or minimum, reflectance value. In practice, the white reference value was not the maximum value even though it has a reflectance of 95%, because specular reflection could occur. We calculated the SNR using the method used in Whetton et al. (2017) [19]. The SNR of each scan was calculated by first correcting the hyperspectral data cube according to Equation (1), after which nonplant pixels were deleted based on the NDVI value [37,38]. We then, per wavelength, divided the mean of the reflectance values over all plant pixels of the image by the standard deviation over all reflectance values of all plant pixels of the image. This led to 224 SNR values, each belonging to one wavelength. Then, the average of these 224 SNR values was taken to yield one SNR value per scan (Figure 3). Scans were taken until each tested setup had hyperspectral data cubes containing at least 30 leek or potato plants.

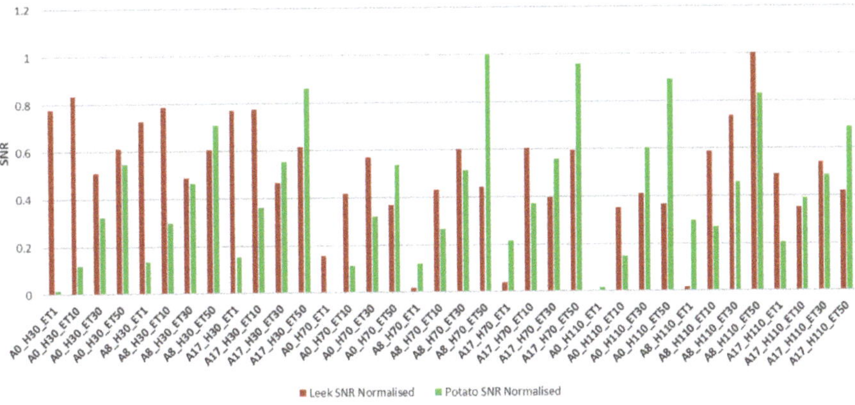

Figure 3. Signal-to-noise ratio (SNR) calculated for the hyperspectral camera for different combinations of setup parameters of height (H), angle (A) and exposure time (ET). Values of SNR were calculated for each wavelength, over all pixels, and then averaged over all wavelengths. Both datasets (leek and potato) have been normalised to be between 0 and 1 using (x−min)/range.

To understand the individual effect of each setup parameter on the resulting SNR, principal component analysis (PCA) was carried out using RStudio (RStudio Inc., USA). The input of the PCA analysis was a matrix containing 36 rows (one for each tested setup) and 4 columns for the height, angle and ET of each setup, with the corresponding SNR value. No normalisation was performed on the SNR data, because tests indicated that each of the available normalisation algorithms in the FactoMineR R package lead to principal components (PCs) that represented less of the variability in the data compared to the PCs obtained without normalisation. The results were represented in PCA factor map plots, also using the FactoMineR R package.

2.2.4. Effect of Artificial Lighting on Hyperspectral Measurements in the Field

To evaluate the effect of artificial light, hyperspectral measurements were taken with the predetermined optimal setup from Section 2, which consisted of an angle of 17°, an ET of 1 ms and a height of approximately 30 cm above a leek canopy. Exact height above crop canopy varied slightly due to the inhomogeneity of the leek plant height within each row and because the field was not perfectly level. Two measurements were taken, one in winter and one in early spring. The first measurement was taken on a day with clear weather, representing sunny conditions, whereas the second measurement was taken under fully overcast, cloudy conditions. The effect of the artificial light was determined by comparing the average reflectance values from crop canopy and white reference target measurements. To compare the shape of the reflectance curve between light on and off scenarios, the reflectance curves were normalised to have values in the 0 to 1 range by using (x-min)/range. These spectra were analysed using Matlab software (The Mathworks, Inc., USA) and ENVI (Harris Geospatial, USA) software.

2.2.5. Effect of Artificial Lighting on Thermal Measurements in the Field

To test the effect of the artificial light on the thermal measurements, a series of images of the leek crop row were taken using the aluminium frame designed for the 3 by 3 m leek plots. These images were taken at 1 Hz. To compare between light on/off treatments, average temperature values were calculated with Flir Tools software (Flir Systems, USA), for each of the images. These values were then plotted using Excel software (Microsoft, USA). The ratio of crop to bare soil is not representative for normal field conditions for images captured at the edge of the crop row. We therefore compared treatments based on images of the middle of the row. It was not possible to reliably remove soil pixels from crop pixels in the thermal images, because there was a lot of overlap between the apparent temperature of the leaves and that of the soil. For this reason, we visually compared thermal patterns on the leaves. These patterns allowed us to see if the top of the crop canopy was heated by the artificial light. We also compared average temperature values over the entire image, since the same spot in the field was measured over different setups (lamps on or lamps off). This means the amount of soil was the same for each setup and it was therefore possible to compare the effect of artificial light on the average apparent temperature between treatments. We followed the development of the crop over several weeks to determine whether any disease was naturally occurring during time of measurement.

3. Results

3.1. Multiple Sensor Setup

The results of the market evaluation of available sensors are presented in Table 1. The combination of most promising sensors [11], namely, a snapshot hyperspectral, thermal and fluorescence sensor was found to cost well over 200.000 euro. Snapshot hyperspectral sensors were available from around 30.000 euro, while pushbroom hyperspectral sensors were available from around 10.000 euro. There were far more pushbroom sensors available than snapshot sensors, and not all suppliers offered snapshot sensors. Thermal cameras found here ranged from 4.000 to 20.000 euro in price. Non-clip-on fluorescence sensors for use in field conditions were available as the CropObserver (Phenovation, The Netherlands), a point-measuring system that works with laser-induced fluorescence, or as the Hyperspec sensor (Headwall, Germany), a camera system that has a video function as well as an imaging function. The point measurement system was available for around 40.000 euro, while the Hyperspec fluorescence sensor cost around 175.000 euro.

Table 1. Available hyperspectral, thermal and fluorescence sensors, with key technical parameters and price ranges. The focus is on sensors that can be used in field conditions.

Manufacturer/ Distributor	Sensor	Spectral Range (nm) or Parameter Measured	Number of Bands	Data Acquisition	On-the-Go Measurement	Sensor Type	Price (excl. VAT, 2018)
Cubert	S185 – FireflEYE SE	450–950	125	S	Yes	HS	€ 39.900
Carbonbee	VNIR	300–1000	256	S	Yes	HS	/
IMEC	VNIR	470–900	150	H	No	HS	€ 15.575
Corning	microHSI 410 SHARK	400–1000	120	P	Yes	HS	$ 30.000
HySpex	VNIR series	400–1000	108–186	P	Yes	HS	/
	ODIN VS-1024	400–2500	427	P	Yes	HS	/
	Mjolnir series	400–1000/2500	200–490	P	Yes	HS	/
Bayspec	OCI-U-2000	600–1000	25	S	Yes	HS	$ 24.980
	OCI-U-1000	600–1000	100	P	Yes	HS	$ 19.980
Senop	Rikola	500–900	50	S	Yes	HS	€ 32.000
Mosaicmill	Rikola	500–900	50	S	Yes	HS	€ 40.500
Resonon	Pika L	400–1000	281	P	Yes	HS	€ 13.640
	Pika XC2	400–1000	447	P	Yes	HS	€ 24.582
Polytec	Nano	400–1000	270	P	Yes	HS	€ 32.860
Specim	FX10e	400–1000	220	P	Yes	HS	€ 11.690
	IQ	400–1000	204	P	Yes	HS	€ 15.950
PhenoVation	CropObserver	/	/	LP	No	PSII efficiency (Chl-Fl)	€ 38.000
Headwall	Hyperspec Fluorescence	670–780 nm	2160	S	Yes	Chl-Fl	€ 174.000
Flir	DUO PRO R 336	7.5–13.5 µm	/	S,V	Yes	Thermal	€ 4.339
	DUO PRO R 640	7.5–13.5 µm	/	S,V	Yes	Thermal	€ 6.342
	A655sc	7.5–14 µm	/	S, V	Yes	Thermal	€ 20.000
Workswell	WIRIS	Emissivity	/	S, V	Yes	Thermal	€ 13.375

HS is hyperspectral, on-the-go measurement is measurement while the sensor is moving on a piece of agricultural machinery, e.g., on a spray boom or a tractor. S is snapshot, P is pushbroom, H is hybrid, HS is hyperspectral, T is thermal, Chl-Fl is chlorophyll fluorescence and PSII is photosystem II.

3.2. Hyperspectral Setup Optimisation

Figure 3 shows the normalised SNR values for a potato and a leek canopy for each of the measurement setups.

Each dataset was normalised (after SNR calculation) using the formula (x-min)/range, which leads to values between 0 and 1. It can be observed that different setups lead to different SNR values. The highest SNR value in the potato crop was obtained with an off-zenith angle of 8°, a height above crop canopy of 70 cm and an ET of 50 ms. A close second was the setup of an 8° angle, a height of 110 cm and an ET of 50 ms. For leek, the best SNR was achieved with an angle of 17°, a height of 70 cm and an exposure time of 50 ms. Examining the SNR values of the potato, a clear trend of stepwise increasing SNR values with increasing ET could be observed. This differed from the leek SNR values, which showed no clear trend. Figure 4 shows the PCA variables factor map plots on the plane formed by PC1 and PC2, and PC1 and PC3, respectively, for hyperspectral measurements of a potato crop. The plot on the plane formed by PC2 and PC3 showed no representation of the SNR vector, so it has been omitted. It can be seen on the axes that PC1 represented 48.67% of the total variance in the data, while the second and third PCs represented 25% of the variance each. The height vector was perpendicular to the SNR vector in the PC1–PC2 plane, indicating that height did not significantly affect SNR for the potato crop. This was confirmed by the plot of PC1 and PC3, where SNR was fully represented by a long vector, as opposed to height, which was not represented in this plane (indicating it is perpendicular to this plane). The angle and SNR showed no relation in the plot of PC1 and PC2, because angle was not represented in this plane. Looking at the projection on the plane of PC1 and PC3, it can be seen that the angle vector had a projection that was almost perpendicular to the SNR vector. This indicates that SNR was not closely related to camera angle for potatoes. The last variable, ET, was almost entirely congruent with the SNR vector in both the plot on the PC1 and PC2 plane and the PC1 and PC3 plane. This indicated that for potato crops, SNR was mostly influenced by ET. Figure 5 shows the PCA variables factor map plots for hyperspectral measurements of leek for the PC1 and PC2 plane, and the PC1 and PC3 plane. The plot on the plane formed by PC2 and PC3 again showed no representation of the SNR vector, so it has been omitted. It can be seen that the first PC represented 36.75% of the total variance in the data, while the second and third PCs represented 25% of the variance each. In these plots, we saw that the relationship between SNR and the other variables was not as clear-cut compared to those of the potato canopy. Comparable to the potato canopy data, the projections of the angle vector were not concurrent with the leek SNR vector on either the PC1 and PC2 plot or the PC1 and PC3 plot. The projection of the height vector in this case seemed to indicate a more significant, negative relation to SNR compared to the potato dataset. ET again seemed to have a significant positive correlation with SNR. This positive correlation with ET was also observed in the SNR values in Figure 3, especially for potato scans. For leek scans, the SNR values are more variable, but still high ETs seemed to correspond to higher SNR values.

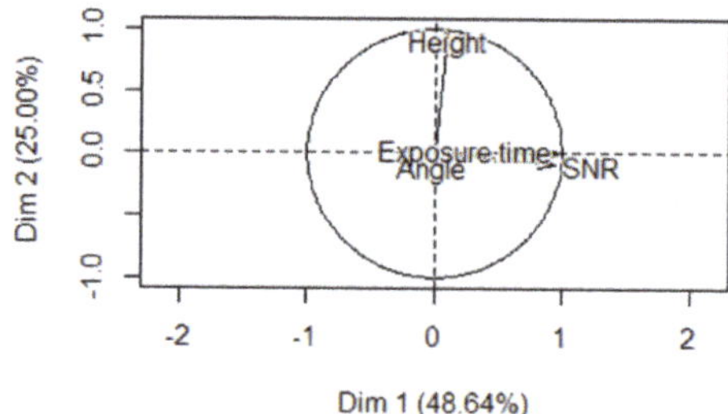

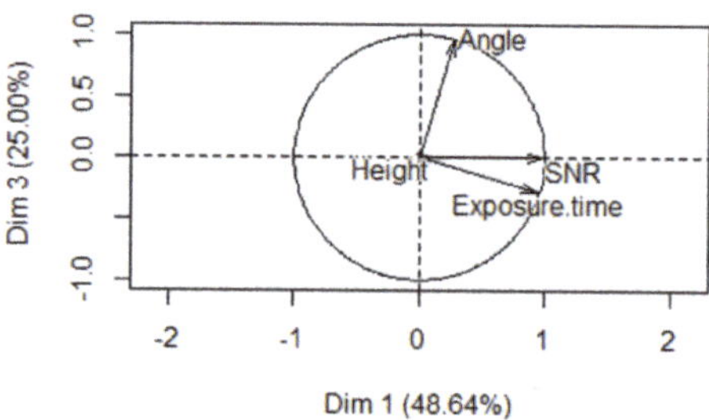

Figure 4. Principal component analysis (PCA) variables factor map plots of a potato canopy, showing the projection on principal components (PC1 and PC2) (top) and PC1 and PC3 (bottom). The projection of each variable vector on the axis formed by the signal-to-noise ratio (SNR) vector gives an indication of the relation of this variable to SNR.

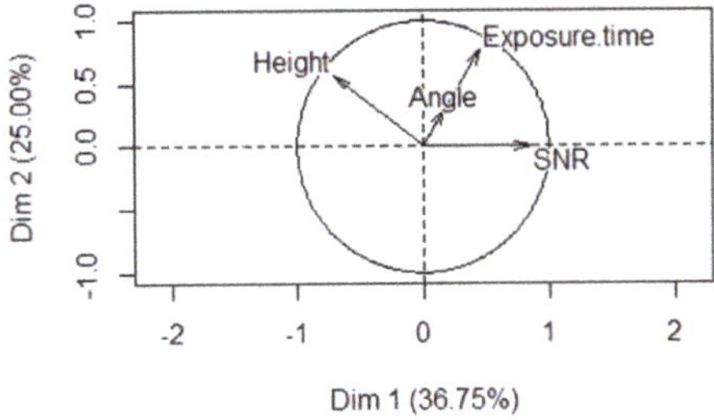

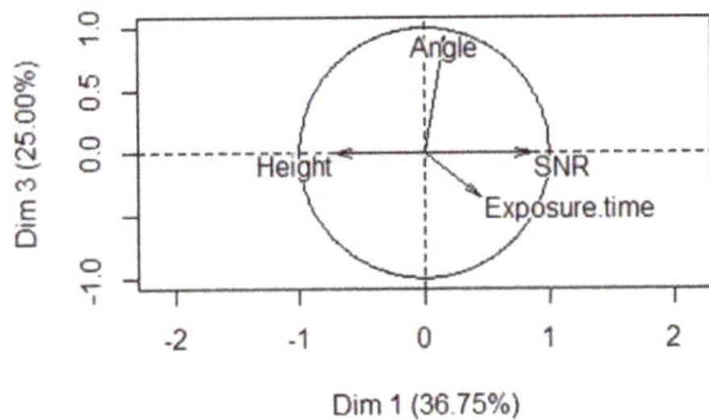

Figure 5. Principal component analysis (PCA) variables factor map plots of a leek canopy, showing the projection on PC1 and PC2 (top) and PC1 and PC3 (bottom). The projection of each variable vector on the axis formed by the signal-to-noise ratio (SNR) vector gives an indication of the relation of this variable to SNR.

3.3. Effect of Artificial Lighting on Hyperspectral Measurements of a Crop Canopy in the Field

Figure 6 shows the average reflectance curves for the scans of a leek row, with artificial light on/off under sunny and cloudy conditions. The curve related to sunny conditions (Figure 6A) showed that although measurements were taken on an exceptionally sunny day (for winter conditions), the lamps still contributed to reflectance. This is in contrast with the results under cloudy conditions (Figure 6B), in which it can be observed that the added illumination apparently decreased reflectance values, especially in the NIR range. The normalized reflectance curves of the on/off treatments were less similar to each other during cloudy conditions compared to sunny conditions (Figure 6C,D). The normalized reflectance curves of the sunny condition experiment appeared very similar in the visible region of the spectrum, while the cloudy conditions curve showed differences, especially in the red colour region of the spectrum (600–680 nm). In the NIR region, the on/off curves of under sunny conditions are inconsistent. During cloudy conditions, the light-on-treatment's reflectance curve is consistently below that of the off-treatment in the NIR region, up to +/-930 nm, after which differences are inconsistent. A depression in the reflectance curve appeared around 950 nm, for cloudy conditions, which did not appear for sunny conditions for both the raw and normalised spectra.

Looking at the increased reflectance of the white reference as a result of switching on the light under sunny conditions (Figure 7A), a mild increase can be seen more or less uniformly over the entire spectrum, with a bigger increase in the middle part of the spectrum compared to the edges at 400 and 1000 nm. It is clear that the halogen lights cause a reflectance pattern that is very similar to the natural light conditions. However, looking at the cloudy white reference reflectance data (Figure 7B), the additional lighting appeared to mainly increase the reflectance values at higher wavelengths, with the largest increase again occurring in the middle part of the spectrum. However, in cloudy conditions, the difference between the light on and light off for white reference curves is much more significant, both in terms of shape of the curve and magnitude of reflectance (Figure 7B). The white reference reflectance curve without artificial light was less bell-shaped and more skewed under cloudy conditions (Figure 7B) compared to that without artificial lights under sunny conditions (Figure 7A), indicating the effect of cloud cover on the reflectance spectrum.

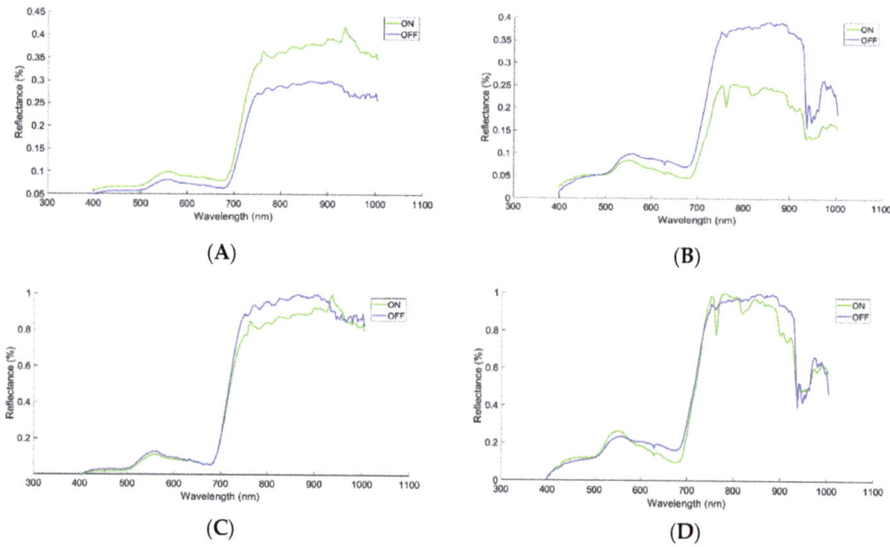

Figure 6. Average reflectance curve of a leek canopy with artificial light (two 500 W halogen lamps) on (dashed blue line) and light off (full green line) scenarios under sunny (**A**) and cloudy (**B**) weather conditions. Normalised spectra plots also shown for sunny (**C**) and cloudy (**D**) conditions.

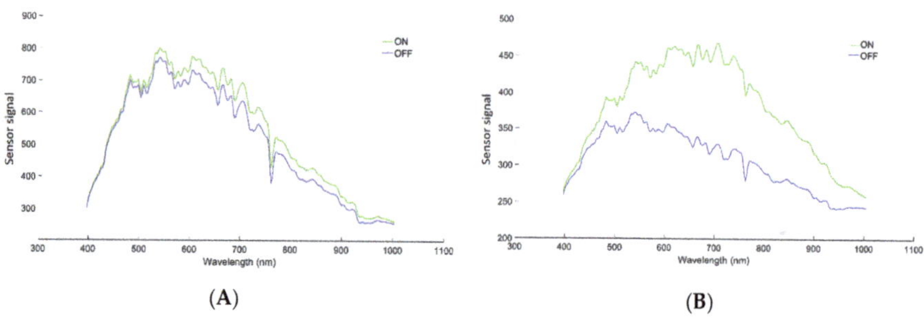

Figure 7. White reference signal curve with artificial light (two 500 W halogen lamps) on (green line) and lights off (blue line) scenarios under sunny (A) and cloudy (B) scanning conditions. Raw sensor signal is shown.

3.4. *Effect of Artificial Lighting on Thermal Measurements of a Crop Canopy in the Field*

Figure 8 compares thermal images measured in conditions with and without artificial light, under sunny (8.1) and cloudy (8.2) conditions.

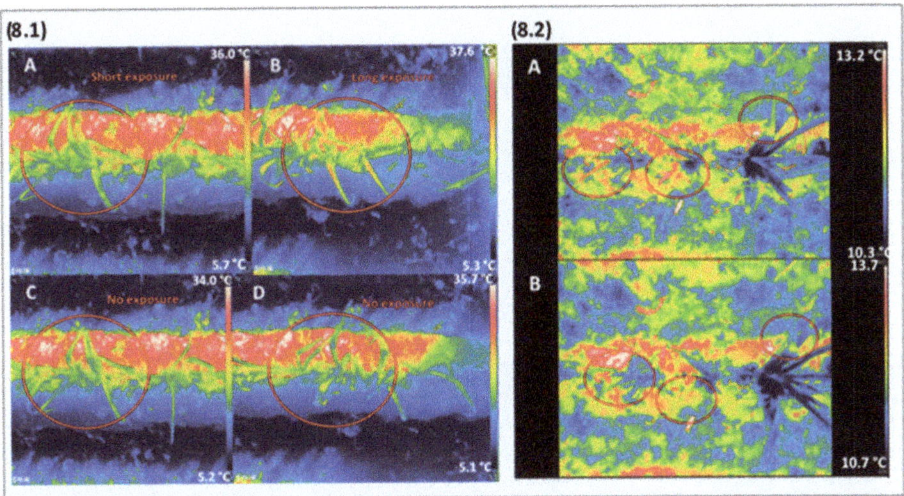

Figure 8. Figure 8.1 shows thermal images of a leek ridge in the middle of the row (Figure 8.1A,C) and at row edge (Figure 8.1B,D). Scans were taken during cloudless, sunny conditions, with additional lighting (Figure 8.1A,B) and in natural light (Figure 8.1C,D). Figure 8.2 shows thermal images taken on a cloudy day, comparing lights on (Figure 8.2A) with lights off (Figure 8.2B). Red circles indicate the areas studied for comparison of hot spot formation. Rightmost red circles of Figure 8.2 (A and B) show the bend of a leaf, which shows up as a hot spot due to physical damage caused by the bending/cracking of the leaf.

Comparing Figure 8.1B,D with Figure 8A,C, it can be seen that the ridge at the edge of the row was slightly slanted, causing it to receive less radiation, which resulted in a decreased average temperature measured in these frames (Figure 8B,D). It can further be observed that the prolonged exposure to the halogen lights at the row edge caused a temperature increase in the top parts of the crop canopy, which is closest to the lamps (Figure 8.1B). This leads to 'hot spots' that were not necessarily related to disease or other types of plant stress, since no disease appeared during the weeks following measurement and the hot spots appeared only after the halogen lamps were turned on. There was no clear formation of hot spots in the middle of the crop row during the light-on treatment (Figure 8A,C). The temperature difference seemed to mainly stem from absorption of radiation by the soil. Figure 8.2 shows the effect of additional lighting during cloudy conditions. This measurement was later in the growing season, when dense weed cover started appearing on the ridges. It is clear from the thermal images that weeds significantly affect measured temperature patterns, as weed temperature was cooler than that of the soil but generally hotter than the leek. No apparent average temperature increase was observed when switching on lights, considering the entire thermal image, with even some minor temperature decrease. The red circled areas on Figure 8.2A show that during this measurement, the artificial light caused hot spots even during movement of the sensor box. This was repeatedly observed in different locations throughout the measured leek rows. These hotspots seemed to be located around sharp bends in the leaves or on lower leaves, which were older and decaying. Looking at healthy leaf tissue (areas outside the red circles), there was no effect of additional light, even on parts of the leek crop that were higher compared to the hotspots. Figure 9 shows the average temperature measured over two rows of leek plants, in conditions with and without artificial lights. There was approximately a 1 °C increase in average temperature due to the addition of artificial light, during sunny conditions. The images taken in the middle of the crop row (image series number 4–8) showed a higher average temperature compared to those at the row edge.

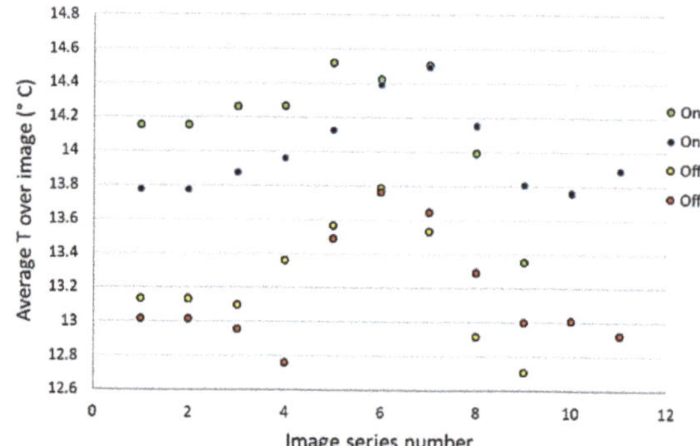

Figure 9. Average T calculated over a series of thermal images captured over two leek crop rows under sunny conditions. Images captured with (on) and without (off) artificial light. Background soil temperature was lower compared to leek temperatures, meaning that lower temperatures correspond to areas with relatively few leek plants, while higher temperature corresponds to the centre of the leek row.

4. Discussion

4.1. Multiple Sensor Setup Selection

The combination of a snapshot hyperspectral, thermal and fluorescence sensor would cost over 200.000 euro. The biggest cost is the snapshot fluorescence camera, at around 175.000 euro. To the best of our knowledge, this is the only sensor of its type available on the market today. The cheapest combination of all three sensors would cost approximately 55.000 euro (Table 1) and consists of a pushbroom hyperspectral sensor, a snapshot thermal camera, and a point measurement fluorescence sensor. It is on the other hand possible to combine a hyperspectral and thermal sensor for under 20.000 euro. The high cost of fluorescence sensors might lead researchers to opt for a cheaper version, such as the handheld MultispeQ clip-on fluorescence sensor (PhotosynQ, USA), which costs around 1000 euro. However, clip-on sensors cannot be used for on-the-go scanning, requiring researchers to invest a lot of time for manual measurements. We therefore believe that for practical applications in precision agriculture, the combination of hyperspectral and thermal sensors provides a good starting point, with relatively low investment costs and less complexity compared to a system including a fluorescence sensor as well. This led to the choice to use a snapshot thermal sensor and a pushbroom hyperspectral sensor for this experiment, both costing no more than 20.000 euro each.

4.2. Hyperspectral Setup Optimisation

The results from Figures 2 and 3 and Figures 4 and 5 confirmed that the method used for optimising the hyperspectral sensor setup for a wheat canopy can likewise be used to optimise the setup for a potato and leek canopy. The most optimal setups in potato and leek consisted of an ET of 50 ms (Figure 3). There was also a clear trend in the potato SNR values indicating a positive correlation with ET, which is in agreement with results found in a wheat canopy [19]. For leek, values appeared more variable. The effect of each variable is shown in the PCA factor map plots of Figures 4 and 5. Both figures indicated that height seemed less important for SNR. The angle also showed no clear correlation to SNR in either of the figures. However, the ET vector was clearly congruent with the SNR vector in the potato PCA factor map plots (Figure 4). Together with the values presented in Figure 3, it is reasonable to conclude that ET was the most important factor affecting SNR for this experiment. This is in agreement with the findings in wheat, where ET was also found to be the most important factor determining SNR values [19]. However, the results in leek were less clear-cut compared to those of potatoes and wheat. This crucial difference means that researchers must always determine the optimal setup for the crop under observation, before measurements. In our results, the optimal setups consisted of an angle of 8°, a height of 70 cm and an ET of 50 ms for potatoes and an angle of 17°, a height of 70 cm and an ET of 50 ms for leek (Figure 3). The optimal setup for leek also registered as the 5th best setup for potatoes. Theoretically, the best measurement setup (of the ones tested in this experiment) for both crops therefore lies at an ET of 50 ms, a height of 70 cm and an off-zenith angle of 8° to 17°. However, some practical constraints have to be considered. First, we observed the occurrence of saturation at ETs of 30 and 50 ms in potato and leek, respectively, similar to the results in wheat [19], but there for a much higher ET of 1000 ms. Only the 1 and 10 ms treatments did not show saturation. It is further important to note that a lower ET results in a higher possible framerate. For some measurement conditions, there is a minimum speed at which the measurement needs to be done (e.g., due to time restrictions, operating speed of a treadmill or driving speed of a tractor). This speed correlates to a minimum framerate necessary to obtain a scan of the full sample, which in turn is associated to a maximum possible ET. From Figure 3, it can be concluded that theoretically the optimal height of scanning of both studied crops seemed to be 70 cm. However, since the variability in potato SNR values and to a lesser extent in leek is mainly caused by ET, the choice of measurement height needs to depend on other factors than the SNR values. We propose that measurement height is determined mainly by pixel resolution and scanning width needed in the experimental context. For example, if the main goal is to scan as large a crop area as possible in a limited amount of time (e.g., farmer field scans), then a height above the crop of 110 m is beneficial. However, if the aim of the experiment is to detect small symptoms (e.g., rust pustules) on the leaves of an experimental leek plot in early stages of infection, a height of 30 cm above the crop is preferable, since this will yield a higher image resolution.

The off-zenith angle shows no clear correlation to SNR like that of the ET, except for leek in the PC1–PC2 plot (Figure 4). The complex interaction between viewing angle and reflectance has been studied for forest canopies [39]. It was found that for white backgrounds, the reflectance decreased with increasing off-zenith measurement angle, while for dark backgrounds the opposite occurred, with increasing reflectance at higher off-zenith viewing angles. In a mixed system such as a crop canopy that contains brighter areas (e.g., phytophthora or other wilting symptoms) and also darker areas (e.g., dark spots or shaded areas), it is difficult to theorise on the effect of viewing angle on reflectance and SNR. Results in pine forests showed that there was a specific angular effect on the reflectance of the red and red edge bands for coniferous trees, possibly due to their canopy structure [40]. Such an angular effect due to canopy structure could contribute to the difference between the results for potato and leek, since angle only seemed to contribute to SNR for leek canopy measurements. This is supported by the clear difference between the oblique growth pattern of a leek canopy, where angular differences are significant between leaves, compared to the relatively homogenous growth pattern of a dense potato canopy, where angular differences might be more easily averaged out. It has further been shown that the angle significantly affects certain canopy measurements, for example of vegetation indices [41]. This is relevant when performing image analysis on, for example, single leaves. In such case, the effect of geometry on reflectance can be modelled to improve data analysis [41–44]. The same applies to satellite imagery, where the modelling of the solar off-zenith angle is crucial [45]. However, for practical applications in proximal crop sensing, such complex crop geography modelling techniques are often omitted, and it is assumed that the effect can be neglected, or vegetation indices are used that are resistant to these effects [41,46]. We therefore propose that the measurement angle needs to be estimated based on the disease of interest. Ideally, scans should be taken perpendicularly to the plane, in which symptoms occur, to maximize the chance of scanning the infected area. If the disease is for example known to manifest on the bottom of the stem, a 0° off-zenith angle will have no chance of detecting it, as opposed to a 17° angle, which makes it possible to scan the lower canopy. Results for powdery mildew detection in grapes also supported the use of higher off-zenith angles [35]. This leads us to advise an off-zenith angle of 17° to detect symptoms on the bottom of the leek stem and lower leaves. Phytophthora damage tends to appear at leaf tips, but also on the base of the leaf where spores are splashed onto the plant from the soil [47]. To detect symptoms on lower leaves of the potato canopy, an angle of 17° is again advised, especially in dry growing conditions that cause the leaves to sag, causing the symptoms to be in the vertical plane perpendicular to the soil.

Although the use of artificial lights has certain advantages (see subsection 3 of the discussion), it also increases the risk of saturation, especially at low measurement heights. The choice of the optimal measurement setup is therefore not only based on the best SNR but is far more complex and should be discussed from practical perspective, taking into account the specific case of each disease or crop under observation. A balance needs to be found between high ET, resulting in better SNR values, while taking into account saturation, pixel size, measurement speed and the structure of the pathosystem. Specifically, for leek, white tip disease (Phytophthora porri) symptoms can easily cause saturation at low scanning heights because of their proximity to artificial lights, even at low ETs. The trade-off between the risk of saturation (at high ETs) and the risk of noisy data (at low ETs) can partly be overcome by spectra preprocessing techniques, which can, to some extent, deal with noisy spectra [48]. It is important to note that the reflectance values in the visible part of the spectrum are much lower compared to the near-infrared part of the plant canopy spectrum [49]. This means that saturation could primarily occur in the near-infrared part of the spectrum, while a low ET and subsequent noisy data could occur more easily in the visible part of the spectrum. Depending on which part of the spectrum is mainly of interest, the saturation/noise trade-off might be different. For proximal disease detection in experimental conditions for leek and potato, we advise a height of 30 cm (instead of the SNR optimum of 70 cm) above the crop canopy, to maintain a high resolution. Because of the risk of saturation at this height, an ET of 1 ms is advised for both leek and potato, instead of the SNR optimum of 50 ms (for both leek and for potato). This needs to be increased if spectra appear "flat", with minor or without features. However, saturation should be absolutely avoided. Since no clear influence was observed for the angle, it is recommended to use the 17° angle because it can measure symptoms on lower canopy parts. This was the theoretical optimal angle for a leek canopy, but for a potato canopy the theoretical optimum was an 8° angle.

4.3. Effect of Artificial Lighting on Hyperspectral Measurements of a Crop Canopy in the Field

To explain the phenomenon that the added light seemed to decrease reflectance of spectra collected from a canopy on a cloudy day (Figure 6B), the spectral profile of the white reference was investigated (Figure 7). At wavelengths up to +/-470 nm, the difference between the light on/off white reference reflectance curves under cloudy conditions (Figure 7B) was minimal, similar to sunny conditions (Figure 7A). At higher wavelengths, the difference increased. Without additional light, the white reference reflectance curve under cloudy conditions was skewed compared to the curve under sunny conditions (due to the effects of cloud cover), leading to an inappropriate correction at high wavelengths. The corrected spectrum of wavelengths > 470 nm was 'stretched' compared to those at lower wavelengths. This is due to lower white reference reflectance values, which increases the final corrected value according to Equation (1). This explains why in Figure 6B, the added light under cloudy conditions caused a decrease in reflectance compared to the light-off scenario. Turning on the artificial light increased the white reference to such a degree that it counteracted the increased reflection from the crop canopy (due to added artificial radiation), resulting in ultimately lower reflectance values. If there would have been no artificial illumination during experiments, the skewed white reference values on cloudy days would cause the final reflectance in the >470 nm range to be relatively stretched (as shown in Figure 6B) compared to the same target measured on a sunny day (Figure 6A), independent of the crop health status. The added light helped to counteract this problem.

After normalisation, the shapes of the light on/off canopy reflectance curves are similar under sunny conditions (Figure 6C), with only a small difference in the NIR part of the spectrum that could be amended by further preprocessing [50]. The difference between light on/off canopy reflectance curves is much more severe under cloudy conditions, especially around 680 nm. This is an important region for disease detection, so it is essential that any change in reflectance is the result of disease, rather than cloud cover variation [11]. The apparent increase in Figure 6D between 600 to 680 nm could be misinterpreted as an increased 'red-orange' colour, which could be interpreted as rust disease symptoms (results not shown). The addition of artificial light helps counteract this effect to some extent, but it is still important to keep this in mind for further data analysis. This area, especially the red colour band at 680 nm, is well documented in literature as being a spectral feature indicating chlorophyll absorption, making it one of the most important features in crop health sensing [19]. The drop in measured reflectance at +/-930 nm appears in both the lights on and lights off curves during cloudy conditions, which suggests that this is a feature, rather than purely the result of variations in solar radiation intensity due to cloud cover (Figure 6D). The fact that this drop occurred only in cloudy conditions lead us to look at the absorbance spectra of cloud cover reported in literature [51]. These authors reported a significant absorption band around 940 nm for stratus clouds, which could possibly explain the decrease in reflectance. This could indicate that the artificial light was not strong enough to compensate for the absorption caused by cloud cover in this spectral region.

Another important aspect of the artificial light is that the angle of reflectance is different compared to that of natural solar radiation [18]. During sunny conditions, the light strikes the crop at a certain angle, depending on the solar radiation. With the artificial light, the light strikes the crop from both sides (because two lamps are used in the present work), at a constant angle to the sensor. This means that the light not only counteracts the effect of cloud cover but also provides a constant source of illumination that helps mitigate the effects of changing solar angle. It is therefore advisable to use as much artificial lights as possible, to reduce the effect of solar radiation variation.

4.4. Effect of Artificial Lighting on Thermal Measurements of a Crop Canopy in the Field

If the crop temperature would increase significantly during sensor movement due to the added radiation, it would be most apparent on the top leaves, which receive most of the radiation. During sunny conditions, hot spot formation was only apparent at the edge of the crop row, where the sensor box stayed stationary for more than 10 seconds (Figure 8.1B). This indicates that even though there was a temperature difference during these measurements, it was mainly caused by absorption of radiation by the soil, not the crop. Otherwise, hot spot formation would occur on the top leaves during measurements (and not only when the box was stationary). Note that this implies that correcting for the temperature difference caused by artificial light is difficult because not every part of the crop row heats up equally. The assumption that bare soil is mainly responsible for the temperature difference between light on/off is also supported by the fact that average measured temperatures varied between the edge and the middle of the crop row, where the soil-to-crop ratio is different (Figure 9). An important observation is that during cloudy conditions (Figure 8.2), there was no apparent temperature increase compared to 1 °C for sunny conditions (Figure 8.1). There was even a small apparent decrease (0.5 °C or less) after turning on the lights for some images during cloudy conditions. The temperature decrease is possibly due to the effect of variations in cloud cover, which are difficult to record during the time of one scan. The lack of temperature difference could also be explained by the dense weed cover that covered the darker soil later in the growing season during the measurements under cloudy conditions. This suggests that it is possible to use halogen lamps in combination with a thermal sensor for weed or crop cover assessment.

During cloudy conditions, hot spot formation in the middle of the crop row was observed during measurements, not only when the sensor box was stationary (Figure 8.2A). The leek crop was older during this measurement, showing signs of wilting on older leaves in places where the leaves were bent or cracked. It was in here that hot spot formation occurred, even when the sensor box passed over the crop in a matter of seconds. This indicates that artificial light interacts differently with diseased or damaged and healthy parts of the crop and that this difference can be observed with thermal cameras. This feature could assist in disease detection using thermal cameras in addition to the fact that these sensors can measure the temperature differences due to evapotranspiration. It is also important to note that the interaction between artificial light and damaged crop areas occurred even over a matter of seconds, which is much faster than the rate at which evapotranspiration changes occur [52]. This could provide new research opportunities, for example by placing thermal cameras with artificial lights at the back of weeders for detecting mechanical damage after passage, which would take longer to show if no artificial light is present.

5. Conclusions

The cost of fluorescence sensors is close to ten times that of a hyperspectral or thermal sensor, leading researchers to favour thermal and hyperspectral imaging. However, very few research groups studied the use of a combination of these two sensors in different applications in crop monitoring and sensing. Our results showed that the setup optimisation method for a pushbroom hyperspectral camera, based on maximum signal-to-noise ratio (SNR), tested in wheat can be extrapolated to other crops. However, different set up parameters should be implemented for different crops to allow successful measurement in practice. The optimal set up parameters found in the present study for potatoes and leek are a camera height of 30 cm, a 17° camera angle and an exposure time (ET) of 1 ms. In line with camera set up results for wheat, ET was the most important parameter affecting SNR, with higher ETs leading to higher SNRs. Measurements in different cropping systems need to be done to determine general applicability of the optimal setup findings. We further concluded that the addition of artificial lights helps counteract the effect of cloud cover on reflectance measurements, aiding disease detection. The temperature difference caused by the additional light appeared to mainly vary with the plant to bare soil ratio. Possible practical applications regarding weed and crop cover assessment with thermal measurements need to be further investigated. Hot spot formation due to artificial light could possibly be used to assess disease stress of certain diseases, due to differential heat absorption of healthy versus diseased crops. A comparison between healthy and diseased plants is needed to confirm the hypothesis that disease detection capabilities increase with the addition of artificial light, for both thermal and the hyperspectral sensing.

Author Contributions: Conceptualization, S.A., J.P. and A.M.M.; methodology, S.A., A.G., J.P. and A.M.M.; validation, S.A., J.P. and A.M.M.; formal analysis, S.A. and S.N.; investigation, S.A.; resources, A.G.; data curation, S.A.; writing—original draft preparation, S.A.; writing—review and editing, S.A., J.P. and A.M.M.; visualization, S.A.; supervision, J.P. and A.M.M.; project administration, A.M.; funding acquisition, A.M.M. All authors have read and agreed to the published version of the manuscript.

Funding: The author(s) disclosed receipt of the following financial support for the research, authorship, and/or publication of this article: This work was supported by the Research Foundation - Flanders (FWO) for Odysseus I SiTeMan Project [Nr. G0F9216N].

Acknowledgments: Apart from the people listed on this paper, a big thanks to the people at Bottelare experimental farm, UGent mechanical workshop and my colleagues who helped me in the field.

Conflicts of Interest: The authors declare no conflict of interest.

References

1. Armengol, J.; Weigand, S.; Von Tiedemann, A.; Kreiter, S.; Duso, C. Education in crop protection: Erasmus Mundus Joint Master Degree–European Master Degree in Plant Health in Sustainable Cropping Systems. *J. Biotechnol.* **2019**, *305*, S8. [CrossRef]
2. Agrios, G.N. *Plant Pathology*, 5th ed.; Elsevier Academic Press: Amsterdam, The Netherlands, 2005; Volume 26.
3. Anderson, P.K.; Cunningham, A.A.; Patel, N.G.; Morales, F.J.; Epstein, P.R.; Daszak, P. Emerging infectious diseases of plants: Pathogen pollution, climate change and agrotechnology drivers. *Trends Ecol. Evol.* **2004**, *19*, 535–544. [CrossRef] [PubMed]
4. Brasier, C.M. The biosecurity threat to the UK and global environment from international trade in plants. *Plant Pathol.* **2008**, *57*, 792–808. [CrossRef]
5. Fisher, M.C.; Henk, D.A.; Briggs, C.J.; Brownstein, J.S.; Madoff, L.C.; McCraw, S.L.; Gurr, S.J. Emerging fungal threats to animal, plant and ecosystem health. *Nature* **2012**, *484*, 186. [CrossRef] [PubMed]

6.   Bebber, D.P.; Holmes, T.; Gurr, S.J. The global spread of crop pests and pathogens. *Glob. Ecol. Biogeogr.* **2014**, *23*, 1398–1407. [CrossRef]

7.   Coakley, S.M. Variation in climate and prediction of disease in plants. *Ann. Rev. Phytopathol.* **1988**, *26*, 163–181. [CrossRef]

8.   Mahlein, A.K.; Kuska, M.T.; Behmann, J.; Polder, G.; Walter, A. Hyperspectral sensors and imaging technologies in phytopathology: State of the art. *Ann. Rev. Phytopathol.* **2018**, *56*, 535–558. [CrossRef]

9.   Rosenzweig, C.; Iglesias, A.; Yang, X.B.; Epstein, P.R.; Chivian, E. Climate change and extreme weather events; implications for food production, plant diseases, and pests. *Glob. Chang. Human Health* **2001**, *2*, 90–104. [CrossRef]

10.   Bohnenkamp, D.; Behmann, J.; Mahlein, A.K. In-Field Detection of Yellow Rust in Wheat on the Ground Canopy and UAV Scale. *Remote Sens.* **2019**, *11*, 2495. [CrossRef]

11.   Mahlein, A.K. Present and Future Trends in Plant Disease Detection. *Plant Dis.* **2016**, *100*, 1–11. [CrossRef]

12.   Nigon, T.J.; Yang, C.; Dias Paiao, G.; Mulla, D.J.; Knight, J.F.; Fernández, F.G. Prediction of Early Season Nitrogen Uptake in Maize Using High-Resolution Aerial Hyperspectral Imagery. *Remote Sens.* **2020**, *12*, 1234. [CrossRef]

13.   Shen, L.; Gao, M.; Yan, J.; Li, Z.-L.; Leng, P.; Yang, Q.; Duan, S.-B. Hyperspectral Estimation of Soil Organic Matter Content using Different Spectral Preprocessing Techniques and PLSR Method. *Remote Sens.* **2020**, *12*, 1206. [CrossRef]

14.   Jin, X.; Li, Z.; Atzberger, C. Editorial for the Special Issue "Estimation of Crop Phenotyping Traits using Unmanned Ground Vehicle and Unmanned Aerial Vehicle Imagery". *Remote Sens.* **2020**, *12*, 940. [CrossRef]

15.   Zhang, J.; Tian, H.; Wang, D.; Li, H.; Mouazen, A.M. A Novel Approach for Estimation of Above-Ground Biomass of Sugar Beet Based on Wavelength Selection and Optimized Support Vector Machine. *Remote Sens.* **2020**, *12*, 620. [CrossRef]

16.   Jiang, Y.; Snider, J.L.; Li, C.; Rains, G.C.; Paterson, A.H. Ground Based Hyperspectral Imaging to Characterize Canopy-Level Photosynthetic Activities. *Remote Sens.* **2020**, *12*, 315. [CrossRef]

17.   Vargas, J.Q.; Bendig, J.; Mac Arthur, A.; Burkart, A.; Julitta, T.; Maseyk, K.; Thomas, R.; Siegmann, B.; Rossini, M.; Celesti, M.; et al. Unmanned Aerial Systems (UAS)-Based Methods for Solar Induced Chlorophyll Fluorescence (SIF) Retrieval with Non-Imaging Spectrometers: State of the Art. *Remote Sens.* **2020**, *12*, 1624. [CrossRef]

18.   Mishra, P.; Asaari, M.S.M.; Herrero-Langreo, A.; Lohumi, S.; Diezma, B.; Scheunders, P. Close range hyperspectral imaging of plants: A review. *Biosyst. Eng.* **2017**, *164*, 49–67. [CrossRef]

19.   Whetton, R.L.; Waine, T.W.; Mouazen, A.M. Optimising configuration of a hyperspectral imager for on-line field measurement of wheat canopy. *Biosyst. Eng.* **2017**, *155*, 84–95. [CrossRef]

20.   Joalland, S.; Screpanti, C.; Liebisch, F.; Varella, H.V.; Gaume, A.; Walter, A. Comparison of visible imaging, thermography and spectrometry methods to evaluate the effect of Heterodera schachtii inoculation on sugar beets. *Plant Methods* **2017**, *13*, 73. [CrossRef]

21.   Thenkabail, P.S.; Lyon, J.G. *Hyperspectral Remote Sensing of Vegetation*; CRC Press: Boca Raton, FL, USA, 2012.

22.   López-Maestresalas, A.; Keresztes, J.C.; Goodarzi, M.; Arazuri, S.; Jarén, C.; Saeys, W. Non-destructive detection of blackspot in potatoes by Vis-NIR and SWIR hyperspectral imaging. *Food Control* **2016**, *70*, 229–241. [CrossRef]

23.   Fitzgerald, G.J.; Rodriguez, D.; Christensen, L.K.; Belford, R.; Sadras, V.O.; Clarke, T.R. Spectral and thermal sensing for nitrogen and water status in rainfed and irrigated wheat environments. *Precision Agric.* **2006**, *7*, 233–248. [CrossRef]

24.   Amigo, J.M. Practical issues of hyperspectral imaging analysis of solid dosage forms. *Anal. Bioanal. Chem.* **2010**, *398*, 93–109. [CrossRef] [PubMed]

25.   Franceschini, M.H.; Bartholomeus, H.; Van Apeldoorn, D.; Suomalainen, J.; Kooistra, L. Intercomparison of unmanned aerial vehicle and ground-based narrow band spectrometers applied to crop trait monitoring in organic potato production. *Sensors* **2017**, *17*, 1428. [CrossRef] [PubMed]

26.   Garzonio, R.; Di Mauro, B.; Colombo, R.; Cogliati, S. Surface reflectance and sun-induced fluorescence spectroscopy measurements using a small hyperspectral UAS. *Remote Sens.* **2017**, *9*, 472. [CrossRef]

27.   Thompson, L.J.; Puntel, L.A. Transforming Unmanned Aerial Vehicle (UAV) and Multispectral Sensor into a Practical Decision Support System for Precision Nitrogen Management in Corn. *Remote Sens.* **2020**, *12*, 1597. [CrossRef]

28. Asner, G.P. Biophysical and biochemical sources of variability in canopy reflectance. *Remote Sens. Environ.* **1998**, *64*, 234–253. [CrossRef]

29. Pinter Jr, P.J.; Jackson, R.D.; Elaine Ezra, C.; Gausman, H.W. Sun-angle and canopy-architecture effects on the spectral reflectance of six wheat cultivars. *Int. J. Remote Sens.* **1985**, *6*, 1813–1825. [CrossRef]

30. Whetton, R.L.; Waine, T.W.; Mouazen, A.M. A Practical Approach to In-Situ Hyperspectral Imaging of Wheat Crop Canopies. In Proceedings of the 13th International Workshop on Advanced Infrared Technology Applications, Pisa, Italy, 29 September–2 October 2015; ISBN 978880250.

31. Whetton, R.L.; Waine, T.W.; Mouazen, A.M. Hyperspectral measurements of yellow rust and fusarium head blight in cereal crops: Part 2: On-line field measurement. *Biosyst. Eng.* **2018**, *167*, 144–158. [CrossRef]

32. Barbedo, J.G.; Tibola, C.S.; Fernandes, J.M. Detecting Fusarium head blight in wheat kernels using hyperspectral imaging. *Biosyst. Eng.* **2015**, *131*, 65–76. [CrossRef]

33. Bousquet, L.; Lachérade, S.; Jacquemoud, S.; Moya, I. Leaf BRDF measurements and model for specular and diffuse components differentiation. *Remote Sens. Environ.* **2005**, *98*, 201–211. [CrossRef]

34. Coops, N.C.; Smith, M.L.; Martin, M.E.; Ollinger, S.V. Prediction of eucalypt foliage nitrogen content from satellite-derived hyperspectral data. *IEEE Trans. Geosci. Remote Sens.* **2003**, *41*, 1338–1346. [CrossRef]

35. Oberti, R.; Marchi, M.; Tirelli, P.; Calcante, A.; Iriti, M.; Borghese, A.N. Automatic detection of powdery mildew on grapevine leaves by image analysis: Optimal view-angle range to increase the sensitivity. *Comput. Electron. Agric.* **2014**, *104*, 1–8. [CrossRef]

36. West, J.S.; Bravo, C.; Oberti, R.; Lemaire, D.; Moshou, D.; McCartney, H.A. The potential of optical canopy measurement for targeted control of field crop diseases. *Annu. Rev. Phytopathol.* **2003**, *41*, 593–614. [CrossRef] [PubMed]

37. Bravo, C.; Moshou, D.; Oberti, R.; West, J.; McCartney, A.; Bodria, L.; Ramon, H. Foliar disease detection in the field using optical sensor fusion. *Agri. Eng. Int. CIGR J. Sci. Res. Dev.* **2004**. Manuscript FP 04 008. Vol. VI.

38. Rouse, J.W., Jr. *Monitoring the Vernal Advancement and Retrogradation (Green Wave Effect) of Natural Vegetation*; Texas A&M Univ., College Station: Texas, TX, USA, 1974.

39. Pisek, J.; Chen, J.M.; Miller, J.R.; Freemantle, J.R.; Peltoniemi, J.I.; Simic, A. Mapping forest background reflectance in a boreal region using multiangle compact airborne spectrographic imager data. *IEEE Trans. Geosci. Remote Sens.* **2009**, *48*, 499–510. [CrossRef]

40. Rautiainen, M.; Lang, M.; Mõttus, M.; Kuusk, A.; Nilson, T.; Kuusk, J.; Lükk, T. Multi-angular reflectance properties of a hemiboreal forest: An analysis using CHRIS PROBA data. *Remote Sens. Environ.* **2008**, *112*, 2627–2642. [CrossRef]

41. Behmann, J.; Mahlein, A.K.; Paulus, S.; Kuhlmann, H.; Oerke, E.C.; Plümer, L. Calibration of hyperspectral close-range pushbroom cameras for plant phenotyping. *ISPRS J. Photogramm. Remote Sens.* **2015**, *106*, 172–182. [CrossRef]

42. Clarke, T.A.; Fryer, J.G. The development of camera calibration methods and models. *Photogramm. Record* **1998**, *16*, 51–66. [CrossRef]

43. Horaud, R.; Mohr, R.; Lorecki, B. On single-scanline camera calibration. *IEEE Trans. Robot. Autom.* **1993**, *9*, 71–75. [CrossRef]

44. Schowengerdt, R.A. *Remote Sensing: Models and Methods for Image Processing*; Elsevier: Amsterdam, The Netherlands, 2006.

45. Grosvenor, D.P.; Wood, R. The effect of solar zenith angle on MODIS cloud optical and microphysical retrievals within marine liquid water clouds. *Atmos. Chem. Phys.* **2014**, *14*, 7291–7321. [CrossRef]

46. Van Beek, J.; Tits, L.; Somers, B.; Coppin, P. Stem water potential monitoring in pear orchards through WorldView-2 multispectral imagery. *Remote Sens.* **2013**, *5*, 6647–6666. [CrossRef]

47. Declercq, B. Integrated Disease Management Based on the Life Cycle of Phytophthora Porri. Ph.D. Dissertation, Ghent University, Ghent, Belgium, 2009.

48. Dasu, T.; Johnson, T. Exploratory Data Mining and Data Cleaning. John Wiley Sons: New York, NY, USA, 2003; Volume 479.

49. Gates, D.M.; Keegan, H.J.; Schleter, J.C.; Weidner, V.R. Spectral properties of plants. *Appl. Opt.* **1965**, *4*, 11–20. [CrossRef]

50. Rinnan, Å.; Van Den Berg, F.; Engelsen, S.B. Review of the most common pre-processing techniques for near-infrared spectra. *TrAC Trends Anal. Chem.* **2009**, *28*, 1201–1222. [CrossRef]

51. Kindel, B.C.; Pilewskie, P.; Schmidt, K.S.; Coddington, O.; King, M.D. Solar spectral absorption by marine stratus clouds: Measurements and modeling. *J. Geophys. Res. Atmos.* **2011**, 116. [CrossRef]
52. Marín, D.; Martín, M.; Serrot, P.H.; Sabater, B. Thermodynamic balance of photosynthesis and transpiration at increasing CO_2 concentrations and rapid light fluctuations. *Biosystems* **2014**, *116*, 21–26. [CrossRef]

 *remote sensing*

Correction

Correction: Xu, M., et al. A Modified Geometrical Optical Model of Row Crops Considering Multiple Scattering Frame. *Remote Sensing* 2020, *12*, 3600

Xu Ma and Yong Liu *

College of Earth and Environmental Sciences, Lanzhou University, Lanzhou 730000, China; max15@lzu.edu.cn
* Correspondence: liuy@lzu.edu.cn; Tel.: +86-139-1940-7135

Received: 16 November 2020; Accepted: 8 December 2020; Published: 11 December 2020

The authors wish to make the following correction to this paper [1].

In the original paper, there was a mistake in Figure 11 as published. During the review process, the reviewer found that correlation coefficient (R) and Root mean square error (RMSE) is confusing in statistics, i.e., regarding the simulation and the measurement, which one is the independent variable on the X-axis and which one is the dependent variable on the Y-axis. Previously, the selection of variables was chaotic, which resulted in inconsistent RMSEs statistical results. Therefore, in the previous revision process, the reviewer recommends using the measurement as values in the X-axis (independent variable) and the simulation as values in the Y-axis (dependent variable) to calculate R and RMSE between reflectance simulated by the row model and field data. Therefore, we recalculated the R and RMSE between reflectance simulated by the row model and field data based on the reviewers' suggestion and then modified the previous confusing statistical results. Finally, to clearly show the statistical results of Figure 11, we added a scatter plot between the sum of the reflectance simulated by the row model and field data in Figure 12 during the modification process. However, confusing statistical results (R and RMSE) in Figure 11 were not deleted in the online version (PDF file). Therefore, these confusing statistical results remain in the published paper, which leads to the difference between the RMSE in Figure 11 and the RMSE in Figure 12. We once again affirm that the statistical indicators (R and RMSE) in Figure 12 in this paper are correct. To avoid ambiguity, we intend to revise Figure 11 (remove R and RMSE in Figure 11). Finally, our paper has no problem with the description of the statistical indicators here, and the authors apologize for any inconvenience caused and state that the scientific conclusions are unaffected.

Replace

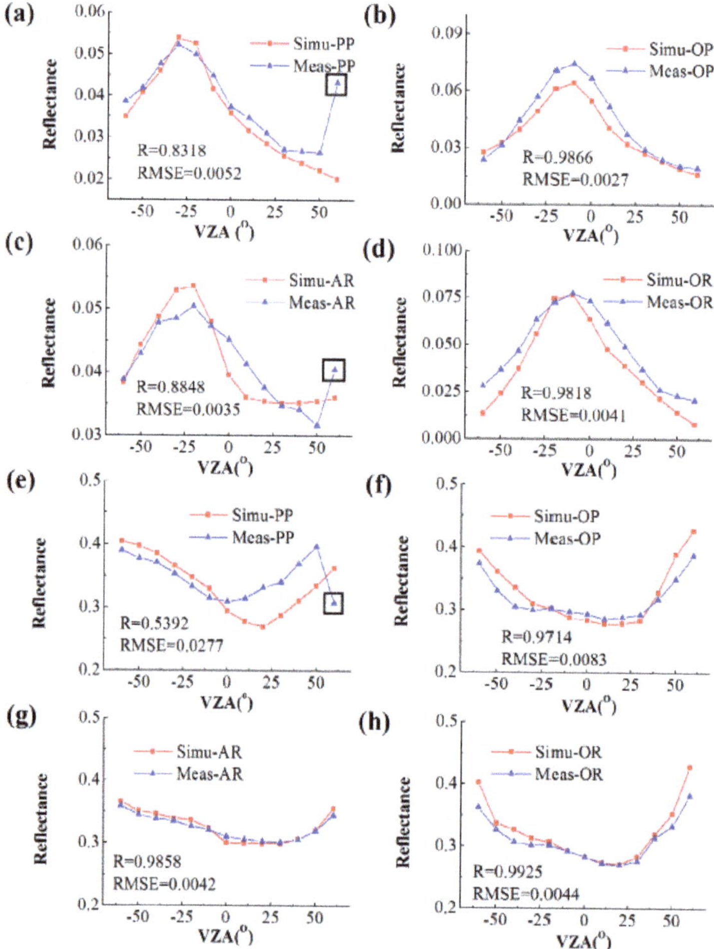

Figure 11. Comparison of the distribution of the sum of the reflectance simulated by the row model and field data in the multiangle observation for the principal plane (PP) mode (**a**,**e**), orthogonal plane (OP) mode (**b**,**f**), along-row plane (AR) mode (**c**,**g**), and orthogonal row plane (OR) mode (**d**,**h**). (**a**–**d**) Red band (670 nm); (**e**–**h**) NIR band (860 nm). VZA is the viewing zenith angle. The black box is the abnormal point in the measurement.

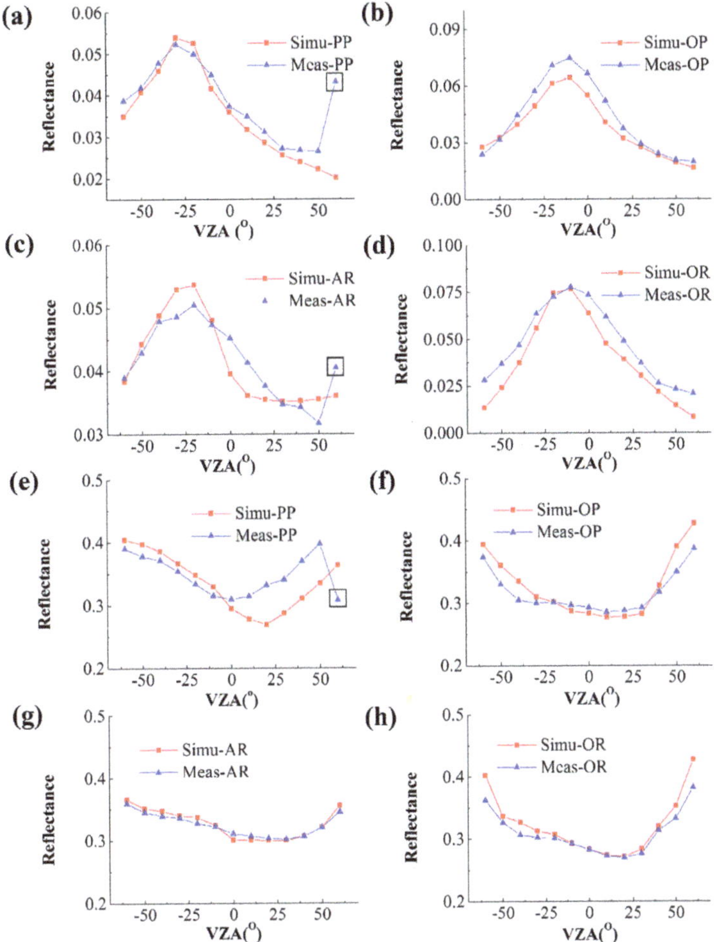

Figure 11. Comparison of the distribution of the sum of the reflectance simulated by the row model and field data in the multiangle observation for the principal plane (PP) mode (**a,e**), orthogonal plane (OP) mode (**b,f**), along-row plane (AR) mode (**c,g**), and orthogonal row plane (OR) mode (**d,h**). (**a–d**) Red band (670 nm); (**e–h**) NIR band (860 nm). VZA is the viewing zenith angle. The black box is the abnormal point in the measurement.

Conflicts of Interest: The authors declare no conflict of interest.

Reference

1. Ma, X.; Liu, Y. A modified geometrical optical model of row crops considering multiple scattering frame. *Remote Sens.* **2020**, *12*, 3600. [CrossRef]

Publisher's Note: MDPI stays neutral with regard to jurisdictional claims in published maps and institutional affiliations.

MDPI
St. Alban-Anlage 66
4052 Basel
Switzerland
Tel. +41 61 683 77 34
Fax +41 61 302 89 18
www.mdpi.com

Remote Sensing Editorial Office
E-mail: remotesensing@mdpi.com
www.mdpi.com/journal/remotesensing